Integrated Forest Biorefineries

RSC Green Chemistry

Series Editors:
James H Clark, *Department of Chemistry, University of York, UK*
George A Kraus, *Department of Chemistry, Iowa State University, Ames, Iowa, USA*
Andrzej Stankiewicz, *Delft University of Technology, The Netherlands*
Peter Siedl, *Federal University of Rio de Janeiro, Brazil*
Yuan Kou, *Peking University, People's Republic of China*

Titles in the Series:

How to obtain future titles on publication:
A standing order plan is available for this series. A standing order will bring delivery of each new volume immediately on publication.

For further information please contact:
Book Sales Department, Royal Society of Chemistry, Thomas Graham House, Science Park, Milton Road, Cambridge, CB4 0WF, UK
Telephone: +44 (0)1223 420066, Fax: +44 (0)1223 420247
Email: booksales@rsc.org
Visit our website at http://www.rsc.org/books

Integrated Forest Biorefineries

Edited by

Lew Christopher
South Dakota School of Mines and Technology, South Dakota, USA
Email: lew.christopher@sdsmt.edu

RSCPublishing

RSC Green Chemistry No. 18

ISBN: 978-1-84973-321-2
ISSN: 1757-7039

A catalogue record for this book is available from the British Library

Published by The Royal Society of Chemistry,
Thomas Graham House, Science Park, Milton Road,
Cambridge CB4 0WF, UK

Registered Charity Number 207890

For further information see our web site at www.rsc.org

Printed and bound in Great Britain by CPI Group (UK) Ltd, Croydon, CR0 4YY, UK

Preface

Biorefineries are our oil refineries of the future where oil is inevitably replaced by plant biomass – the most abundant and renewable resource on earth. *Integrated Forest Biorefineries: Challenges and Opportunities* describes how science, engineering and technology advancements could enhance the value extracted from lignocellulosic fibers by employing a biorefinery approach to the biomass conversion processes. This reference book documents recent progress and suggests future prospects for research and development of integrated forest biorefineries (IFBRs) as the path forward for the pulp, paper and other fiber-processing industries. Due to the strong and increasing off-shore competition and global movement toward and incentives for green fuels and chemicals, the pulp and paper industry needs to create additional revenues and diversify their products and markets to remain competitive. The focus of the book is on the transformation of pulp and paper mills into IFBRs that would require the development of advanced bio-based processes to bring about economic, environmental and social benefits to industry and society. The major research needs, technological challenges, potential products derived from the IFBR platforms, and models for the complex utilization of forest biomass for biofuels, biochemicals, market pulp and paper products are critically reviewed and discussed. The book summarizes the latest R&D of renewable energy production from forest biomass and related issues of feedstock supply, biomass and industrial sustainability, life cycle analysis, economic and policy regulations.

The book is organized in 11 chapters and starts with an extensive introductory overview of the current state of forest biorefining, IFBR research goals, production platforms, technological hurdles and perspectives for future developments. It sets the stage for the following chapters by presenting the current technical obstacles and providing justification for the need to evolve the pulp and paper mills into technologically-advanced IFBRs that would benefit

RSC Green Chemistry No. 18
Integrated Forest Biorefineries
Edited by Lew Christopher
© The Royal Society of Chemistry 2013
Published by the Royal Society of Chemistry, www.rsc.org

from product and market diversification and a higher degree of economic independence. The next four chapters deal with economic, environmental and industrial sustainability issues related to policy regulations, biomass production, supply chains, product development and market conditions for the forest products industry to achieve the goals of industrial manufacturing, economic growth and environmental benefits in a socially responsible manner. The subsequent three chapters focus on the selective removal of hemicellulose, lignin and extractives from woody biomass and their potential utilization through economic modeling and process integration to obtain biofuels and value-added bioproducts. The last three chapters review the latest developments in utilization of woody biomass for production of hydrocarbon fuels through gasification and pyrolysis, biohydrogen by anaerobic dark fermentation, and bio-based composite materials. The economics of described IFBR technologies and their environmental impact have been emphasized in most of the chapters. Practical examples and case studies are presented, where relevant and applicable, and product and process details are presented in sufficient detail to allow the reader a better understanding of research accomplishments. Altogether, the book is believed to fill a critical gap in our systematic knowledge about emerging perspectives on IFBR development, capturing the diversity and complexity of the on-going biorefinery research for sustainable production of bioenergy and bioproducts - a challenge and an opportunity that we currently face in our endeavors to transition to a bio-based economy and society.

The book is aimed at both industrialists and academics from diverse science and engineering backgrounds including chemical and biotechnology companies, governmental and professional bodies, and scholarly societies. I sincerely hope that this book will be a useful reference and source of valuable information for a wide audience of industry and academic researchers, instructors, students, business leaders, decision- and policy-makers, consultants and other professionals in the field of renewable energy. I would like to sincerely thank all the authors who, as renowned experts in their field, have made valuable contributions to this book. Although the contributing authors represent different science and engineering disciplines and different industry, federal and academic institutions, they all share common vision and interests in the area of forest biorefineries. Their opinions and views expressed in the book should not necessarily represent those of the organizations they work for. My sincere gratitude also goes to the editorial and production staff at RSC Publishing who all have worked efficiently and diligently under tight deadlines to ensure that the high standards of RSC have been maintained in the book.

Lew P. Christopher
Rapid City, South Dakota
September, 2012

Contents

RSC Green Chemistry No. 18
Integrated Forest Biorefineries
Edited by Lew Christopher
© The Royal Society of Chemistry 2013
Published by the Royal Society of Chemistry, www.rsc.org

Chapter 8   Lignin Recovery and Lignin-Based Products   180
Göran Gellerstedt, Per Tomani, Peter Axegård and
Birgit Backlund

**Chapter 9   Integrated Forest Biorefineries: Gasification and Pyrolysis for
            Fuel and Power Production   211**
            *Sushil Adhikari, Suchithra Thangalazhy-Gopakumar and
            Steven Taylor*

CHAPTER 1
Integrated Forest Biorefineries: Current State and Development Potential

LEW P. CHRISTOPHER

Center for Bioprocessing Research and Development and Department of
Civil and Environmental Engineering, South Dakota School of Mines and
Technology, Rapid City 57701, South Dakota, USA
Email: lew.christopher@sdsmt.edu; Tel.: +1-605-394-3385

1.1   Introduction

The motivation for development and use of biofuels is currently driven by
several important factors: 1) diminishing reserves of readily recoverable oil;
2) increasing demand and prices of petroleum-derived fuel; 3) concerns over
increasing greenhouse gas emissions and global climate change; 4) growing
food needs; and 5) desire for energy independence and security. The total
energy world consumption in 2005 was 488 EJ with U.S. consumption of 22%
of the world total, which is expected to surpass 650 EJ by 2025 and grow
approximately 40% over the next 25 years. Future oil supply is not unlimited or
assured as currently available petroleum fuel reserves are estimated to become
nearly depleted within 40 years. Crude oil prices have risen from less than $20/
barrel in the 1990s to nearly $100/barrel in 2007 with a current annual volatility
of crude oil prices exceeding 30%.[1] Government-controlled national oil companies and organizations, many in countries that are unstable or prone to
conflict, command and control more than 75% of the world's known oil

RSC Green Chemistry No. 18
Integrated Forest Biorefineries
Edited by Lew Christopher
© The Royal Society of Chemistry 2013
Published by the Royal Society of Chemistry, www.rsc.org

reserves and global oil production. The U.S. imports 10 million barrels of oil per day of the existing oil reserves of 1.3 trillion barrels.

In 2004, fossil fuels accounted for 86% of the U.S. total energy consumption, with an additional 8% from nuclear power and only 6% from renewable sources, including 3% of biomass-derived biofuels. Biomass, however, is the single renewable resource on earth, reproduced at 60 billion tons per year (as organically-bound carbon), that has the potential to supplant the use of liquid transportation fuels and help create a more stable energy future. In general, around 30% of the world's primary energy is derived from biomass with around 430 g carbon produced per square meter of land per year. The U.S. and other regions of the world have abundant biomass resources which are much more evenly distributed and accessible throughout the planet than the oil reserves. In their "billion ton vision", the U.S. Department of Energy (DOE) reported that nearly 1.3 billion dry tons of biomass could become available to produce biofuels and displace more than 30% of the nation's consumption of liquid transportation fuels.[2] However, the biomass share of the U.S. energy supply in 2004 was less than 3% of the total, compared to 40% and 23%, derived from petroleum and coal, respectively. Although biomass ranks well below petroleum, natural gas, and coal and is about one-half of nuclear, it surpasses hydroelectric and other renewable sources, and in 2009, the share of biomass in the total U.S. energy consumption exceeded 4% for the first time.

While ethanol production from corn and sugarcane (first generation biofuels) is a well-established process, cellulosic ethanol (second generation biofuels) is yet to be commercialized. The DOE Roadmap envisages large-scale production of second generation biofuels to become a nation-wide reality beyond 2020.[3] The cellulosic biomass base is composed of a wide variety of forestry and agricultural resources that include forest thinning, wood mill residues, logging residues, paper waste, tree trimmings, grass clippings, energy crops such as switchgrass and miscanthus, sugarcane waste (bagasse), wheat straw, rice straw, corn stover and corn cobs.

According to the "billion ton vision" of DOE, two-thirds of the biomass resources in the U.S. represent agricultural waste, whereas about a third is forest-based. Forests cover 30% of the earth (about 3.9 billion hectares) and play a major role in preservation of biodiversity, soil conservation, and pre-vention of climate change serving as a major carbon dioxide sink.[4] The forest resources are sustainable and provide long-term economic benefits to more than 1.6 billion people with a market of forest products estimated at $327 billion per year. Wood is used to produce heat for domestic and industrial purposes, and due to its strength properties, it is the primary construction material in more than 145 countries around the world. Wood is also used as a chemical feedstock to produce charcoal, tar, pitch, pulp fibers, paper, *etc.* Therefore, the wood-based value chain includes wood products, paper pro-ducts, energy, and wood-derived chemicals. Due to the wood's increasing use and demand, approximately 130 000 km^2 of forest are lost as deforestation every year. In addition, large areas of forest lands are littered with an unnatural accumulation of stunted, overcrowded trees and woody debris. Decades of fire suppression have disrupted the natural fire cycle of U.S. forests. Fires on these overstocked stands are more intense and harder to control than forest fires in previous decades, and they often result in catastrophic crown fires that kill large areas of forestlands. An estimated 8.4 billion dry tons of material needs to be removed from the national forests to reduce the risk of fire hazard, insect infestation, and disease. This vast source of biomass is available for production of wood products, chemicals, and energy. The biomass-derived energy use is projected to grow 35% from 2010 to 2024.

To combat the above processes, sustainable forest management practices need to include activities such as forest carbon credits, afforestation, reforestation, and the active support of government, forest companies and landowners. Government policies, carbon credits and carbon markets can be used as a tool to offset greenhouse gas (GHG) emissions. "Carbon credit" is a generic term for any tradable certificate or permit representing the right to emit one ton of carbon dioxide or the mass of another greenhouse gas with a carbon dioxide equivalent to one ton of carbon dioxide. Since GHG mitigation projects generate credits, this approach can be used to finance carbon reduction schemes between trading partners and around the world though carbon markets.[5] Plant-derived biofuels as a carbon-neutral technology have to achieve at least 60% lower emissions than petroleum fuel based on lifecycle studies that include all emissions resulting from making the fuel from the field to the tank. Meeting these goals will require significant and rapid advances in biomass feedstock and conversion technologies; availability of large volumes of sustainable biomass feedstock; demonstration and deployment of large scale, integrated biofuels production facilities; and development of an adequate biofuels infrastructure.

The use of non-food cellulosic biomass to produce biofuels presents a solution to the growing food vs. fuel debate - the dilemma regarding the risk of diverting farmland or crops for biofuels production in detriment of the food supply on a global scale, thereby having potential adverse impacts on food price, land use change, carbon and energy balance. For example, the share of corn destined to ethanol production in the U.S. reached 25% in 2007 and according to a recent study for the International Centre for Trade and Sustainable Development, the market-driven expansion of ethanol in the U.S. increased corn prices by 21% in 2009, in comparison with what prices would have been, had ethanol production been frozen at the 2004 level.[6] However, the United Nation's Organization for Economic Cooperation and Development (OECD) released a report in 2008 that provided estimates according to which up to 12% of the global coarse grain production and 14% of global vegetable oil production could be used for biofuels without having any significant impact on food prices.[7]

The Energy Independence and Security Act (EISA) of 2007 established a life-cycle GHG standard of 20% emission reduction for corn-based ethanol, based on the 2005 emission level, which would discourage the use of coal for process heat and limit further expansion of corn-based ethanol plants. The life-cycle analysis (LCA) studies mandated by EISA, dealing with data quality, allocation, system boundaries and sensitivity analysis, can profoundly shape the conclusions of biofuels production and land use change analysis. Using GHG and energy as functional units, the biofuel LCA accounts for the inputs and outputs of biofuel production, characterized in terms of energy requirements and yields, economic costs and benefits, and environmental costs and values. Most LCA results for lignocellulosic crops conclude that biofuels can supplement energy demands and mitigate GHG emissions to the atmosphere[8] because fermentation-derived ethanol is already part of the global carbon cycle. Also, blending oxygenates such as ethanol and methyl tertiary butyl ether (MTBE) are well recognized for causing reduced carbon monoxide levels by improving the overall combustion of the fuel. However, global equilibrium economic models have estimated that indirect land-use change associated with an increased use of corn-based ethanol could potentially double GHG emissions associated with that fuel pathway in the next 30 years.[9] This model is based on projections which suggest that biofuels would cause farmers to convert forests and grasslands to new agricultural lands that would otherwise be conserved. Measuring what really changes as a result of bioenergy policies and crops compared to what is expected to occur in their absence is an important challenge that must be addressed to improve land-use accounting. To protect ecosystems and improve livelihoods through more sustainable land-use practices, the forces that actually drive deforestation should be better understood. From this perspective, provided environmental preservation concerns are met, deforestation could be minimized, if much underutilized land is used for biomass production.

EISA contains a number of provisions to increase energy efficiency and the availability and use of renewable energy in the U.S. One key provision of EISA is the setting of a revised Renewable Fuels Standard (RFS). The revised RFS mandates the use of 36 billion gallons per year (BGY) of renewable fuels by 2022. The revised RFS has specific fuel allocations for 2022 that include use of: 16 BGY of cellulosic biofuels; 4 BGY of advanced biofuels; 1 BGY of biomass-based biodiesel; and 15 BGY of conventional biofuels (*e.g.*, corn starch-based ethanol). This potential resource is more than sufficient to provide feedstock to produce the required 20 BGY of cellulosic biofuels by 2022 – the year in which the revised RFS mandates the use of 36 BGY of renewable fuels. By 2030 the target is to replace 30% of the transportation fuel supply with biofuels, equal to 60 billion gallons of ethanol, which would require the use of approximately 0.75 billion tons of biomass.

In 2009, the U.S. produced 10.94 billion gallons of ethanol, and together with Brazil, both countries accounted for nearly 90% of the world's production. In 2010, the ethanol production in the U.S. reached 13.30 billion gallons which exceeded the RFS mandate for the previous year – 2010, whereas in 2011, 13.95 billion gallons of ethanol were produced. In U.S. alone, the number of ethanol biorefineries increased from 110 (as of January 2007) to 209 (as of January 2012). For the same period, the ethanol production capacity increased by more than 60% – from 5.5 BGY to 14.9 BGY (http://www.ethanolrfa.org/pages/statistics/). The recent increase in ethanol production was driven by a combination of high crude oil prices, RFS for domestic renewable fuel consumption, tax credits for ethanol blenders, and large net exports in 2010 and 2011.

However, ethanol production in the U.S. is still mainly corn-based, therefore, breakthrough technologies are needed to make cellulosic ethanol cost-competitive with corn-based ethanol. Although significant progress has been recently made towards commercialization of cellulosic ethanol, there are still economic, social and environmental challenges that need to be addressed. These include significant and rapid advances in biomass feedstock and conversion technologies; availability of large volumes of sustainable biomass feedstock; demonstration and deployment of large scale, integrated biofuels production facilities; and development of an adequate biofuels infrastructure. A minimum profitable ethanol selling price of $2.50/gallon can compete on an energy-adjusted basis with gasoline derived from oil costing $75–$80/barrel. At the lower oil prices ($45–$50/barrel), cellulosic technology may not be as competitive and could require policy supports and regulatory mandates to drive the market. The biofuels and bioproducts strategies need to be based on a thorough assessment of opportunities and costs associated with the upward pressure on food prices, intensified competition for land and water, and deforestation. As the feedstock costs comprise more than 20% of the production costs, it has now been widely recognized that biomass waste such as agricultural and forest waste can provide a cost-effective alternative to improve the economic viability of bioethanol production.[10] Despite technology advancements and declining processing costs for biofuels production, the profit margins for ethanol plants have been shrinking due to increasing feedstock costs and soaring prices of agricultural commodities. Costs and subsidies for biofuels are partly compensated by the expected economic, environmental and social benefits including increased energy security and reduced dependence on imported fossil-based fuels; diversification of energy and chemicals supply and markets; reduction of GHG emissions to mitigate climate change; job creation opportunities in rural areas; and overall improvement of quality of human health and life.

The supply and demand forces of market fundamentals have contributed to volatility in oil prices in recent years, and by transitioning toward higher energy efficiency and additional domestic sources of renewable fuels, such as biofuels, there is high potential to reduce U.S. market uncertainty and increase energy security. The depleting oil reserves and the use of traditional fuel with the associated logistics issues can be offset by deploying biorefineries that integrate various conversion technologies to derive energy and chemicals from locally available biomass resources. Furthermore, co-products such as corn gluten feed and meal, corn oil, glycerin, natural plastics, fibers, cosmetics, liquid detergents and other bioproducts, will increase with biofuel production and improve profitability. Currently, however, of the 100 million metric tons of chemicals produced annually in the U.S., only about 10% are biobased.[11]

1.2 Integrated Forest Biorefineries

A biorefinery is a facility that integrates biomass conversion processes and equipment for sustainable processing of biomass into a spectrum of value-added bio-based products (food, feed, chemicals, materials) and bioenergy (biofuels, power and/or heat). The biorefinery is analogous to today's petroleum refinery, which generates multiple fuels and products from petroleum. However, in contrast to the petroleum-based products, the biorefinery products are non-toxic, biodegradable, reusable and recyclable. The biorefinery takes advantage of the various components in biomass and their intermediates, therefore maximizing the value derived from the biomass feedstock. It employs a multidisciplinary approach that integrates physical and mechanical methods, chemical and biological conversion, catalysis and biocatalysis to obtain high-efficiency, low-cost, and low-energy consumption. Biorefineries are continuously evolving as new advancements in research of biomass feedstock, related processes and products become available for sustainable energy production – a challenge and an opportunity that we currently face in our endeavors to transition to a biobased economy and society.

An Integrated Forest Biorefinery (IFBR) is a biorefinery that can process forest-based biomass such as wood and forestry residues to bioenergy and bioproducts including cellulosic fibers for pulp and paper production (Figure 1.1). As lignocellulose consists of four major components – cellulose, hemicellulose, lignin and extractives (Table 1.1) – the IFBR has four production

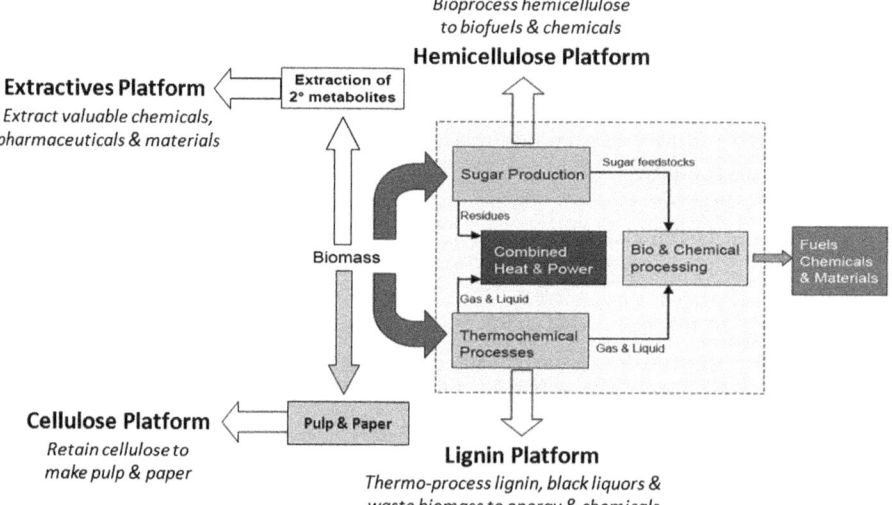

Figure 1.1 The IFBR concept.

Table 1.1   Major components of wood.

Wood Components	*Content (%)*
Cellulose	40–50
Hemicellulose	15–35
Lignin	20–30
Extractives	1–5

platforms that can be used in an integrated manner for production of biofuels and high-value bioproducts. A unique feature of the IFBR is that the cellulose platform is predominantly dedicated to production of pulp and paper rather than cellulosic ethanol. The existing prototype of the future IFBR are the pulp and paper mills, in particular the chemical pulp mills. The pulp and paper industry has the world's largest non-food biomass collection system that provides a primary source of cellulosic feedstocks. The U.S. paper and forest products industry made a commitment to increase the development of biomass fuels with the strategic goal of evolving existing pulp and paper mills into forest biorefineries that export substantial amounts of renewable, sustainable energy and chemical products while continuing to meet the growing demand for traditional pulp, paper and wood products. The pulp and paper mills are most suited for biorefinery large-scale developments as they are located near the forest and agricultural residuals and have existing infrastructure to transport the raw materials and finished products. In the U.S. alone, mills collect and utilize over 120 million dry tons of wood per year as a raw material and produce power from biomass of which nearly 60% is derived from wood residuals and spent liquors. Furthermore, pulp and paper mills also have a highly trained workforce capable of operating energy and biorefinery systems. The U.S. pulp and paper industry is the world's largest manufacturer of forest products[12] that employs nearly 1.3 million people with a payroll of over $50 billion per year. In 2010, the U.S. produced 76 million tons of paper and paperboard and nearly 50 million tons of wood pulp. However, the U.S. pulp and paper and other fiber processing industries need to create additional revenues and diversify their products and markets to remain competitive. This would enhance the profitability of these facilities thereby providing a higher degree of technological and market flexibility and economic independence. There are a number of reasons that necessitate the conversion of pulp and paper mills into IFBRs: 1) unstable, fluctuating oil prices and uncertainties about oil reserves; 2) strong, increasing off-shore competition; 3) global warming and increasing GHG emissions; 4) global movement toward and incentives for green fuels and chemicals.

For the implementation of the IFBR concept, new and more advanced bioprocessing and conversion technologies, that are both cost efficient and environmentally friendly, are needed to bring about the expected economic, environmental and social benefits. This transformation and technology upgrade would only be possible through development of sound economic studies in partnership with government and industry. The IFBR research needs are grouped in three focus areas:[13]

1) Sustainable Forest Productivity: Development of fast-growing biomass plantations designed to produce economic, high-quality feedstocks for bioenergy and bioproducts;
2) Extracting Value Prior to Pulping: Extraction of hemicellulose from wood prior to pulping to produce commercially viable chemical and liquid fuel products;

3) New Value Streams from Residuals and Spent Pulping Liquors: Use of thermo-chemical conversion technologies for production of fuels and chemicals and physico-chemical extraction processes to recover high-value materials from residuals and spent liquors.

As indicated earlier, the cellulose platform in the IFBR is reserved mainly for production of cellulosic fibers for pulp and paper, which is the core business of the pulp and paper industry. This is in contrast to a lignocellulosic feedstock biorefinery such as ethanol biorefinery where cellulose is hydrolyzed and fermented to ethanol and biochemicals. In the IFBR scenario, cellulose from pulp and paper mill waste such as paper mill sludge, that is not used in paper production because of inferior quality, can be converted to ethanol and other biofuels.[14] In the U.S., the pulp and paper industry generates about 5 million tons of paper sludge per year which has low value as a waste product but is a promising feedstock for ethanol production with more than 15% return on investment that requires no pretreatment as it is already delignified. In the past decade, ethanol production from cellulosic biomass has been extensively researched.[15] There are ongoing efforts to reduce production costs of the entire biomass to ethanol process: from feedstock production, harvesting and transportation to enzymatic hydrolysis, ethanol fermentation and product recovery. Process and cost improvements are focused on feedstock pretreatment efficiency, enzyme costs for cellulose hydrolysis and strain improvement for ethanol fermentation.[16] Simultaneous saccharification and fermentation (SSF) and consolidated bioprocessing (CBP) are well recognized strategies for integration of process steps with outstanding potential for cost reductions of up to 50%.[17]

The wood delignification (pulping) methods are well established.[18] Chemical (cellulose) pulp can be manufactured from wood using mechanical, semi-chemical or fully chemical methods (kraft and sulfite processes). Over the past 60 years the kraft process has proven as the most versatile and economical pulping process which is now the predominant method for production of cellulose pulp. For example, in 2000, 131 million tons of cellulose pulp were produced of which 117 million tons was kraft pulp (89% of total). In this chapter, the cellulose platform for pulp and paper production will not be discussed as this technology is well established and commercialized. However, emphasis has been given on the hemicellulose, lignin and extractives platforms for bioenergy and biochemicals production that relate to the IFBR focus areas of "Extracting Value Prior to Pulping" and "New Value Streams from Residuals and Spent Pulping Liquors".

1.2.1 Hemicellulose Platform

1.2.1.1 Hemicellulose Composition and Structure

Hemicellulose, the second most abundant polysaccharide after cellulose, are amorphous heterogeneous polymers comprising 15–35% of lignocellulosic biomass with a degree of polymerization (DP) of 80–200.[19] Hemicellulose forms an interface in the cell wall matrix with binding properties mediated by covalent and noncovalent interactions with lignin, cellulose and other polysaccharides.[20] The close association between the biopolymers in plant biomass is realized via chemical bonds, predominantly between lignin and hemicelluloses, in lignin–carbohydrate complexes (LCCs) that include benzyl-ether, benzyl-ester and phenyl-glycoside types of linkages. The composition and structure of hemicellulose (heteropolymer) are more complicated than that of cellulose (homopolymer) and can vary quantitatively and qualitatively in various woody species.[21] Due to the lower DP, the chemical and thermal stability of hemicelluloses is lower and their alkali solubility - higher than that of cellulose. The building blocks of hemicellulose (polyoses) include pentoses (D-xylose and L-arabinose) and hexoses (D-glucose, D-galactose and D-mannose). Sugar acids (acetic, 4-*O*-methyl glucuronic acid, ferulic/coumaric acids) make up the remainder of the hemicellulose structure. Xylans and glucomannans are the two predominant types of hemicellulose in hardwoods and softwoods, respectively, and their composition and proportion varies by species. Typically, softwoods have more mannose and galactose and less xylose and acetyl groups than hardwood. The hardwood xylans as complex heteropolysaccharides, comprising β-1,4-linked D-xylopyranose units, are highly substituted (Figure 1.2). The xylopyranose unit of the xylan main chain can be substituted at the C2 and/or C3 positions with acetic acid (at both C2 and C3 position in hardwoods), 4-*O*-methylglucuronic acid (at C2 position in both hardwoods and softwoods), and arabinose (at C3 position in softwoods). Arabinose may be further esterified by phenolic acids which crosslink xylan and lignin in LCCs in the cell wall matrix. The uronic acid groups in hardwood xylans are not evenly distributed, with one

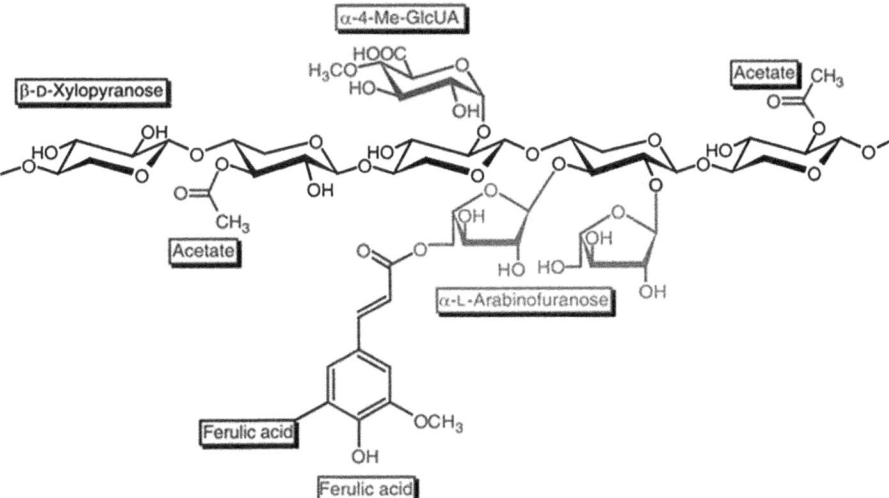

Figure 1.2   Chemical structure of xylan.

uronic group for every ten xylose units. In softwoods, every eight xylose residues are substituted with arabinose by α-1,3-glycosidic linkages whereas the ratio of xylose to glucoronic acid is 4:1.[18] The galactoglucomannans can be classified into two fractions with different galactose contents – galactose-poor fraction and galactose-rich fraction with a corresponding galactose/glucose/mannose ratio of 0.1/1/3, and 1/1/3, respectively, and acetyl content of 6% in both fractions. The softwood xylan is a linear polymer of D-xylopyranose units slightly branched with 1-2 side chains of arabinofuranose and glucoronic acid per molecule. The degree of substitution of hardwood xylan with acetyl groups can vary from 8% to 17% corresponding to 3.5-7 acetyl groups per 10 xylose units, and on average every second xylose unit is acetylated.

1.2.1.2   Fate of Hemicellulose during Pulping

In the kraft process, hemicellulose is partially depolymerized, debranched and solubilized in the cooking liquor.[22] The two important reactions of carbohydrate degradation during the kraft cooking are the "peeling" reaction and the ß-elimination of 4-*O*-methyl-D-glucuronic acid. In the peeling reaction, a stepwise depolymerization of the carbohydrate occurs at the reducing end sites of the polymer chain. The reaction generates a monosaccharide that undergoes a benzilic acid rearrangement to form an isosaccharinic acid. The reaction also forms a new reducing end on the remainder of the polymer, which can undergo further peeling reactions. The peeling reaction continues in carbohydrates until the introduction of a carboxyl group at the reducing end which protects the carbohydrate against further peeling. The carbohydrate material lost in the peeling reaction is converted to various hydroxyl acids that consume alkali and reduce the effective concentration of the pulping liquor. Subsequent losses of hemicellulose occur during the heating period of the kraft cook thereby about 40% of xylan is lost and glucuronic acid is converted to hexenuronic acid by ß-elimination of methanol. By the end of the cook, 60–70% of the glucuronosyl and 10% of the arabinosyl substituents in softwood xylan are removed. Due to a pH drop in the pulping liquor caused by debranched acetyl residues towards the end of pulping, part of dissolved xylan, lignin and lignin-xylan complexes are reprecipitated back onto the fiber surface. The extent of this readsorption depends on the alkaline cooking conditions and wood species, however, the reprecipitated xylan has a low molecular weight without side-chain groups and a high degree of crystallinity.[23] For instance, half the xylan content of pine kraft pulp is estimated as relocated xylan whereas up to 14% of birchwood xylan can be reprecipitated during kraft pulping. During acid sulfite pulping, redeposition of xylans onto the fiber surface has not been observed. The possible reasons for this would be that the harsh cooking conditions and presence of acid-resistant residual acetyl and 4-*O*-methylglucuronic acid groups act as barriers against the adsorption and intercrystallization of xylan onto the cellulose micromolecules. Significant amounts of xylans are hydrolyzed and solubilized in the sulfite pulping process. For instance, in sulfite cooking of birch only 45% of the original xylan remains in pulp after 20 min and its original DP of 200 is reduced to less than 100. The bonds between the pentose units (arabinose and xylose) are hydrolyzed much more rapidly than the gly-copyranosidic bonds. However, the glucuronic acid-xylose and xylose-acetic acid linkages are relatively more resistant to the acid hydrolysis conditions and little cellulose is lost in the sulfite cook.[24] The degradation products of the hemicellulose acid hydrolysis appear in the cooking liquor in the following approximate order: arabinose > galactose > xylose > mannose > glucose > acetic acid > glucuronic acid. In the above order, glucose is derived mostly from the glucomannan rather than cellulose polymer. Thus, the residual xylan in sulfite pulps is less accessible since it is localized mainly in the secondary cell walls, although the xylan distribution across the cell wall is more uniform than in kraft pulps.

1.2.1.3 Hemicellulose Extraction

Kraft pulping is a low-yield process (yield is typically less than 50%) since half the hemicellulose (or up to 15% of the wood weight) and almost all of the lignin components of the wood are dissolved into the spent pulping liquor. This "black liquor" is processed downstream in a recovery boiler which burns the organics in the black liquor to produce process steam and recovers the inor-ganic cooking chemicals for re-use. A typical kraft pulp mill can process about 600 000 metric tons of wood per year of which about 15% (90 000 metric tons) is released into the spent liquors as degraded hemicellulose. Because lignin has a relatively high heating value (26.9 MJ/kg), it is cost-effective to recoup the heating value by combustion. The hemicelluloses, however, have a low heating value (13.6 MJ/kg) and remain underutilized through the incineration process as hemicellulose contributes about a quarter of the total energy recovered in the recovery furnace. Therefore, extracting the hemicellulose from wood chips prior to pulping could be used to produce higher value chemicals and polymers and enhance the profitability and competitiveness of the paper mills. Selective removal of the hemicelluloses prior to pulping can be accomplished without degrading the wood fibers. Hemicellulose extraction of up to 10% of wood weight has been shown to have no impact on the pulping process and strength properties of the resulting pulps. In fact, kraft mills may benefit from decreased cooking times and increased pulp throughput of up to 20%, especially for mills with limited capacity of their recovery boilers. Different process options for hemicellulose extraction before pulping have been described.[25–27] One pro-posed method of hemicellulose removal uses green liquor as a pretreatment chemical.[27,28] Green liquor is an alkaline aqueous solution generated in the pulping recovery process which is comprised of sodium hydroxide, sodium carbonate and sodium sulfide. Under alkaline extraction conditions, xylan in wood chips is dissolved in oligomeric form while glucomannans are degraded by the peeling reaction,[21] therefore this method is applicable to hardwoods. Extraction of hardwood chips using alkaline chemicals to extract up to 10% of hemicellulose results in a final liquor that is near-neutral pH, preserving the pulp yield and paper quality.[29] According to another hemicellulose extraction method, known as "hot water extraction", wood chips are treated in absence of mineral acids or bases at 160–170 °C with water only, replacing alternative costly pre-treatment methods.[30] Extraction using hot water generates acidic conditions (pH of 3.5) due to release of acetyl groups from wood hemicellulose. Under these conditions, hemicellulose is autohydrolyzed to generate a wood hydrolyzate, containing monosugars and acetic acid, that is then subjected to multiple separation and fractionation steps to produce commercial chemicals and fermentation products. The extracted woody biomass has a higher energy content and contains fewer easily degradable components which allows for its efficient processing and conversion to pulp and paper, wood pellets, fiberboard and nanocellulose.[31,32] Other methods for hemicellulose extraction, that have been researched with a varying degree of success, include microwave-assisted extraction, use of supercritical carbon dioxide, ionic liquids, near critical water, ammonium hydroxide, *etc.* Based on the "billion ton vision" of DOE, nearly 400 million tons of hemicellulose are available in the U.S. for bioprocessing to fuels and chemicals. In addition, every year approximately 15 million tons of hemicellulose are produced by the pulp and paper industry alone, and according to preliminary results, this can yield in excess of 2 billion gallons of ethanol and 600 million gallons of acetic acid with a net cash flow of $3.3 billion.

1.2.1.4 Bioproducts from Hemicellulose

1.2.1.4.1 Enzymes. Due to their complex structure, the complete breakdown of naturally occurring branched hemicelluloses requires the concerted action of several enzymes with different functions. These are classified in two groups, hydrolases and esterases, based on the nature of linkages that they can cleave. The glycosyl hydrolases are involved in the enzymatic hydrolysis of the glycosidic bonds of hemicellulose. Of major importance are the endo-β-1,4-xylanases or 1,4-β-D-xylan xylanohydrolases (3.2.1.8) that can randomly hydrolyze internal xylosidic linkages on the backbone of xylan polysaccharide. The main products formed from xylan hydrolysis by xylanase are xylobiose, xylotriose and substituted xylooligosaccharides depending on the mode of action of the particular enzyme (Table 1.2). The xylooligomers liberated by xylanase are

Table 1.2 Xylan-degrading enzymes.

Enzyme	Mode of Action
Endo-xylanase	Hydrolyses interior β-1,4-xylose bonds of xylan backbone
Exo-xylanase	Releases xylobiose from xylan backbone
β-Xylosidase	Releases xylose from xylobiose
α-Arabinofuranosidase	Hydrolyses α-arabino-furanose from xylan
α-Glucurosidase	Releases glucuronic acid from glucuronoxylans
Acetyl xylan esterase	Hydrolyses acetyl ester bonds in acetyl xylans
Ferulic acid esterase	Hydrolyses feruloyl ester bonds in xylans
ρ-Coumaric acid esterase	Hydrolyses ρ-coumaryl ester bonds in xylans

converted to xylose by 1,4-ß-D-xylosidase (EC 3.2.1.37). The so-called accessory enzymes such as acetyl xylan esterases, phenolic acid esterases, arabinofuranosidases and glucuronidases cleave side groups from the xylan backbone. All xylanolytic enzymes act synergistically in xylan hydrolysis. Xylanases can be classified structurally into two major groups, family 10 and family 11. Family 10 enzymes have a relatively high molecular weight whereas family 11 xylanases are relatively low molecular weight with low or high pI values. The release of reducing sugars from xylan however has not been shown to correlate to the family belonging of enzyme. The enzyme-substrate interaction is dependent on substrate specificity and kinetic properties of enzyme and can be influenced by pH, presence of xylan binding domain and ionic strength of protein and xylan molecule. Since xylan is negatively charged due to the presence of glucuronic acid side-chain groups, the efficiency of binding of enzyme to xylan is affected by the pH of reaction and pI of protein. For instance, if pH is below the pI value, the enzyme can be completely bound to the polysaccharide. The xylanolytic enzyme system of a variety of microorganisms have been extensively investigated and several exhaustive reviews have appeared.[33–35]

The optimization of fermentation techniques and isolation of more efficient microbial strains has led to a significant increase in the production rates of xylanase. Fungal systems are excellent xylanase producers, but often co-secrete cellulases which can adversely affect pulp quality. One way of overcoming this is by using suitable separation methods to purify xylanases from contaminating cellulase activity. This approach, however, is expensive and impractical. By applying appropriate screening methods and selection of growth conditions, it is possible to isolate naturally occurring microorganisms which produce totally cellulase-free xylanases or contain negligible cellulase activity. Alternatively, genetically engineered organisms could be used to produce exclusively xylanase. Most xylanases studied are active in slightly acidic conditions (pH 4-6) and temperatures between 40 and 60 °C. The current trend is, however, to produce enzymes with improved thermostability and activity in alkaline conditions to fit operations at harsh industrial conditions. ß-Xylanases are produced by many microorganisms on xylan-rich substrates.[35]

Currently, most commercial enzymes are mainly produced in a conventional submerged fermentation process, which is an inherently expensive operation best suited for high value antibiotics and other pharmaceutical products. Solid substrate fermentation is an economically viable alternative for enzyme production which offers numerous advantages over the submerged fermentation systems as many enzymes and other biochemicals can be produced by solid state fermentation at a fraction of the cost for submerged fermentation.[36] The solid state fermentation allows the direct use of *in-situ* enzymes (*i.e.* xylanase for pulp pretreatment and bleaching) without their prior downstream processing. The substrate (*i.e.* paper pulp which contains xylan), which is initially used as a carbon source for enzyme production, subsequently becomes the target substrate of enzyme (xylanase) action. This approach could certainly improve the economics and enhance the efficiency of the biobleaching technology due to the operational simplicity of solid state fermentation, high volumetric productivity and concentration of enzyme and production of substrate-specific enzymes in a water-restricted environment.[37] Advantages include high concentration of the product and simple fermentation equipment as well as low effluent generation and low requirements for aeration and agitation during enzyme production.[38] Due to the considerably lower production costs, the *in-situ* xylanase has been shown to be more cost-efficient when compared to commercial liquid products.

Spent sulfite liquor (SSL) is derived from the delignification of wood chips in an aqueous solution of acid bisulphites with an excess of SO_2, resulting in the solubilisation of lignin and leaving the wood cellulose largely undegraded.[39] The solids of the resultant black liquor contain 50 to 65% lignosulphonates, 15 to 22% sugars and 2 to 5% volatile acids such as acetic acid. The sugars found in the SSL include xylose, mannose, galactose, arabinose and glucose with xylose concentrations of 70–85% of the total sugars (Table 1.3). Following concentration, SSL becomes a concentrated waste with a high BOD and COD (> 1000 g/L) levels and needs treatment prior to disposal. The utilization and recovery of the valuable organics in this effluent would, therefore, be more desirable than its simple discharge. The microbial utilisation of SSL has been studied for production of various metabolites such as lactic acid, single-cell yeast protein[39] and ethanol.[40] Recently, the use of this inexpensive carbon source as inducer of xylanase activity has also been demonstrated.[41] Potential advantages include reduced xylanase production costs and development of effluent-free technology that impact positively on the environment. However, the xylose present in SSL is difficult to ferment due to the presence of inhibitory compounds such as acetic acid (> 10 g/L) and polyphenols (> 10 g/L), therethore strain improvement through genetic engineering and microbial adaptation on SSL have been employed.[42] Currently, Tembec Inc., Temiscaming, Quebec, an acid bisulfite dissolving pulp manufacturer, produces 14 million liters per year of industrial alcohol by fermenting hexose sugars in SSL.[43]

Xylan-degrading enzymes, and in particular xylanases, have a great potential in industrial processes such as saccharification of lignocellulosic biomass to fermentable sugars for production of biofuels and biochemicals, bread making, clarification of beer and juices, enzymatic retting of flax, surface softening and smoothing of jute-cotton blended fabrics.[44] Nevertheless, the most important application of these enzymes to date is their use in the pulp and paper industry. Xylanases have been reported to enhance inter-fiber bonding through fibrillation without reducing pulp viscosity. Xylanase-treated pulps have shown improved beatability and brightness stability. When applied together with

Table 1.3 Sugar composition of spent sulfite liquors from hardwoods (g/L).

Sugars	Aspen	Birch	Eucalyptus
Arabinose	1.5	–	0.9
Galactose	–	0.6	2.4
Glucose	0.5	1.1	2.2
Mannose	3.1	6.4	0.1
Xylose	24.3	21.1	23.6

cellulases, xylanases can improve the drainage rates of recycled fibers and can facilitate the release of toners from office waste and the following flotation and washing steps. The xylanase production on large scale constitutes approximately 50% of the total enzyme market and the demand for xylanases grows about 25% per year, with a major application in bleaching of paper pulps.

The use of xylanases at pulp and paper mills to facilitate bleaching (biobleaching) and improve fiber properties is one the most important large-scale biotechnological applications of recent years.[45,46] The enzymatic improvement in pulp bleachability depends on a number of factors such as the wood source, pulping and bleaching processes as well as properties and substrate specificity of the enzyme. Factors such as inhibitory effect of residual pulping and bleaching chemicals in pulp as well as degradation end products on xylanase efficiency, presence of xylan-lignin and xylan-cellulose bonds may as well impact on extent of xylan hydrolysis and pulp bleachability. Restrictions in the enzymatic removal of xylan from pulp have been assigned to retarded accessibility and chemical modification of residual hemicellulose. Accessibility problems arise from the fact that chemical pulping and bleaching apparently remove the more accessible portion of xylan from the cell walls, leaving the remaining part in locations, that are less accessible to xylanase. Xylanases should contain no or very low cellulase activity as cellulases prove detrimental to yield and strength properties of pulp. The bleaching efficiency of xylanase is measured either as the reduction in the amount of chemicals used for bleaching of pulp or the brightness gain induced by the enzyme. As the biobleaching effect is dependent on the amount of enzyme used, the enzyme production costs should be kept as low as possible to ensure a cost-effective bleaching process. The major benefits from the enzyme bleaching are: 1) Reduced bleaching costs; 2) Reduced chemical consumption; 3) Increased pulp throughput; and 4) Reduced pollution.[47] A few hypotheses exist to explain the phenomenon of xylanase-aided bleaching of pulp, although the exact mechanism is not completely understood. It should be noted that the proposed mechanisms for biobleaching are not mutually exclusive and more than one model can be involved depending on pulp type, on one side, and substrate specificity of xylanase to a specific xylan type in pulp, on another.[48]

The initial model proposed suggested that xylanases attack and hydrolyze mainly xylan redeposited on the fiber surface thereby enabling the bleaching chemicals a better and smoother access to residual lignin.[49] During kraft pulping, pulp xylan is first solubilized and later on part of it is redeposited back onto the pulp fibres. Xylanase acts on these reprecipitated xylans by partially hydrolysing them to facilitate extraction of lignin during pulp bleaching (Figure 1.3). The second hypothesis suggests that xylanases can partly hydrolyze xylan that is involved in lignin-xylan complexes thereby reducing the size of these complexes and improving their mobility and extractability from the cell walls. Indirect evidence does exist that lignin-carbohydrate bonds are formed during biosynthesis and aging of wood as well as during kraft pulping and that xylose is released as the main sugar component of isolated lignin-carbohydrate complexes. The biobleaching effect appeared to be accompanied by a decrease in the

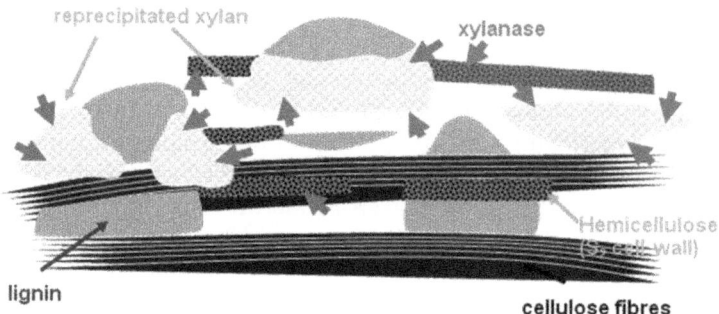

Figure 1.3   Mechanism of xylanase-aided bleaching of paper pulps.

DP of xylan and a slight reduction in xylan content. It has also been reported that xylan-chromophore associations can be generated during alkaline pulping which contribute to pulp color and brightness reversion of pulps. A direct brightening effect has been observed following xylanase pretreatment of pulp. This could be due to a direct removal of lignin fragments involved in lignin-xylan complexes and/or removal of xylan derived chromophore structures. This hypothesis is supported by the recent findings that during the kraft cook the methylglucuronic acid of xylan can be modified to hexenuronic acid giving rise to double bond chromophore-type formations. Xylanases may also be able to disrupt, to an extent, the physical interlinking between xylan and cellulose within the fiber matrix thereby improving the fiber swelling and generating macropores to facilitate lignin removal. The biobleaching effect observed with some hardwood sulfite pulps may also be caused by improved pulp porosity. This suggestion is based on the fact that in acid sulfite pulps, in contrast to kraft pulps, xylan is not reprecipitated on the fiber surface but is largely entrapped across the fiber the cell walls.

Another enzyme of industrial importance, that can be produced on hemicellulose-based substrates such as xylan and xylose, is glucose (xylose) isomerase (EC 5.3.1.5). This enzyme is used industrially to convert glucose to fructose in the manufacture of high-fructose corn syrups, HFCS.[50] HFCS is produced by milling corn to produce corn starch which is first treated with alpha-amylase to produce shorter chains oligosaccharides and then with glucoamylase to produce glucose. Finally, xylose isomerase (also known as glucose isomerase) converts glucose to a mixture of about 42% fructose and 50–52% glucose (HFCS-42) with some other sugars mixed in. This 42–43% fructose-glucose mixture is then subjected to a liquid chromatography step, where the fructose is enriched to about 90% and then back-blended with 42% fructose to achieve a 55% fructose final product (HFCS-55). While the relatively inexpensive alpha-amylase and glucoamylase enzymes are added directly to the slurry and used only once, the more costly xylose isomerase is packed into columns and used repeatedly until it loses its activity. Thus, production of HFCS using xylose isomerase is the major application of immobilized enzyme technology.[51]

The most widely used varieties of high-fructose corn syrup are HFCS-55 (mostly used in soft drinks) and HFCS-42 (used in many foods and baked goods). In the U.S., HFCS is among the sweeteners that have primarily replaced sucrose. Factors for this include governmental production quotas of domestic sugar, subsidies of U.S. corn, and an import tariff on foreign sugar, all of which combine to raise the price of sucrose to levels above those of the rest of the world, making HFCS less costly for many sweetener applications. Pure fructose is the sweetest of all naturally occurring carbohydrates and 1.73 times as sweet as sucrose with 44% less calories than sucrose.[52] Fructose has the lowest glycemic index (GI of 23) of all the natural sugars and may be used in moderation by diabetics. In comparison, ordinary table sugar (sucrose) has a GI of 65 and honey has a GI of 55. Per relative sweetness, HFCS-55 (containing 55% fructose and 45% glucose) is comparable to sucrose. Sweetness is measured against sucrose as a reference with a sweetness index of 1.0. However, compared to sucrose, HFCS is less expensive, has better solubility and stability in solution, easier transportation and use than sucrose. HFCS represents at least 40% of all sweeteners added to foods, beverages and soft drinks. From 1970 to 1999 the HFCS production in the U.S. increased 10-fold and currently, together with sucrose, dominates the industrial sugar market in the U.S.[53] The average American consumed approximately 17.1 kg of HFCS in 2008 versus 21.2 kg of sucrose. In Japan, HFCS consumption accounts for one quarter of total sweetener consumption. The world market for HFCS was 5 million tons in 2004. In 2010, HFCS accounted for 37% of the caloric sweetener market in the U.S. At wholesale, HFCS is most often priced at $0.05-0.20 per pound lower than refined cane and beet sugar. However, the increase in the HFCS consumption has coincided with the increase in incidence of obesity, diabetes, and other cardiovascular diseases and metabolic syndromes. There are also major concerns about the mercury contamination of HFCS during production and its toxicity to honey bees with possible contribution to colony collapse disorder of honey bees.[51]

1.2.1.4.2 Xylan. Hemicelluloses such as xylan need to have a certain degree of purity before they can be utilized in any process of industrial importance. Their isolation however is restricted as they form hydrogen bonds with cellulose, covalent bonds with lignin (mainly ether linkages), acetic and hydroxycinnamic acids (ester linkages). Xylan source and recovery process (extraction) directly impact the physical and chemical properties of the recovered polysaccharide and determine its applicability. Xylans can be extracted from lignocellulosic materials or partially delignified pulps. Xylan fractionation from lignified materials yields polysaccharides with major proportions of lignin, whereas higher purity xylans are obtained when isolated from pulps, especially bleached pulps.[54] However, the properties of xylans have not been fully characterized, defined and exploited. Although annual plants have been proven a rich source of xylan, because of the difficulties in extraction and purification of xylans and hemicellulose in general, an efficient

Table 1.4   Major large-scale applications of xylan.

Pulp and Paper Industry	Pharmaceutical Industry	Chemical Industry	Food Industry	Fermentation Industry
• Beater additive– improved swelling, porosity, drainage, strength • Fiber coating • Wood resin stabilizer	• Anticoagulant • Anti-cancer agent • Cholesterol-reducing agent • Wound treatment agent • HIV inhibitor • Tabletting material • Hydrogels • Dietary fiber	• Furfural • Thermoplastic material • Polypropylene filler • Paint formulations • Gel-forming material • Chiral polymer building blocks	• Xylose • Xylitol • Biodegradable polymers – plastics, films, beverage packaging, coatings with increased hydrophobility and water resistance (acetyl xylans)	• Enzymes– xylanase, xylose isomerise • Biopolymers– polyhydroxy-alkanoates

isolation process has never been realized.[55] Furthermore, the great variety of xylan structures makes their individual use difficult as a better understanding of their physico-chemical and functional properties is needed. Xylans from hardwoods have been isolated using a combination of alkali and steam,[56] aqueous ammonia[57] and steam explosion.[58] Barium hydroxide was used as a selective extraction chemical for fractionation of arabinoxylans.[59] Xylan extractability is related to their interaction with other cell wall constituents such as lignin. Table 1.4 summarizes some of the most important current and potential applications of xylan in the pulp and paper, pharmaceutical, chemical, food and fermentation industries.[60]

1.2.1.4.3 Ethanol.   For the economic production of ethanol from lignocellulosics, the fermentation of both hexose and pentose sugars is an economic necessity. *Saccharomyces cerevisiae* is used universally for industrial ethanol production because of its ability to produce high concentrations of ethanol with high inherent ethanol tolerance. However, native *S. cerevisiae* cannot ferment xylose, hence, engineering *S. cerevisiae* for xylose utilization has focused on adapting xylose metabolic pathway from xylose-utilizing yeasts. This can be achieved by introducing pentose-utilizing capability into efficient ethanol producers such as *Saccaromyces* and *Zymomonas*.[61] Genes encoding for xylose reductase, xylose isomerase (xi), xylulokinase (xk), transaldolase (tal) and transketolase (tkl) are inserted to enable the pentose-phosphate pathway through which xylose can enter the glycolysis pathway of glucose fermentation to ethanol (Figure 1.4). Employing this strategy for *Z. mobilis*, 85% of the theoretical ethanol yield on xylose was attained.[62] In attempts to generate *S. cerevisiae* strains that are able to ferment D-xylose, the XYL1 and XYL2 genes of *Pichia stipitis* coding for the xylose reductase (xr) and xylose dehydrogenase (xdh), respectively, were introduced into

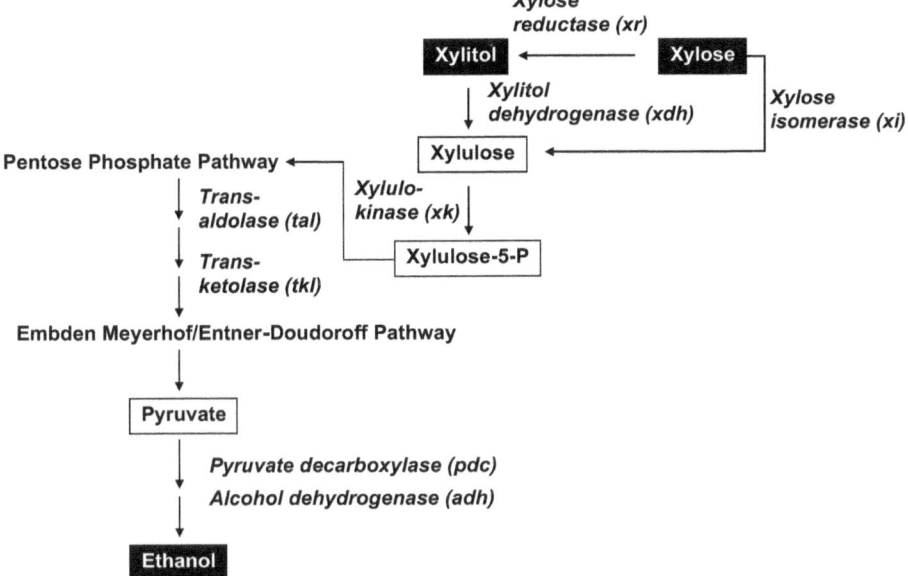

Figure 1.4   Fermentation pathways of xylose to xylitol and ethanol.

S. cerevisiae by means of genetic engineering.[63] Although *P. stipitis* ferments pentose and hexose sugars (xylose, glucose, mannose, galactose and celliobiose) and produces ethanol at faster rates and yields than other pentose-fermenting yeasts, it is not as ethanol and inhibitor tolerant as the traditional ethanol producing yeasts.[64] The maximum ethanol concentration achieved with *P. stipitis* is 6.5% as compared to 18% with *S. cerevisiae*. The ethanol production rate of *P. stipitis* on glucose is also lower than that of *S. cerevisiae*. Nearly all reported xylose isomerase-based pathways in *S. cerevisiae* suffer from poor ethanol productivity, low xylose consumption rates and poor cell growth compared with the oxidoreductase pathway. To increase the xylose isomerase activity before expressing the mutant enzyme in *S. cerevisiae*, directed evolution can be used. This approach improved the aerobic growth rate by 61-fold and both ethanol production and xylose consumption rates in *S. cerevisiae* by nearly 8-fold. Moreover, the mutant enzyme, which had 77% higher activity than the native enzyme, enabled ethanol production under oxygen-limited fermentation conditions.[65]

It is well known that bacterial fermentations of xylose to ethanol are associated with low ethanol yields, slow fermentation rates, byproduct formation (acids) which requires additional product separation, contamination problems due to the neutral pH requirements for bacterial growth, bacterial sensitivity to inhibitors, and intolerance to high ethanol concentrations.[66] To overcome these problems, the carbon flow in bacterial species is diverted from native fermentation products to ethanol by introducing pyruvate decarboxylase (pdc) and alcohol dehydrogenase (adh). PDC catalyzes the nonoxidative decarboxylation of pyruvate to produce acetaldehyde and carbon dioxide, whereas ADH catalyzes the reduction of acetaldehyde to ethanol during fermentation. Several Gram-negative bacteria such as *Escherichia coli*, *Z. mobilis*, *Klebsiella oxytoca*, *Klebsiella planticola* and *Enterobacter cloacae* have been engineered for ethanol production.[66–70] On mixed sugars (xylose and glucose), a recombinant *E. coli* produced 103–106% of the theoretical yield of ethanol.[71] Recombinant Gram-negative *E. coli* KO11[72] and Gram-positive *Clostridium cellulolyticum*[73] were constructed to produce ethanol from acid hydrolysates of hemicellulose and lignocellulosic biomass, respectively. The ethanol production by the recombinant *C. thermocellum* increased 53%. It should however be noted that the most common problems with recombinant microorganisms that still need to be resolved are their instability, slower production rates and reduced robustness compared to the wild strains.[63]

Ethanol fermentation can also be accomplished with existing industrial xylose isomerase (xi) products that convert xylose to xylulose (see Figure 1.4), which is a fermentable sugar by *S. cerevisiae*. However, there is a mismatch between the optimum pH and temperature of xylose isomerase (pH 7.5, 55 °C) and those for ethanol fermentation (pH 5, 30 °C), which makes this approach economically unviable. The major challenges in bioethanol production are the ethanol and inhibitor tolerance of fermenting microorganisms, fermentation rates and low sugar concentrations. There is a need to develop stable and robust ethanologenic microorganisms capable of tolerating high ethanol and inhibitor concentrations that can convert high levels of sugar concentrations with a solids content of 20% or higher. The economics of bioethanol production would be significantly improved if these microorganisms could produce cellulose and xylan degrading enzymes and ferment mixed sugars. The ultimate goal is to develop a consolidated bioprocessing (CBP) technology that integrates enzyme production, cellulosic biomass hydrolysis and fermentation of mixed sugars in a single step.[74] The key objectives of this goal are to: 1) perform metabolic engineering of native cellulolytic organisms such as *C. thermocellum* and increase the product yield and titer,[75,76] or alternatively clone the cellulolytic abilities that are lacking in native highly efficient ethanologens such as *S. cerevisiae*; 2) improve the microbial tolerance to substrate and product inhibition.[77,78] The functional expression of cellobiohydrolase (CBH) in *S. cerevisiae* increased 20-120-fold.[79] Despite these advances of genetic engineering in yeasts, their cellulose-degrading abilities are still much lower than those of *C. thermocellum* and *T. reesei*.[80]

1.2.1.4.4 Organic Acids. The U.S. DOE has identified twelve building block chemicals that can be produced from sugars via biological or chemical conversions.[81] The twelve building blocks can be subsequently converted to a number of high-value bio-based chemicals or materials. Building block chemicals, as considered for this analysis, are molecules with multiple functional groups that possess the potential to be transformed into new families of useful molecules (Table 1.5). The synthesis for each of the top building blocks

Table 1.5   U.S. DOE building block chemicals.

1. 1,4-succinic, fumaric and malic acids
2. 2,5-furan dicarboxylic acid
3. 3-hydroxy propionic acid
4. aspartic acid
5. glucaric acid
6. glutamic acid
7. itaconic acid
8. levulinic acid
9. 3-hydroxybutyrolactone
10. glycerol
11. sorbitol
12. xylitol/arabinitol

and their derivatives was examined as a two-part pathway: 1) transformation of sugars to the building blocks; and 2) conversion of the building blocks to secondary chemicals or families of derivatives. A second-tier group of building blocks was also identified as viable candidates. These include gluconic acid, lactic acid, malonic acid, propionic acid, the triacids (citric and aconitic); xylonic acid, acetoin, furfural, levoglucosan, lysine, serine and threonine. Most of the organic acids identified as building block chemicals from the first- and second-tier group can be produced from lignocellulosic biomass including biomass waste – that offers a less expensive alternative in replacement of their petrochemical counterparts, provides a sustainable way of waste disposal and generates income to the industrial and rural sectors.[82]

In the last decade, microbially-produced organic acids[83,84] find increased use in the food industry and as raw materials for manufacture of biodegradable polymers.[85] For instance, lactic acid, with a current price of $1.5/kg for a 88% purity food-grade product, has the potential of becoming a very large volume, commodity-chemical intermediate produced from renewable carbohydrates for use as feedstocks for biodegradable polymers (polylactic acid, PLA), oxygenated chemicals, environmentally friendly "green" solvents (ethyl lactate), plant growth regulators, and specialty chemical intermediates (acrylic acid). Currently, lactic acid is the most promising starting material for chemical synthesis and transformations due to the presence of the adjacent and highly reactive carboxylic and hydroxyl groups. Lactic acid in food products usually serves either as a pH regulator or as a flavoring agent and is used in a wide range of food applications such as yougurt, bakery products, beverages, meat products, confectionery, dairy products, salads, dressings, ready meals, *etc.*[84] Organic acids find application in food preservation because of their effects on bacteria.[86] The non-dissociated (non-ionized) organic acids can penetrate the bacteria cell wall and disrupt the normal physiology of certain types of bacteria such as *E. coli*, Salmonella and Campylobacter species that are pH-sensitive and cannot tolerate a wide internal and external pH gradient. Upon passive diffusion of organic acids into the bacteria, where the pH is near or above neutrality, the acids dissociate and the cations lower the bacteria internal pH, leading to situations that impair or stop the growth of bacteria. Furthermore, the anions of the dissociated organic acids accumulate within the bacteria and disrupt their metabolic functions leading to osmotic pressure increase that is incompatible with the bacterial survival. The world production of lactic acid in 2009 was 258 000 tons with a projected growth of 329 000 tons by year 2015 and 367 000 tons – by 2017. Primary growth catalysts include sustained demand from end-use industries, heightened R&D activity and emergence of new applications. Lactic acid consumption in chemical applications, which include PLA polymer and new "green solvents", such as ethyl lactate, is expected to expand 19 % per year.[87] Lactic acid based biodegradable polymers is a niche area rapidly gaining ground, and is poised to become one of the most promising end-use applications for lactic acid. In addition, growing environmental concerns arising from use of plastics will further clear the path for biopolymers in the long run. The recent announcements of plant expansions and building of new development-scale plants for production of lactic acid and/or polymer intermediates by major U.S. companies such as Cargill, Chronopol, A.E. Staley, and Archer Daniels Midland (ADM) attest to this potential. Major international manufacturers of fermentative lactic acid include Purac (Netherlands), Galactic (Belgium), and several Chinese companies. In late 1997, Cargill joined forces with Dow Chemical and established a Cargill-Dow PLA polymer venture, NatureWorks LLC, which exists today as a stand-alone company. NatureWorks LLC has constructed a major lactic acid facility in Blair, Nebraska, which began operations in 2002 with a capacity of 180 000 metric tons of lactic acid per year.[88]

Levulinic acid can be produced from renewable resources including hemicellulose by the Biofine process through a two-step, acid-catalyzed, high-temperature reactions.[89] The process involves acid hydrolysis of the biomass polysaccharides to monosugars at 210–220 °C in the first step, followed by their degradation to levulinic acid and tars (Biofine char) at 190–200 °C – in the second step. The by-products that are formed during the process are formic acid (from the cellulose fraction of biomass) and furfural – from the hemicellulose fraction. If the two fractions are separated, as in the case of IFBR, the cellulose fraction is used for pulp and paper production whereas the hemicellulose fraction can be converted into furfural. As xylose and other 5-carbon sugars in the hemicellulose fraction are recalcitrant to ethanol fermentation, their utilization for production of furfural could be a viable alternative. Furfural can then be used as a commodity chemical or converted to levulinic acid. The carbon-rich Biofine char is potentially an ideal feedstock for a gasification reactor which allows it to be converted to a high energy "syngas" that can either be used as source of other chemicals via a Fischer Tropsch conversion process or burned as fuel gas in a boiler or gas turbine for energy (energy content of 12 000 BTU per pound or 25 000 KJ/kg). A Biofine plant with a capacity of 300 tons per day could produce 13 000 tons of furfural per year from lignocellulosic biomass containing 25% hemicellulose by mass. This represents 2.9% of the global furfural consumption in 2004. Using only heat and pressure in a carefully controlled chemical environment, the Biofine conversion avoids the many challenges facing other biomass conversion technologies. The Biofine process enables the use of a broader range of lig-nocellulosic feedstock including low-value forest residues, whole tree chips, agricultural residues, food wastes, recycled paper, sewage, paper mill sludge and municipal solid waste. Levulinic acid at 77% of the maximum theoretical yield was produced on paper mill sludge that contained 57 wt% cellulose and 8 wt% hemicellulose.[90] The feedstock flexibility is one of the greatest strengths of this process in the marketplace of biomass conversion technologies. By far the largest potential market for levulinic acid is in the production of oxygenated fuels for both transportation (gasoline and diesel) and energy generation (heating oil and gas turbine fuels). Levulinic acid can be converted into the oxygenated gasoline fuel additive methyltetrahydrofuran (MTHF). MTHF can also be produced directly from the pentose fraction in hemicellulose via fur-fural.[91] MTHF has several attractive properties as a gasoline fuel additive:[92] 1) can be mixed with gasoline in amounts of up 30% with no adverse impact on engine performance; 2) has good anti-knock properties (motor octane value of 80); 2) has energy density of 90% of that of gasoline; 3) has relatively low volatility (R.V.P = 3 psi), approximately that of gasoline; 4) can serve as co-solvent for ethanol in gasoline mixtures (MTHF significantly reduces the vapor pressure of ethanol when co-blended in gasoline). The Biofine process has been commercialized and is currently one of the most advanced lignocellulose-processing technology available. Table 1.6 summarizes the most prominent applications of organic acids that can be derived from hemicellulose.

1.2.1.4.5 Furfural. Furfural represents a renewable building block chemi-cal which is currently regaining attention as a biobased alternative for the production of industrial and household chemicals[100] – from antacids and fer-tilizers to plastics and paints (Figure 1.5). It is the first and most important product derived from hemicellulose on an industrial scale. About 10% of the mass of xylan-rich plant residues such as agriwaste and forest waste can be recovered as furfural.[101] Furfural is usually obtained through dehydration of pentoses, particularly xylose, or hemicelluloses, at high temperature (200–250 °C) or in the presence of mineral acids as catalysts. The applications of furfural are grouped as follows:

1. A sustainable substitute for petroleum-based building blocks used in production of fine chemicals and plastics.[102] Furfural is used as a feed-stock for production of furfuryl alcohol, furan, methylfuran, tetra-hydrofuran and furoic acid. The vast majority of furfural (more than 60%) is converted into furfuryl alcohol, a well-established industrial commodity, which has found growing applications as a source of a variety of materials with notable recent progress. Furoic acid can be obtained via oxidation[103] and furan – via palladium catalyzed vapor phase decarbonylation.[104] Tetrahydrofurfuryl alcohol is a widely used precursor for specialty chemicals and as a binder in catalyst for the new pebble bed reactors. Another furfural derivative, tetrahydro-2-furanmethanol, is

Table 1.6 Applications of organic acids.

Organic Acid	Applications
Citric Acid[93]	70% of total production used in confectionary and beverage products, 30% – in pharmaceuticals (anticoagulant blood preservative, antioxidant) and metal cleaning; selling price decreased with market shift from pharmaceuticals to food applications (879 000 t produced in 2002).
Lactic Acid[94]	Acidulant, flavor enhancer, food preservative, feedstock for calcium stearoyl-2-lactylates (baking), ethyl lactate (biodegradable solvent) and polylactic acid plastics (100% biodegradable) for packaging, consumer goods, biopolymers (approved by FDA); estimated U.S. consumption of 30 million lb with 6% growth pa; potential demand of 5.5 billion lb as a very large volume-commodity chemical.
Itaconic Acid[95]	Feedstock for syntheses of polymers for use in carpet backing; paper coating N-substituted pyrrolidinones for use in detergents and shampoos; cements comprising copolymers of acrylic and itaconic acid.
Aspartic Acid[96]	For synthesis of aspartame; monomer for manufacture of polyesters and polyamides; polyaspartic acid as substitute for EDTA with potential market of $450 million per year.
Fumaric Acid[97]	For manufacture of synthetic resins, biodegradable polymers; intermediate in chemical and biological synthesis.
Malic Acid[98]	Acidulant in food products; citric acid replacement; raw material for manufacture of biodegradable polymers; for treatment of hyperammonemia and liver dysfunction; component for aminoacid infusions.
Succinic Acid[99]	Used as acidulant, pH modifier, flavoring and antimicrobial agent, ion chelator in electroplating to prevent metal corrosion, surfactant, detergent, foaming agent; for production of antibiotics, amino acids and pharmaceuticals; 270 000 t in 2004; U.S. domestic market estimated at $1.3 billion per year with 6–10% annual growth.
Levulinic Acid[90]	For synthesis of methyl tetrahydrofuran (gasoline extender), diphenolic acid (for epoxy resins), tetrahydrofuran (solvent), 1,4-butanediol (polymer intermediate), succinic acid (specialty chemical), delta-aminolevulinic acid (active chemical in herbicides and pesticides), sodium levulinate (antifreeze ingredient), ethyl levulinate (diesel oxygenate).

being developed as a solvent for cleaning electronic components, as chemical coupling agent in organic syntheses and for making vinyl resin, dyes and rubber.[105]

2. Intermediate in the conversion of biomass to alkane-based liquid fuels (biomass-to-liquid, BTL). Polyoses and/or pentosans are first converted to furfural by acid-catalyzed dehydration, followed by aldol condensation and hydrogenation over solid base catalysts to obtain a C7-C15 fraction of hydrocarbon fuel.[106] Currently, the typical process for BTL production is gasification following by Fischer–Tropsch process, a technology which is feasible economically only on large scale.

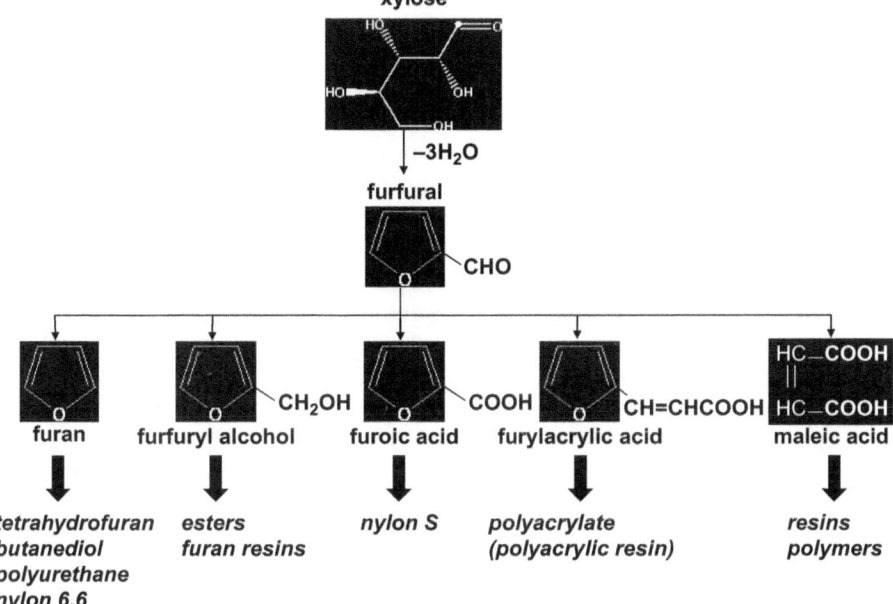

Figure 1.5   Furfural-based applications.

3. Solvent in the refining of lubricating oils, diesel fuels and vegetable oils. Furfural has been widely utilized as an extractant due to its broad solubility in ethanol, ether and water, for separation of saturated from unsaturated compounds and selective extraction of aromatics from hydrocarbon oils in petroleum refining.[107–109]
4. Fungicide/Pesticide/Nematicide: Furfural is a new pesticidal active ingredient intended for use as a fumigant to control root-infesting plant parasitic nematodes and fungal plant diseases in greenhouse soil used for growing ornamentals and other non-food commodities.[110,111] The technical formulation contains 99.7% furfural and is for use in formulating end-use products. The end-use product contains 90% furfural in a liquid formulation.[112] Furfuryl alcohol can be further hydrogenated to tetrahydrofurfuryl alcohol (THFA). THFA is used as a nonhazardous solvent in agricultural formulations and as an adjuvant to help herbicides penetrate the leaf structure. Plant parasitic nematodes cause an estimated annual loss to agriculture of $35 billion worldwide. Furfural derivatives have been used as fungicides and herbicides.[113]

China is the biggest supplier of furfural and accounts for around half of the global capacity. The world production of furfural in 2005 was about 250 000 t/year. Currently, the well-established acid-based technologies produce some 300 000 tons of furfural per year at a stable price of $1000/t ($1/kg). Provided the production cost of furfural can be further reduced, new application opportunities would rise, such as for drug and specialty chemical manufacture, replacement of phenol in foundry resins, and specialized polymers.[105]

1.2.1.4.6 Xylitol. Xylitol is a five-carbon naturally occurring sugar alcohol found in the fibers of fruits, vegetables and beech wood.[114] However, because of the small amounts, the quantitative extraction of xylitol from natural sources is economically unfeasible. Xylitol is currently produced chemically on a large scale by hydrogenation of xylose, which converts the sugar aldehyde into a primary alcohol.[115] Hydrogenation is carried out at high pressures (up to 50 atm), high temperature (80–140 °C) using expensive catalysts (Nickel Raney) and expensive purification processes.[116] The xylitol yields are low – on average 50–60% from xylan. Xylitol price is about \$7/kg[117] which is comparatively higher than that of natural sweeteners.

The drawbacks of the chemical process can be overcome by using a biological route of xylitol production that is carried out by microorganisms at low temperature (30–35 °C). The microbial conversion employs naturally fermenting yeasts (*Candida*) such as *C. tropicalis and C. guillermondi* that yield 65–90% of the theoretical maximum from xylan. Alternatively, recombinant strains containing a xylose reductase gene (*i.e.* recombinant *S. cerevisiae*) can be used with a very high production yield of 95% from the theoretical maximum. The microbially produced xylitol requires less purification than the chemical process.[118] Hemicellulosic hydrolyzates from hardwoods and agricultural residues are used as feedstock for xylitol production. Hemicellulose hydrolyzates from aspen,[119] poplar,[120] eucalyptus,[121] rice straw,[122] barley bran[123] and corn cobs[124] have been studied for xylitol production. Due to the presence of inhibitory compounds, pretreatment of these hydrolyzates is usually required to detoxify them prior to microbial conversion.[125] Various approaches are being considered to remove fermentation inhibitors or minimize their formation, such as neutralization,[121] use of activated charcoal,[126] overliming,[127] ion exchange resins,[128] solvent extraction,[129] intracellular acidification,[130] yeast strain variation,[131] laccase[132] recombinant strains[133] and adaptation of the microbial strains.[124] Although the biotechnological production of xylitol has made significant progress over the past decade, further process and product optimization is necessary to make this technology compete with the chemical production method on large scale. Milestone on the way of commercialization include better understanding of xylose metabolism into xylitol (Figure 1.4), product inhibition, air supply, lag phase of xylitol formation, *etc.* New and more efficient methods for xylitol production using genetic engineering, immobilized cells, mixed cultures and enzymatic biocatalysis are currently being researched.[134]

The primary interest in xylitol is based on its properties as an alternative sweetener. It is as sweet as sucrose, twice as sweet as sorbitol, and nearly three times sweeter than mannitol. Xylitol is a non-fermentable sugar alcohol with dental health benefits in caries prevention, showing superior performance to other polyols (polyalcohols). Xylitol inhibits the microbial deterioration of tooth enamel as it is not utilized by the acid-forming oral bacteria.[135] Due to its anti-cariogenic and anti-plaque action, xylitol is used around the world as a sweetener in chewing gums, pastilles, and oral hygiene products such as toothpaste, fluoride tablets and mouthwashes. Its plaque-reducing effect is manifested by attracting and starving harmful micro-organisms because cariogenic bacteria prefer fermentable six-carbon sugars as opposed to the non-fermentable xylitol.[136] More than 10% of its use is in sugar-free chewing gums which have a world market of more than $12 million per annum.

Possessing approximately 40% less food energy, xylitol is a low-calorie alternative to table sugar. Absorbed more slowly than sugar, it does not contribute to high blood sugar levels or the resulting hyperglycemia caused by insufficient insulin response. Because xylitol does not depend on insulin to enter the glycogenolytic metabolic pathways, it is used for treatment of diabetes. Its glycemic index is approximately ten-fold lower than that of sucrose and more than two times lower than fructose. This characteristic has also proven beneficial for people suffering from metabolic syndrome, a common disorder that includes insulin resistance, hypertension, hypercholesterolemia, and an increased risk for blood clots.

Xylitol also has potential as a treatment for osteoporosis – it prevents weakening of bones and improves bone density.[137] Studies have shown that the xylitol-containing chewing gum can help prevent upper air and ear infections.[138] When bacteria enter the body, they adhere to the tissues using a variety of sugar complexes. The open nature of xylitol and its ability to form many different sugar-like structures appears to interfere with the ability of many bacteria to adhere which was attributed to the increased effectiveness of endogenous (naturally present in the body) antimicrobial factors.[139] Xylitol is also one of the building block chemicals that can be used in production of ethylene glycol, propylene glycol, lactic acid, xylaric acid, and for synthesis of unsaturated polyester resins, antifreeze, *etc.* (Figure 1.6).

1.2.1.5 Hemicellulose-Based Biorefinery

A simplified diagram of IFBR based on generation of multiple products from hemicellulose is presented in Figure 1.7. The flowchart illustrates the enormous potential that the IFBR hemicellulose platform has to produce an array of biofuels and high-value products in an integrated, cost-efficient and environmentally-friendly way.[140] There are however technological challenges that need to be overcome to make these processes economically viable.[141] These challenges are related to optimization of process conditions to maximize the biorefinery-derived value such as: 1) improvements in extraction efficiency of hemicellulose for minimal sugar degradation while preserving the pulp and paper properties; and 2) improvements in pentose fermentation and tolerance of microbial producers to inhibitors present in the hemicellulose extracts and hydrolyzates. The schematic in Figure 1.7 only depicts some of the possible scenarios of processes and products for the IFBR development. It would be

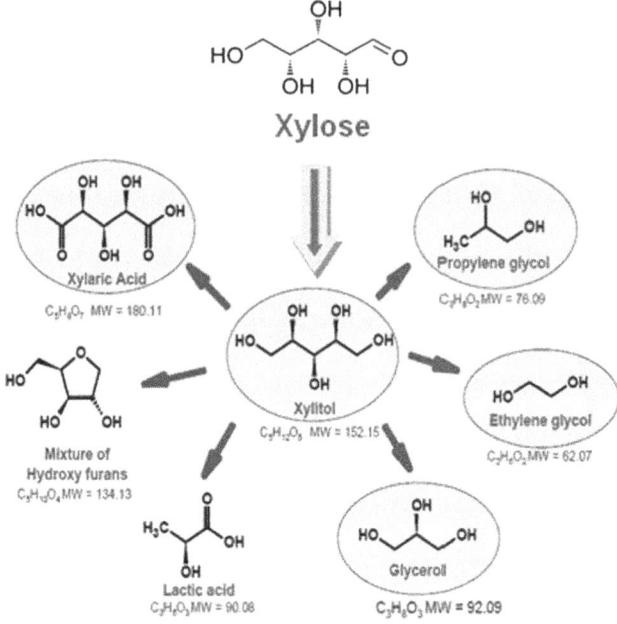

Figure 1.6   Xylitol as a building block chemical.

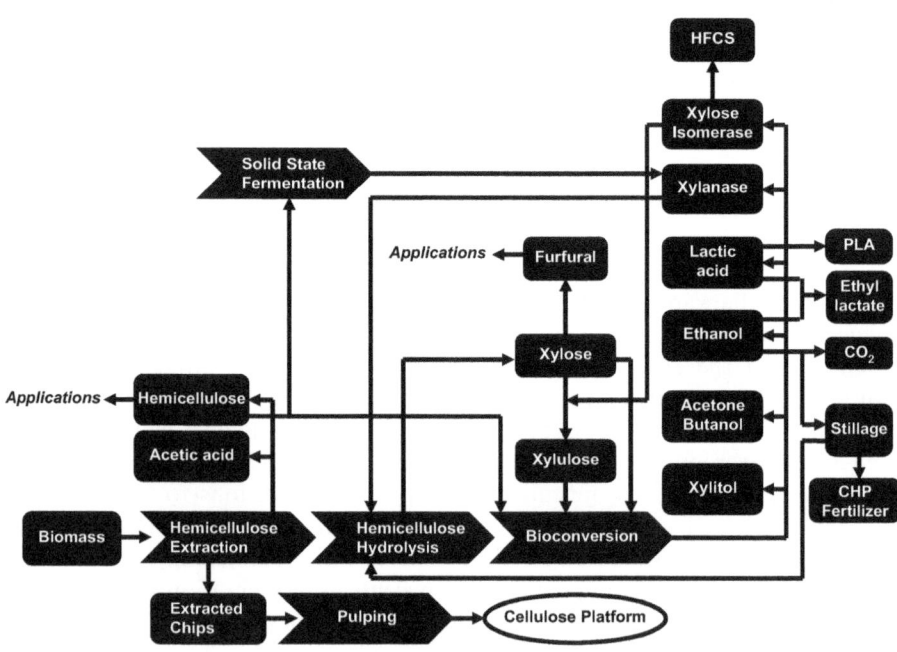

Figure 1.7   Hemicellulose platform of IFBR.

impractical and probably impossible to present all pathways of the hemicellulose platform. Moreover, the IFBR development is a dynamic, technology- and market-driven process of continuous improvements and adjustments. The strive for more cost-effective manufacturing technologies, robust supply chains, new markets, commercial opportunities and economic benefits will determine the best products and direct the necessary changes of the transformation process to a IFBR for a given mill. Recent studies of the economic and commercial potential of IFBR have suggested that the longer-term competitive advantage of implementing IFBR is more likely related to the supply chain and manufacturing flexibility of pulp and paper mills than technology.[142] Therefore, mills would need to first establish a viable market for a specific product before investing in the implementation of a particular process technology, meaning that product design and marketing should precede process design.

1.2.2 Lignin Platform

1.2.2.1 Lignin Composition and Structure

After cellulose, lignin is the second most abundant organic polymer on earth. Softwoods generally contain more lignin (25–35%) than hardwoods (20–25%). Lignins are complex, three-dimentional biopolymers consisting of phenylpropanoid units containing both aromatic and aliphatic groups.[143] The phenilpropane units (C9 or C6C3), known as monolignols or lignin precursors, are linked together through C-C and C-O-C bonds and have different amounts of methoxy groups (Me). The dominant bond is the β-O-4 linkage. Three types of monolignols have been identified: *p*-coumaroyl alcohol, confineryl alcohol and synapyl alcohol (Figure 1.8). Hardwood lignins are synthesized from mainly sinapyl and coniferyl precursors in proportions from 1:1 to 1:3. Softwood lignins contain up to 95% coniferyl units with small quantities of *p*-coumarolyl monolignols (up to 5%).[18,144,145] The lignin macromolecule also contains a variety of functional groups that have an impact on its reactivity such as methoxyl groups, phenolic hydroxyl groups, and few terminal aldehyde groups. Only a small proportion of the phenolics hydroxyl groups are free since most are occupied in linkages to neighboring phenylpropane linkages. Carbonyl and alcoholic hydroxyl groups are incorporated into the lignin structure during enzymatic dehydrogenation. Lignin is more concentrated in the middle lamella and primary cell wall. Lignins surround the cellulose-hemicellulose matrix to provide stiffness to the cell walls and glue the cells together. Lignin as a hydrophobic polymer serves as a barrier against water penetration and is resistant toward degradation by most microorganisms except white-rot fungi and some bacteria.[146]

1.2.2.2 Fate of Lignin during Pulping

In the pulp and paper industry, lignin is separated from other components of wood (delignification) in a process called chemical pulping using two principal

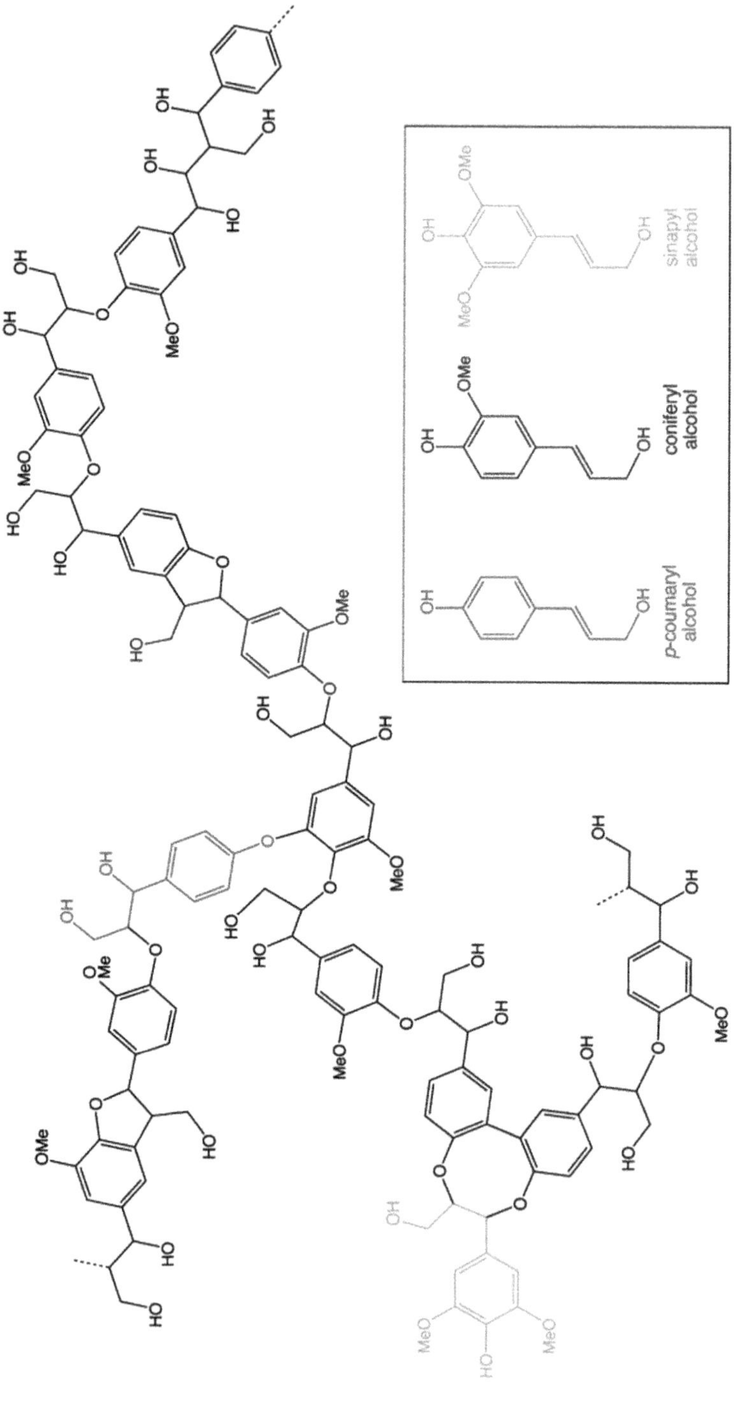

Figure 1.8 Structural model of lignin.

methods – alkaline (kraft, soda) and acidic (sulfite) pulping processes. The removal of lignin allows individual fibers to be freed from the wood matrix with mild mechanical treatment. Pulping must be able to remove lignin from fibers through chemical degradation, while minimizing damage to the cellulosic portion of the fibers to maintain strength. The kraft pulping process accounts for 98% of chemical pulp production in the U. S. and 92% of chemical pulp production in the world. Kraft pulping is an alkaline process that uses sodium hydroxide and sodium sulfide (white liquor) at 160–180 °C and pH above 12. The hydroxide and hydrosulfide anions react with lignin during the kraft cook. The alkaline attack causes fragmentation of native lignin into lower molecular weight segments. These lignin fragments are reacted with the thiol groups generated by the sodium sulfide to form thiolignins which are soluble in strong alkali and can be precipitated upon acidification. The lignin solubility in the cooking liquor is increased by cleavage of the linkages holding the phenyl-propane units together, thereby generating free phenolic hydroxyl groups.[18] The most prevalent degradation reactions occurring during kraft pulping include the cleavage of α-aryl ether and β-aryl ether bonds which increases the amount of phenolics hydroxyl groups. The formation of quinine and catechol structure and carboxyl groups is increased due to the oxidative conditions during delignification. The carbon-carbon linkages, however, are more stable and tend to survive the kraft pulping process. In addition to the native alkali-stable LCCs surviving the cook, alkali-stable LCCs may be formed during the cook. The lignin fragment of the native LCCs is believed to be linked exclusively with hemicelluloses, while the lignin fragment of the LCCs that are formed during pulping may be more frequently linked to cellulose.[147] The alkali-stable linkages between lignin and carbohydrates in the LCCs survive the kraft cook and have been suggested to be present in kraft pulps. Overall, about 85–95% of the lignin originally present in wood is dissolved in the cooking liquor.[148] The black liquor is the waste liquor that is released after the kraft pulping process is completed. It contains most of the cooking chemicals and the dissolved wood substances including lignin, organic acids (acetic acid, formic acid, saccharinic acid), hemicellulose (xylan) and other compounds. In the sulfite process, a mixture of sulfonic acid and a bisulfite salt is used to degrade and solubilize lignin in the form of lignosulfonates. The water-soluble lignosulfonates are released with the waste pulping liquor (sulfite spent liquor) after the sulfite cook. The lignosulfonates are normally mixtures, with wide molecular weight profiles, which contain 70–75% lignin and up to 30% impurities such as carbohydrates, ash, and other inorganic materials. Incorporation of the sulfur into biorefinery products could be a serious environmental problem and its removal would add to the overall production costs. In soda pulping, the oldest chemical pulping method, delignification occurs by the action of sodium hydroxide which causes fragmentation of native lignin and its dissolution in the cooking liquor. Compared to the kraft and sulfite lignin, the soda lignin is sulfur-free and chemically less modified. In the Organosolv pulping, a mixture of organic solvents with or without water is used as the cooking liquor. The Organosolv lignin is produced by a selective solubilisation in the solvent mixture. The Organosolv process makes it possible to obtain a lignin product with a higher homogeneity which is less modified compared to other lignins. Overall, the chemical pulping solubilizes 40–50% of the wood dry weight – mainly lignin (on average, 90% of original lignin), up to 50% hemicellulose, up to 20% cellulose and most of the extractives.[149]

1.2.2.3 Bioproducts from Lignin

1.2.2.3.1 Applications of Lignin as a Polymer. It has been estimated that the current lignin production in the existing pulp and paper industry is between 50 and 60 million tons per year of which 98% is burned as a low cost fuel in the chemical recovery boiler and only 2% is used commercially. The commercial lignin is mainly lignosulphonates originating from sulfite pulping (about 1 million tons/year) and less than 100 000 tons/year of kraft lignin.[150] In addition, a major fraction of the kraft lignin is converted into a water-soluble sulfonated lignin that competes with the lignosulfonates in some applications. As a result, the existing lignin products are based on the low-value lignosulphonates used for dispersing and binding applications, and the lignin markets are stagnated at $300 million and very low growth rates.[151] One of the major reasons for the lignin market stagnation is the heterogeneous nature of lignin products having non-uniform and non-standard quality and properties. For example, the molecular weight of kraft lignin extracted from the black liquor can range from 200 to 200 000 grams per mole. The lignosulfonates have a very high polydispersity of 4.2 to 7 and degree of sulfonization of up to 0.5 per phenylpropanoid unit which corresponds to up to 8% sulfur content.[152] In comparison, kraft lignin has a polydispersity of 2.5 to 3.5 and contains 1–3% sulfur. The heterogeneity of lignin is a result of the biomass source, pulping method and recovery process used that impact on the lignin composition, size, properties and reactivity. Lignins possess unique chemical reactivity because of their heterogeity, presence of different functional groups and limited accessibility of reactive site at *o*- and *p*-positions, which makes it difficult to control a specific reaction in a desirable direction.[153] Impurities of organic and inorganic nature in lignins and difficulties in recovering lignins from product streams also contribute to the challenges associated with lignin markets.[154] A recent report from DOE estimates that 0.225 billion tons of lignin (biorefinery lignin) can be produced from processing 750 million tons of biomass feedstock for biofuels production.[155] According to this report, opportunities for commercial applications of lignin can be grouped into: power, fuel and syngas products; macromolecules; and low molecular weight aromatic, phenolic and/or miscellaneous monomer compounds. About 75% of the commercial uses of lignin as dispersants, emulsifiers, binders, and sequestrants are based on its polymer and polyelectrolyte properties. The large-scale challenge for application of industrial lignins is to find methods for their utilization which will bring profits high enough to justify the development and commercialization of lignin technologies. Some opportunities for lignin utilization in an IFBR are shown in Figure 1.9.

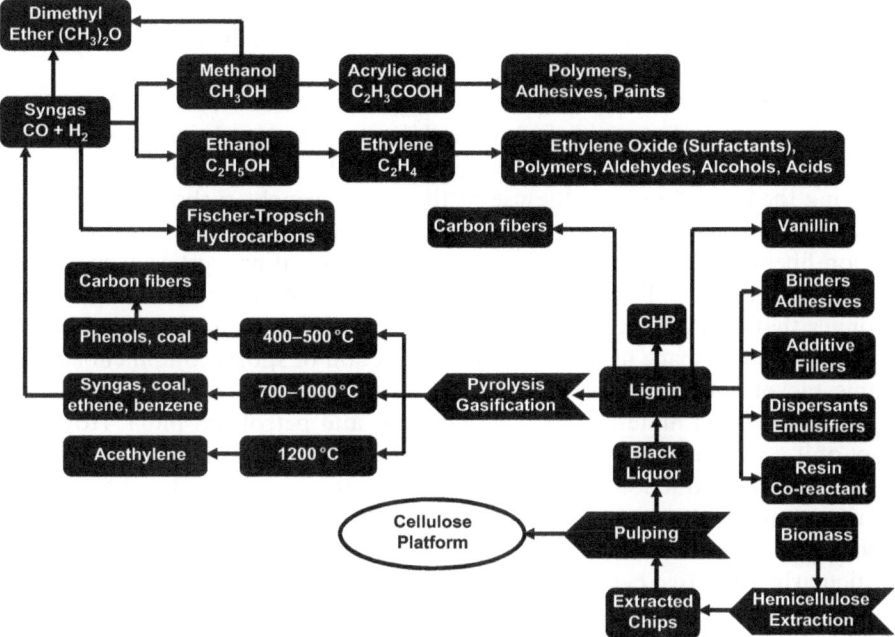

Figure 1.9   Lignin platform of IFBR.

Kraft lignin precipitates upon acidification of the black liquor using carbon dioxide or acid available at the mill. Apart from providing lignin for various applications, its precipitation from the black liquor would alleviate the load on recovery-limited boilers which would in turn increase the pulp production capacity. To date, most of the kraft lignin applications are low-tonnage or pilot scale products. Kraft lignins have been used to produce binders, resins including ion-exchange resins, carriers for fertilizers and pesticides, thermo-plastic polymers, dyes and pigment chemicals, mineral technology, asphalt, lead storage batteries, phenolic resins, activated carbon and carbon fibers.[156] Thermoplastic blends of lignin and lignin-derivatives in synthetic polymers were shown to be subject to property engineering via lignin content and lignin modification.[157] Both unmodified or chemical modified lignin have been used as a filler or in blends with other natural and/or synthetic polymers[158,159] Phase-compatabilizing lignin modifications are revealed for the incorporation of lig-nin into such thermoplastic polymers as polyolefins, polyacrylics, PVC and others.[160] The first prototype printed circuit board for the electronics industry was formulated by IBM with a 50% lignin-containing epoxy resin.[157] As kraft lignins are insoluble in water, advanced modifications are needed to make products such as asphalt emulsifiers.[150] The wet strength of kraftliner has been reported to increase by laccase-aided grafting of lignin model com-pounds.[161,162] An increased wet strength in kraftliner could therefore be facilitated by use of black liquor lignin derivatives with a high free phenolic content. The use of kraft lignin derivatives in this application could become a large scale business as 25–30% of the world paper production is corrugated board material including testliner and kraftliner.[12]

Another potentially large market for lignin is the low cost production of carbon nanofibers for use in the automobile and light truck industry. As currently more than half the vehicle weight is due to its metal construction, by lowering the overall weight, the use of the lighter weight carbon fiber composites could dramatically decrease the vehicle fuel consumption. Furthermore, carbon fibers are known for their superior mechanical properties as measured by their tensile index. For example, the strongest carbon fiber is about five times stronger than steel. Carbon fibers, due to their strength and light weight, are currently used in space technology and production of sports equipment such as tennis rackets. Today, more than 90% of the carbon fibers are made from petroleum-derived materials: polyacrylonitrile and petroleum pitch. However, to permit economic use of carbon fiber composites in commercial production vehicles, fiber production will need to be substantially increased and fiber price decreased from the current $25/kg to $7/kg or less. To achieve this cost objective, high-volume, renewable or recycled raw materials such as lignin are particularly attractive because the cost of these materials is inherently both low and insensitive to changes in petroleum price. Sufficient fiber to provide 10 to 100 kg for each of the 13 million cars and light trucks produced annually in the U.S. will require an increase of 5 to 50-fold in worldwide carbon fiber production. The volume of lignin produced and burned by the domestic pulp and paper industry is about 1 000 times greater than the worldwide carbon fiber production of about 28 000 tons per year.[163] It has been estimated that 20% of the lignin potentially available in the U.S. could produce enough carbon fiber to displace all the steel used in domestic passenger vehicles. Kraft lignin, organosolv lignin and steam exploded lignin have successfully been used to produce carbon fiber.[164,165] The technical challenges for low cost carbon fiber production from lignin, as for other lignin applications, include lignin impurities and lignin molecular weight polydispersity. A large commercial worldwide market for carbon fibers currently exists at $125 million with a projection of $3.6 billion in 2014.

A potential advent for lignin as a low cost additive would be its use in polyurethane formulations to displace petroleum-derived compounds[27] thereby improving the thermal and mechanical properties of kraft lignin urethanes.[166] However, sulfur-free lignins such as Organosolv or soda lignin should be used to avoid any environmental problems caused by the release of sulfur-containing gases. The global market for polyurethanes was estimated at 13.65 million tons in 2010 with a revenue of $33 billion and is expected to reach close to 18 million tons and $55 billion by 2016. Besides its use in polyurethanes and polyesters, technical lignins are also of interest in phenolic and epoxy resins where lignin, due to its phenol-like structure, can replace phonolic compounds in the synthesis of phenol-formaldehyde (PF) resins.[167,168] The PF resins were the first completely synthetic commercial polymer. Kraft lignin can be used to displace up to 70% of the phenol required for PF resins.

The lignosulfonate characteristics are different from those of kraft lignins. Due to the presence of sulfonic, carboxylic and hydroxyl groups, lignosulfonates possess unique colloidal properties which allow their application as dispersing agents for oil well drilling products, detergents, stabilizers, binders, surfactants, adhesives, cement additives, battery expanders, for use in dyestuffs, pesticides particleboards and animal feeds. Their usage as dust binding liquor on gravel roads was one of the first large products based on lignosulphonates. Molding resins take advantage of the their water adsorption, dispersing, adhesion and lubricating properties. Methods of separating and modifying lignosulphonates have been developed and a wide range of applications such as concrete additives, feed binders and surface modification additive in lead acid batteries have been found.[150] Lignosulfonates have been extensively studied for their adhesive properties with poor reproducibility of the bonding effects, due to variable properties of lignin from various sources.[167] The lignosulfonates were shown to reduce the amount of water to produce a more uniform concrete product with improved durability, density and strength. Due to their dispersing properties, they reduce the amount of water needed in the manufacture of bricks and ceramic products. The lignosulfonate market is currently dominated by Borregaard LignoTech with a capacity of 500 000 tons per year. Tembec of Canada sells about 75 000 tons/year on 50% dry weigh basis of liquid and powdered sulfite spent liquor and products.

The sorption properties of lignins open up new avenues for their utilization in medicinal products.[157] It is known that dietary fiber, mainly composed of cellulose and lignin is resistant to hydrolysis by the digestive enzymes of humans and animals. Research indicates that lignin, a major dietary insoluble fiber source, may alter the fate and metabolism of soluble fibers.[169,170] Digestibility of dietary fiber and crude protein in animals was inhibited with cellulose and lignin being the major determinants for changes in digestibility.[171] Lignin was found to have a strong negative influence on fiber digestion and was undigested in both the small and large bowel of humans.[172] Lignosulfonates are used in animal feeds as a pellet binder and to increase viscosity of molasses for easier transportation. Kraft lignins and lignosulfonates possess certain antibiotic activity due to the presence of phenolic and carboxyl groups. Lignin extracts from corn stover residues generated during ethanol production were shown to exhibit antimicrobial activities against pathogenic bacteria and yeast which could be an application in antimicrobial packaging.[173] Lignin acted as an antioxidant against oxidative agents and had a DNA-protective effect in mice cells.[174] Organosolv lignins from hybrid poplar with more phenolic and less aliphatic hydroxyl groups, low molecular weight and narrow polydispersity were reported to have high antioxidant activity.[175] Research is underway to demonstrate the use of lignin nanotubes as carriers of cancer-fighting drugs (http://news.ufl.edu/2012/03/29/nanotech/).

1.2.2.3.2 Application of Lignin Degradation Products. Combustion, gasification, pyrolysis or liquefaction can degrade lignin to a different extent, depending on the severity of the process conditions, from partial depolymerization to low molecular weight lignin fractions (pyrolysis at 400–650 °C in absence of air) to fully oxidized end products of lignin – carbon dioxide and water (combustion at up to 2000 °C in excess of air). These processes convert lignin to power, liquid fuel and syngas products. The choice will depend on the process economics, lignin availability (in the form of waste cooking liquor, forest waste, *etc.*) and the desired end product.[176–180] As lignin has a heating value of nearly 27 MJ/kg or 17 000 Btu/lb and contributes as much as 40% of the energy content of lignocellulosic biomass, lignin combustion from the black liquor is widely practiced today in paper mills to produce process heat, power, steam and to recover pulping chemicals. Black liquor is normally concentrated via multiple effect evaporators to 40–50% solids and then burned for its heating value (12 000 to 13 000 Btu/dry lb). However, the value of lignin realized through heat and power only reflects the price of fossil fuel. It has been estimated that a 30 tons per day lignin-to-fuel plant would require an installed capital cost of $10 million and would have a payback period of 3–4 years. Kraft pulp mills (more than 90% of the world production) have increasingly experiencing bottleneck problems in their recovery boiler as the installed capacity becomes too small after increase of the fiber lines capacity. Only in the U.S., black liquor is produced at about 80 million tons a day as dry solids which represents 40% of the global production rate per day. Debottlenecking of recovery boilers by partial lignin precipitation or gasification of the black liquor are new strategies already contemplated by some pulp mills. Compared to a conventional recovery boiler, a black liquor gasifier can increase the total energy efficiency of a chemical pulp mill and produce a synthesis gas for production of fuels and chemicals.

Gasification of lignin (black liquor) is carried out at 750–900 °C to convert it to a gaseous fuel (syngas) through partial oxidation (1.5–1.8 kg air per kg lignin vs 6–7 kg air/kg lignin in case of combustion). In case of black liquor gasification, concentration of black liquor to 70–80% solids precedes its gasification (Figure 1.10). In addition to syngas, a mixture of carbon monoxide and hydrogen, the gas stream contains water vapors, carbon dioxide, nitrogen, ammonium, hydrogen sulfide, hydrogen chloride, and methane.[181] The presence of contaminants in syngas can cause various problems such as GHG emissions (nitrogen and chlorine compounds), corrosion (sulfur and chlorine compounds, alkali metals), deactivation of catalyst (sulfur and chlorine compounds), water and air pollution and clogging of equipment (tars and particulates). Cyclones, barrier filters, electrostatic precipitators, venture scrubbers and catalytic cracking are used for gas cleanup.[176] Thereafter, syngas can be used for heat and power applications, production of hydrogen employing water-gas-shift technology,[182] methanol, ethanol and Fischer-Tropsch (FT) hydrocarbons (wax, diesel, gasoline and naphta).[183] The power applications include the use of syngas in spark ignition gas engines, in a gas turbine to produce electricity, or in a boiler to produce heat by combustion.[184] Hydrogen can be used for production of chemicals and fertilizers, in fuel cells for electricity generation, for refinery hydrotreating and as a transportation

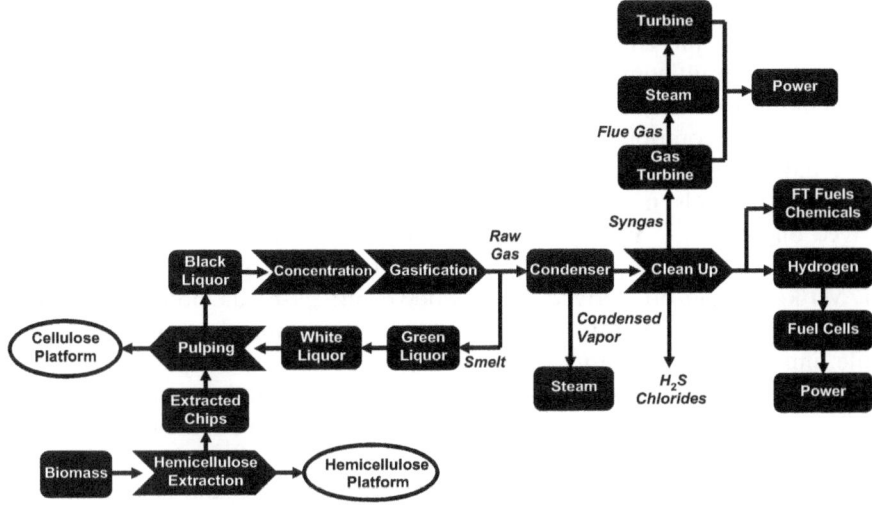

Figure 1.10 Black liquor gasification.

fuel. Methanol is a starting chemical for production of chemicals such as formaldehyde, methyl acetate, acetic acid, ethylene propylene and fuels such as dimethyl ether (DME). Methanol and ethanol can be converted to acrylic acid and ethylene, respectively, and from there – to a variety of synthetic products including polymers, adhesives, surfactants, paints, *etc.* (see Figure 1.9). Technologies to produce DME and FT chemicals are well established. In the FT, the carbon monoxide in the syngas adsorbs on the catalyst surface which induces the hydrocarbon chain reaction.[185] This reaction is a sequence of hydrogenations and carbon monoxide additions occurring repeatedly until the chain growth ends by desorbing the hydrocarbon product from the catalyst surface. The product composition is influenced by the nature and reactivity of the catalyst, the hydrogen to carbon monoxide ratio and operating conditions (temperature of 260 °C and pressure of 400 to 2 600 psi). The technical needs for FT synthesis include economical purification of syngas streams and catalyst and process improvements to reduce unwanted products such as methane and higher molecular weight products such as waxes.[186,187] Catalyst and process improvements are still needed to scale up the thermochemical conversion of syngas to ethanol and other alcohols. It has been estimated that syngas production accounts for 50% to 75% of the total cost of the end product such as hydrogen, alcohols, DME *etc.* To reduce costs, the biomass-to-fuels process should be optimized in order to obtain the highest yield, least cost configuration.[188] The gasification economics of different lignin sources could differ, however, it has been assumed that one ton of lignin can yield 55 gallons of ethanol and 19 gallons of 3 to 5 carbon mixed alcohols.[155] It is expected that costs will decrease as gasification technology matures and with increasing scale of production.[189]

Pyrolysis of lignocellulosic biomass produces three fractions: liquid (bio-oil), gaseous (hydrogen, methane, carbon dioxide, carbon monoxide, *etc*) and solid (bio-char). Typically, fast pyrolysis (650 °C for less than 5 s) results in 60–70% bio-oil, 15–25% bio-char and 10–20% gases.

Bio-oil are multicomponent mixtures derived from depolymerization and fragmentation reactions that, depending on the biomass source and pyrolysis method, can be composed of more than 300 organic compounds that belong to acids, aldehydes, ketones, alcohols, esters, anhydrosugars, furans, phenols, guaiacols, syringols, nitrogen compounds, hemicellulose-, cellulose- and lignin-derived oligomers.[190] Pyrolysis has the advantages of: 1) reduced size of biomass (for easier transportation) with increased energy density of bio-oil (21 MJ/kg of bio-oil derived from wood) and gases as potential biofuels; 2) lower process temperature in comparison to combustion and gasification while limiting gas pollutants such as dioxins; 3) process simplicity despite its chemistry complexity; 4) economically feasible technology for small-scale application (50–100 tons/day) of portable units distributed close to the biomass source with potential for job creation in rural areas. Bio-oil can be used in three ways: 1) directly for combustion in boilers, diesel engines and gas turbines for CHP generation; 2) upgraded via gasification and hydroprocessing to FT chemicals, fuels and power; or 3) as a source of valuable chemical compounds (Figure 1.11).

Rapid heating of fast pyrolysis reactor and rapid cooling of pyrolysis vapors are two important parameters that can maximize the bio-oil yield. Bio-char can be utilized as solid fuel (18 MJ/kg), for production of activated carbon and other chemicals, for carbon sequestration, bioremediation and for improving soil functions such as soil erosion, water and nutrients retention, *etc.*[191,192]

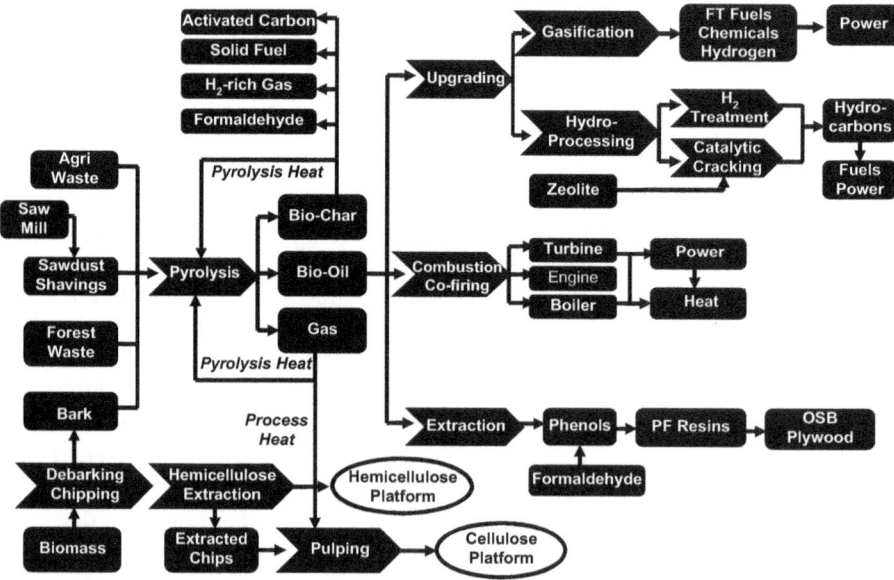

Figure 1.11   Lignin and biomass pyrolysis.

The gaseous fraction containing syngas and other low molecular hydrocarbons can be used to provide heat to the pyrolysis unit or other process streams in the IFBR. Biochar production is maximized by flash pyrolysis that involves heating of biomass under moderate to high pressure in a retort with yields of 60% biochar and 40% volatiles (bio-oil and syngas) whereas slow pyrolysis typically yields up to 35% biochar.[193] A number of pyrolysis reactors have been developed for the bio-oil production, including bubbling fluidized bed, entrained bed, circulating fluidized bed, rotating cone, screw pyrolysis reactor, vacuum pyrolysis reactor, *etc.* BTG (www.btgworld.com) and Dynamotive (www.dynamotive.com) have already established demonstration plants for biomass fast pyrolysis, suggesting that the fast pyrolysis is near commercial. However, there are still some technical hurdles that need to be overcome before large-scale implementation of pyrolysis. Challenges with the use of bio-oil are related to its high acidity that can cause corrosion problems, high viscosity that makes it difficult to transport in pipes; high inorganic and suspended char content that can cause erosion and equipment plugging; high water content that leads to low homogeneity; high oxygen content that can reduce stability and heat value; and thermal instability leading to decomposition into less useful products.[194] Pyrolysis research including bio-oil production and characterization, reactor design for pyrolysis, pyrolysis parameters, reactions and kinetic mechanisms, *etc.* have been recently reviewed in several publications.[193,195–199] Current technology developments focus on stabilization of pyrolysis oil, catalyst improvements and bio-oil compatibility with existing petroleum infrastructure and technology.

Fast pyrolysis of lignin[200] begins with its thermal softening at around 200 °C and yields bio-oil liquid products that contain a high molecular weight fraction (pyrolytic lignin), monomeric phenolic compounds and low degradation compounds[201] such as methanol, hydroxyacetaldehyde and acetic acid. The pyrolytic lignins account for 13.5–27.7 wt% of crude bio-oils[202] with average molecular weight of 650–1300 g/mol[203] and typically characterized by dimeric and monomeric biphenyl, phenyl coumaran, diphenyl ethers, stilbene and resinol structures.[204] Therefore, pyrolysis may be useful as a technology for the controlled molecular weight reduction of lignin that can offer some unique possibilities for conversion to useful aromatics. The lignin-derived phenolic compounds can be recovered from bio-oils[205] and used to displace phenol in production of PF resins.[206] Work at the National Renewable Energy Laboratory (NREL) has focused on a solvent-aided recovery of phenolic stream from bio-oil.[207] To increase the formation of monomeric phenolic compounds in fast pyrolysis, biomass can be impregnated with alkali[208,209] or use catalytic cracking of pyrolytic lignin.[210] The major obstacle in the extraction of valuable products from bio-oil is their low concentrations that renders current recovery technically difficult and economically unattractive.

The first industrial use of lignin was in production of vanillin which today is based on nitrobenzene oxidation of lignosulfonates under alkaline conditions.[211] Vanillin is a phenolics aldehyde used in the food industry as a flavoring agent.[212] The ice cream and chocolate industries together comprise 75% of the market for vanillin with smaller amounts being used in confections and baked goods. The global market for vanillin is estimated to be between 15–16 000 tons worth of $180 million per year. Vanillin and related phenols can also be produced by microbial degradation of lignin.[213] White-rot fungi are believed to degrade wood lignin by excreting extracellular oxidative enzymes – laccases and peroxidases.[214,215] Due to size limitation and redox potential incompatibility of these enzymes, lignin oxidation is mediated by small molecular compounds, mediators, that are capable of penetrating the cell walls to access lignin.[216] Recent research suggests that bacteria, in particular soil- and, often aromatic-degrading bacterial species, are also capable of breaking down lignin.[217] *Sphingobium sp.* SYK-6[218] can metabolize lignin ß-aryl ether to vanillin Degradation of Kraft lignin by *Bacillus* sp. and *Aneurinibacillus aneurinilyticus* produced ferulic acid – a compound used as an anti-oxidant food additive.[219] *Thermotoga fusca* and *Streptomyces viridosporus* metabolized lignin to a water-soluble intermediate polyphenolic polymeric lignin (APPL) that can be precipitated by acidification[220,221] Phenolics have been extensively studied for their antimicrobial, anti-oxidative, and anticorrosive properties.[222] Investigations revealed that phenolic fractions from kraft and sulphite spent liquors can provide a source of antioxidants with very strong radical scavenging activity.[223] Phenols have some pharmacological properties that could suppress the activation and expression of the HIV-1 gene.[224]

1.2.3 Extractives Platform

1.2.3.1 Composition, Classification and Properties of Extractives

Wood extractives, also called extraneous compounds or secondary metabolites, are low molecular weight, nonpolymeric chemical compounds that are not considered essential to the cell wall structure and can be extracted from wood with various neutral solvents such as organic solvents or water. The wood extractives vary significantly within the same tree family, genera, species, from heartwood to sapwood and from season to season. Their content is normally less than 5% and may be concentrated in higher quantities in bark, tree branches, roots and wounded wood. They impact color, odor, taste, decay resistance to wood and may be toxic to fungi, bacteria and termites. For example, heartwood extractives retard wood decay, rosin is exuded by conifers to protect wounded tissues and as protection from insects, bark compounds minimize animal invasions, *etc.* Some of these phytochemicals may be also toxic to the producing plant[225] and may cause some undesirable effects such as brightness reversion in pulps, pitch problems in papermaking, inhibition of concrete and glue setting, *etc.* It should be noted that the extract composition is a function of the solvent used. Furthermore, the volatile components in the wood resin cannot be measured as they will evaporate during the solvent recovery and extract dessication. No single solvent can extract all extractives present in wood. For the pulp and paper industry, the most important class of extractives is the wood resins - lipophilic compounds that are soluble in solvents such as hexane, acetone, ethanol, benzene, dichloromethane and diethylether. Due to environmental and health concerns for the use of aromatic and chlorinated solvents, determination of extractives in wood and pulp is currently standardized using acetone as a solvent. Acetone extracts more wood substance as dry weight compared to dichloromethane which was previously used as a standard solvent in the pulp and paper industry. More recently, supercritical fluid extraction with carbon dioxide was used.[226,227] The advantages of using this method include reduced use of organic solvents, higher extraction yield in comparison to acetone and dichloromethane, and shorter analysis time. Wood resins may contain fats and fatty acids, steryl esters and sterols, terpenoids and waxes. Chemically they are grouped into: terpenes, phenolic and aliphatic compounds.[228]

Terpenes are cyclyc compounds composed of isoprene units (C_5) such as monoterpenes (2 isoprene units, C_{10}), sesquiterpenes (3 isoprene units, C_{15}), diterpenes (4 isoprene units, C_{20}), triterpenes (six isoprene units, C_{30}) and tetra-terpenes (eight isoprene units, C_{40}). Terpenes are contained in high quantities in the resin ducts of softwoods such as pine. Turpentine consists of volitile oils such as α- and ß-pinene (monoterpenes) used in household pine oil cleaners with pleasant aroma. The softwood monoterpenes and sesquiterpenes provide the typical pine forest odor and are collected in the turpentine fraction from the digester. Resin acids such as abietic and pimaric acid are oxygenated diterpenes that are used in rozin sizing to control the water absorption in paper products. Resin acids can be allergenic and toxic. Together with the fatty acids, they are separated from the black liquor as tall oil soap. Oxygenated terpenes such as resin acids are also called terpenoids. Some terpenoids may be allergenic and should be washed from pulps intended for use in hygienic products. Of commercial significance is the triterpenoid betulin, $C_{30}H_{50}O_2$, contained in the bark, pulp and pitch deposits of European birch (*Betula alba*). Although betulin is strictly a bark component (25–35%) that gives the typical white look of birch trees, it is introduced in the pulp and paper mill through the residual bark of incompletely debarked birchwood chips. Betulinic acid has been explored as a potential treatment for skin cancer for more than five years. Pure betulin, its derivatives and other extractives from birch bark are also tested for their effectiveness in treating HIV and respiratory syncytial virus (RSV), viruses that can cause severe cold-like symptoms and pneumonia. Betulin and some of its derivatives have shown strong antifeedant properties towards many important pest species. Phytol, acyclic diterpene (C_{20}) is found in leaves of woods and plants and comprises more than 30% of the chlorophyll molecule. Extracts from stevoside, a terpene glycoside, have been used as sweeteners[229] as stevoside is more than 300 times sweeter than sucrose.[230] The red and yellow plant pigments contain carotenoids – tetraterpenes (C_{40}) used in food coloring, cosmetics (lotions, powders, lipsticks) and vitamin synthesis.[231] Rubber, a polymeric isoprene containing up to 6 000 isoprene units, is used for its elastic properties.[232] The global natural rubber production is likely to rise 7.8% to 11.8 million tons in 2012 against 10.9 million tons in 2011, while consumption may touch 11.7 million tons in 2012. Sterols contain a tetracyclic ring and relate to triterpenes with sitosterol being the main wood sterol commonly occurring in extractives from conifers including tall oil. Sterol glycosides can be used for heart treatment as they have a strong effect on the heart muscle. Sitosterol is structurally close to cholesterol – the main sterol in humans. Recent dietary recommendations emphasize the possibility of lowering LDL cholesterol levels through consumption of products enriched with plant sterols and stanols as these are not synthesized in humans.[233]

Phenolic compounds contain one or more aromatic rings with a varying amount of hydroxyl groups. They are mainly found in the bark of many wood sciences and are common in heartwood. For example, suberin is a typical component of bark and can be extracted from the cork tissue of birch bark. The suberin structure is not completely determined but resembles that of lignin. The potential applications of suberinic acid are for production of coatings, surfactants and lubricants. Phenolic compounds are removed from wood during pulping and are present in the spent liquors. Some phenolics like dihydroquercetin can interfere with sulfite pulping. Resistance to decay in Scots pine is due to pinosylvins[234] – a dimeric phenol belonging to the class of stilbenes. Stilbenes are commonly found in the heartwood of pine species and can be hydroxylated, methylated or glycosylated.

Flavonoids, tannins and lignans are common classes of phenolics compounds. Over 4000 different flavonoids have been isolated. Their main function is to provide resistance of trees to insects and microbial degradation.[235] Polyflavonoids are used to convert animal hides to leather. Some flavonoids such as quercetin, chatechin and xanthohumol are potent antioxidant with potential health benefits including reduced risk of cancer, heart disease, asthma, and stroke. Flavonoids act as antioxidants by neutralizing oxidizing free radicals, including the superoxide and hydroxyl radicals, which are formed in the human body by the reduction of oxygen and may cause cancer and coronary heart disease and accelerate human aging.[236,237] The flavonoid compounds catechin, epicatechin, and quercetin were identified and quantified in spruce, pine, and fir species with white spruce bark containing the most abundant source of catechin (3600 ± 100 ppm). Naringenin is a flavonoid that was isolated from the bark of *Choerospondias axillarisis* and is considered to have a bioactive effect on human health as antioxidant, free radical scavenger, anti-inflammatory, carbohydrate metabolism promoter and immune system modulator. Certain flavonoids have antihistamine, antimicrobial, memory- and even mood-enhancing properties. Procyanidin, extracted from *Pinus radiata* bark, is a flavonoid-based antioxidant that has been approved for human nutrition.[238] Another flavonoid, anthocyanin, a water-soluble vacuolar pigment and a powerful antioxidant that occurs in all tissues of higher plants, can protect eyes from cataracts.

Tannins have molecular weights ranging from 500 to over 3000 (gallic acid esters) and up to 20 000 (proanthocyanidins) and are considered to have anti-feedant properties. They can bind to and precipitate proteins, amino acids and alkaloids. Wood tannins from oak are used in tanning animal hides into leather. Tannins have been separated in two classes: hydrolyzable and condensed tannins. The hydrolyzable tannins are further divided into gallotannins and ellagitannins, with gallic acid and ellagic acid, respectively, as essential components. The hydrolyzable tannins are mixtures of simple phenols and glucose esters of gallic and digallic acids.[239] They have been used to displace up to 50% of phenol in the manufacture of PF resins.[240] Ellagic acid has antitumor activity and is used as a sedative and tranquilizer.[241] Bark contains 10–12% tannins that are used as pharmaceuticals (anti-diarrheal agents), corrosion inhibitors and adhesives. More than 90% of the commercial production of tannins is for condensed tannins that are used in a number of applications including PF resins, tyre cord adhesives, foundry core binders and wood preservatives.[242]

Lignans are one of the major classes of phytoestrogens, derived from phenylalanine via dimerization of substituted cinnamic alcohols (monolignols) to a dibenzylbutane structure. Lignans act as antioxidants and display a range of biological activities including enzyme inhibition, fish toxicisty, growth inhibition, insect antifeedant properties.[243] Lignans are capable of binding to estrogen receptors and interfering with the cancer-promoting effects of estrogen on breast tissue. For example, podophyllotoxin is currently studied for its possible effects on breast, prostate and colon cancer. Hydroxymatairesinol was detected as the major lignan constituent in knots of Norwegian spruce with approximate content of 5.5 wt%.[244] Hydroxymatairesinol has anti-cancer properties used in treatments of breast, colon and prostate cancer, cardiovascular diseases and as a dietary supplement. Recent research has uncovered the naturally occurring existence of hydroxymatairesinol as the dominant lignan in wheat, triticale, barley, corn, amaranth, millet and oat bran.

Aliphatic compounds of wood extractives include fatty acids, fatty alcohols and hydrocarbons (alkanes). The fatty acids are mainly present as esters with glycerol. Triglycerides are saponified during kraft pulping to produce soaps such as sodium soaps (liquid) and potassium soaps (solid). The dominating fatty acids are the unsaturated C_{18} fatty acid – oleic, linoleic and linolenic acids that constitute 75–85% of the total amount of fatty acids. Alkanes can also accumulate in wood tissues – *n*-heptane is a component of turpentine from *Pinus sabiniana*.[245] The content of nitrogen compounds in wood is less than 0.1% and is due to presence of amino acids and proteins involved in cell wall biosynthesis[246] and alkaloids.

Alkaloids are a group of naturally occurring chemical compounds that contain basic nitrogen atoms and are found in higher concentrations in bark, roots and leaves. Alkaloids can be purified from crude extracts by acid-base extraction and have pharmacological effects, can act as poisons and are used as medications, as recreational drugs, or in entheogenic rituals. Cocaine, caffeine and nicotine are classified as alkaloids.[247] Reserpine is an alkaloid first isolated from *Rauwolfia* species of South American evergreen trees and shrubs. It is used as an antihypertensive drug and to treat disorders including schizophrenia.[248] Another alkaloid isolated from the bark of the South American plant *Chondrodendron tomentosum*, tubocurarine, is used adjunctively in anesthesia to provide skeletal muscle relaxation during surgery or mechanical ventilation. Quinine, an antimalarial alkaloid, has been isolated from the tropical rain forest – the most dense and biogenetically diverse forest areas of the world that is largely unexplored and is currently an untapped source of valuable, biologically active natural products including pharmaceuticals and neutraceuticals. Some of these products still have superior attributes over synthetically derived drugs and/or are more economically extracted from its natural sources.[249] Taxol, originally isolated from the bark of Pacific yew, *Taxus brevifolia*,[250] is the most effective antitumor agent developed in the past three decades. It has been used for effective treatment of a variety of cancers including refractory ovarian cancer, breast cancer, non-small cell lung cancer, AIDS related Kaposi's sarcoma, head and neck carcinoma and other cancer types.[251] In attempts to make taxol less expensive and more widely available via industrial fermentation, recent research has focused on taxol-producing endophytic fungi.[252]

With respect to their influence on the pulping process, extractives are classified in two groups: saponifiables such as fatty and resin acids, glycerides and some steryl esters that form soluble soaps with alkali; and unsaponifiables (also called neutrals) such as waxes, fatty alcohols, sterols, terpene alcohols, *etc.* that do not form soaps and can cause pitch problems. Fatty acids (30–60%), resin acids (40–60%) and unsaponifiables (5–10%) constitute the tall oil fraction of black liquor (Figure 1.12) which is formed during kraft pulping by saponification of softwood extractives-based fats and waxes to sodium salts of fatty and resin acids.[253] They contain a polar hydrophilic (carboxylic) end and a nonpolar, hydrophobic (hydrocarbon) end and associate to form a micellar colloid called micelle. The electric charge of the polar carboxyl groups stabilizes the micelle and prevents their agglomeration. The neutrals (unsaponifiables) such as sterols and waxes can be solubilised within the hydrocarbon nonpolar interior of the micelle. However, as the cationic strength (sodium cations) increase, the negative charge on the outer surface of the micelle decrease and is eventually neutralized which causes aggregation and precipitation of the micelles as raw rosin soap from the black liquor.[254] The raw rosin soap is

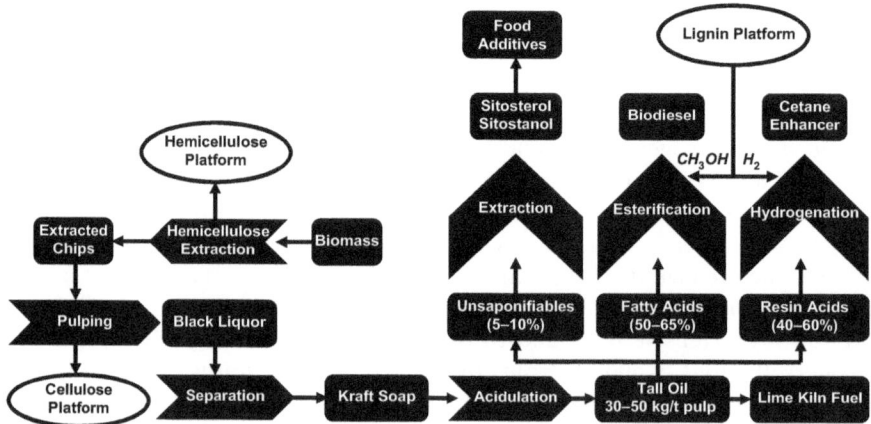

Figure 1.12   Extractives platform: Tall oil recovery and utilization possibilities.

removed from the black liquor during its evaporation (from 15% solids to about 25-30%) by skimming the surface with an average yield of 30–50 kg/ ton pulp produced from highly resinous softwood specie like southern pines. Thereafter the soap is allowed to settle to release any retrained black liquor and acidified with sulphuric acid to produce the crude tall oil. The crude tall oil is normally fractionated by vacuum distillation to several commercial fractions and sold for soaps, rosin size, adhesives, rubbers, inks and other products. Globally, about 2 million tons/year of tall oil are refined whereas in the U.S. alone more than 700 million liters of tall oil and turpentine could be produced. The unsaponifiable fraction contains mainly sterols such as sitosterol and sitostanol which, upon further isolation and purification, can be used as cholesterol-lowering food additives.[233] The resin acids can be utilized together with the fatty acids in biodiesel production, or used for production of cetane enhancers by catalytic hydrogenation and cracking (Figure 1.12).[255] Cetane enhancers are compounds added to diesel fuel oil to raise the fuel's measured cetane level. The addition of cetane enhancers to diesel fuel is one recognized retrofit technology used to reduce diesel engine emissions. In a IFBR scenario, the hydrogen and methanol required in the production of cetane enhancers and biodiesel, respectively, could be generated and supplied from the lignin platform via the syngas route as discussed earlier.

1.2.3.2 Fate of Extractives during Pulping and Bleaching

Most of the wood extractives are removed during the kraft cooking process. However, some sterols and alcohols, formed from steryl esters and wastes, have low solubility in the cooking liquor and may be carried out entrapped in the micelles with the brownstock pulp. Sulfite pulps normally contain more extractives than kraft pulps as fatty and resin acids are insoluble in the acid cooking. These extractives can agglomerate to form particles, often referred to as pitch, which may deposit on the equipment causing operating difficulties, or may appear in the finished pulp and paper as contaminating particles. Pitch deposits may cause spots, specks, streaks, breakage or holes in the paper that altogether result in significant downtime for cleaning in the papermaking process. Extractives can lower the surface energy of paper leading to decreased wettability with a negative impact on products such as tissue paper. Extractives can also decrease the bonding of toner particles in printing papers, or decrease the bonding between individual fibers in the paper thereby reducing the paper strength.[228] Oxidation of wood extractives, especially unsaturated fatty acids and their esters, can result in formation of volatile, off-flavor compounds that, if used as food wrapping paper, can affect the food odor or taste.[256] Extractives can react with the bleaching chemicals to form oxidized and modified extractives that can impact brightness stability and cause brightness reversion.[257] For example, oxidized lignans have colored structures, whereas chlorinated extractives liberate hydrochloric acid that can cause formation of colored compounds from polysaccharides.[258,259] To avoid this problem, removal of extractives prior to bleaching becomes necessary. An alkali pretreatment on pine bisulfite pulp prior to hydrogen peroxide bleaching was shown to remove 86% of extractives.[260] Hot water treatment[261] as well as pressing and washing[262] have been proposed as effective methods for dissolution of the lipophilic extractives. Anionic surfactants (naphthalene sulphonate or lignosulphonate) and non-anionic surfactants (fatty alcohol ethoxylate or alkyl phenol ethoxylate) have been reported to reduce 30 % of the dichloromethane extractives in peroxide bleaching of CTMP aspen pulp.[263] Pitch in mechanical pulps acts as polyanions to cause anionic trash. Traditional methods to control pitch problems include wood seasoning before pulping or adsorption and dispersion of pitch particles with chemicals such as talc during the pulping and papermaking process. Methods for pitch control using resin-degrading fungi and lipase enzymes have also been developed.[264,265]

1.3   Concluding Remarks

The main technological challenges that need to be overcome to maximize the fiber value in a future IFBR are related to process improvements in extraction of hemicellulose for minimal sugar degradation and preservation of pulp and paper properties; fermentation of mixed sugars and tolerance of microbial producers to inhibitors and ethanol; pyrolysis and gasification efficiency. Process integration in a IFBR would reduce the number of process steps and reuse the process streams in a waste-free and closed cycle operation mode thereby reducing the overall energy demands. Waste heat could be utilized to integrate other manufacturing opportunities whereas additional energy requirements could be met by combustion (or another thermal process such as gasification or pyrolysis) of waste biomass. Process integration tools can be employed to identify products that can be economically produced by a pulp and paper mill. It is anticipated that, in addition to pulp and paper products, the future IFBR will extract/generate significant amounts of hemicellulose and lignin, respectively, and create new market opportunities for production of chemicals, polymers, new wood composites, liquid fuels, ethanol, *etc*. Hemicellulose pre-extraction from wood chips prior to pulping, or lignin precipitation from black liquor provide opportunities for higher paper production by 20% (hemicellulose extraction), or 15% (lignin precipitation). These new biorefinery technologies could be used as a recovery debottlenecking tool at existing mills.[266] Additional benefits from hemicellulose extraction include reduced alkali consumption and black liquor load at increased delignification rate. Debottlenecking through black liquor gasification is another technological opportunity that is expected to reach commercial readiness by 2015 and is currently actively being pursued by the industry seeking federal and state grants, loans and investors funding. Black liquor gasification would allow replacing aged, low-efficient and high-maintenance recovery boilers, that are currently still in use at pulp mills, offering the potential for overall cost reductions, more efficient energy recovery, emissions reduction, improved safety, separation of pulping chemicals to maximize pulp yield through modified pulping technologies at internal rate of return on incremental capital investment of 14–18% assuming $50/bbl of oil.[267,268] To supplement the bioenergy and bioproducts derived from hemicellulose and lignin, the use of bark and foliage extracts from wood would enable the IFBR to enter the markets for functional food additives, neutraceuticals and pharmaceuticals that have shown a steady 15% growth per year over the past decade. Extractives usually comprise a minor proportion of wood biomass, however, for a large scale IFBR operation, they can be a potential source of high-value co-pro-ducts.[269] A simplified flow diagram of the IFBR with the four production platforms is displayed in Figure 1.13.

The key to a successful, convergent value chain is the identification of potential mill-specific products and processes that can be implemented and integrated at the IFBR. The following factors will have to be considered when identifying the IFBR technology pathway and specific product mix: 1) mill-specific factors – location, existing technology, wood species, production levels; 2) market analysis; 3) product design analysis – price competitiveness, supply chain analysis, life-cycle assessment; 4) process design analysis of emerging, cost-effective technologies; 5) process flexibility; 6) investment risk; 7) speed of new product to market. It should be noted that the flexibility in product diversification and risk minimization would decrease with increase in the IFBR scale of production. The distinction between main products and by-products will have a major influence on the functioning and organization of the forest products markets.[270] Depending on the reliance of forest resources for product and revenue diversification and on the demand for new forest products on the market, the competition for and prices of forest-based biomass and materials may increase. Following the development of a technology strategy, pulp and paper mills need to formulate their business strategy: 1) formulate their competitive and marketing strategy and long-term objectives of a phased approach

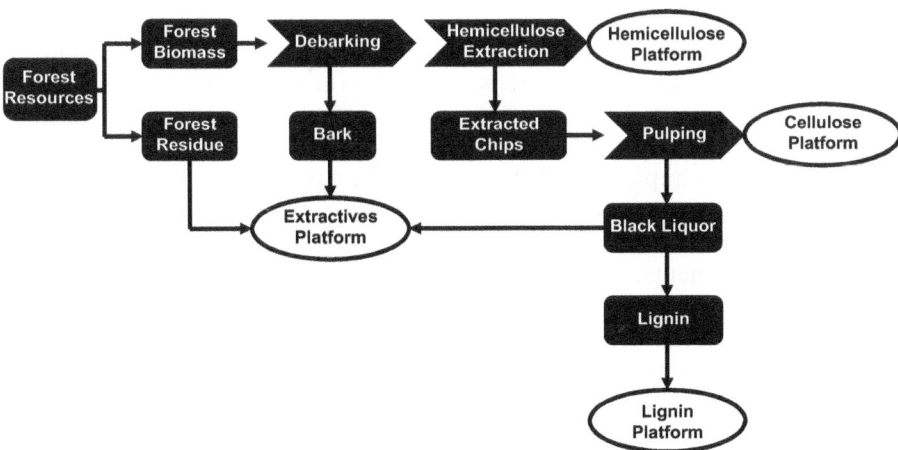

Figure 1.13   IFBR production platforms.

for incremental implementation of IFBR; 2) identify and secure supply chains and marketplace; 3) form strategic alliances with companies in other sectors with established marketplace; 4) invest in innovation and R&D to verify and optimize the products and technology choice and explore alternative platforms and concepts. The techno-economic and commercial risks of the technology and business strategy of the IFBR must be identified and mitigated.[271] To accomplish the IFBR implementation plan, government support will be of vital importance. The IFBR economic and operational feasibility may be evaluated using a demonstration or pilot plant at an existing pulp mill, or a small kraft mill that has been closed. The IFBR options could add another 30–50% of profitable revenue with 25–40% return on investment for the pulp and paper sector.

Biotechnology will continue to play a major role in the future R&D to provide cost efficient and environmentally friendly solutions to increased forest productivity and product diversification. The forest biotechnology research is expected to revolutionize advancements in tree genetic engineering for improved resistance to insects, pathogens and environmental stressors, leading to development of fast-growing biomass plantations designed to produce economic, high-quality material for building, construction, pulp, paper, bioenergy and bioproducts. For instance, Weyerhaeuser envisions a yearly production of over 100 million seedlings of Douglas-fir and loblolly pine developed through the company's biotechnology program.[272] A number of microbial and enzymatic processes have been already developed and some of them commercialized in the pulp and paper industry. Enzymatic and microbial processes include biopulping, biobleaching, biorefining, bioremediation, enzymatic grafting of paper, enzymatic delignification, pitch control, enzymatic control of fines and drainage properties, enzymatic control of tissue softness, enzymatic deinking, enzymatic depithing, *etc.*[273–276]

Additional jobs, tax benefits and air emissions reductions may be generated to address the societal need for utilizing renewable resources rather than fossil fuels in the production of commodity products, liquid fuels and electricity. The socio-economic challenges need to address: 1) the complex systems of policies and regulations in different countries and make them more compatible; 2) the environmental impact of biomass removal; 3) the pressure from environmental groups on policy makers; and 4) the unstable commodity and fuel prices. Barriers, that can impact on IFBR development projects, include regulatory uncertainties, opposition from local communities due to environmental or other local concerns, difficulties securing permits for new technologies or power purchasing agreements, negative effect from new company failures, *etc.* Policy integration and policy interventions should be carefully considered prior to introduction to avoid undesirable effects on markets and prices of raw materials and end products. The successful implementation of the IFBR will likely require the establishment of strategic collaborations and partnerships with government and industry experts. Different technological, value-chain, commercial and financial partnerships and strategic alliances, depending on the business model and strategy of the IFBR, will have to be created in order to enable risk mitigation and value creation for a successful transformation into a IFBR.[277] New ways for integration and cooperation with other industries such as the forest, agricultural and chemical sectors must be identified. The IFBRs are a pivotal milestone in our efforts to implement sustainable, low footprint, environmentally beneficial technologies that meet the global demands for energy and bioproducts through the intelligent use of our renewable resources toward a bio-based economy.

Acknowledgements

Financial support by the Center for Bioprocessing Research and Development (CBRD) at the South Dakota School of Mines & Technology (SDSM&T), the South Dakota Board of Reagents (SD BOR), the South Dakota Governor's Office for Economic Development (SD GOED), and the U.S. Air Force Research Laboratory (AFRL) is gratefully acknowledged. I am thankful to D. Christopher and K. M. Christopher for the excellent editorial assistance with the manuscript.

References

1. J. L. Smith, World oil: Market or mayhem? *J. Econ. Persp.*, 2009, **23**, 145–164.
2. R. D. Perlack, L. L. Wright, A. F. Turhollow, R. L. Graham, B. J. Stokes and D. C. Erbach, Biomass as feedstock for a bioenergy and bioproducts industry: The technical feasibility of a billion-ton annual supply, *U.S. Department of Energy*, 2005. Available at: http://www.osti.gov/bridge.
3. U.S. DOE, *Breaking the Biological Barriers to Cellulosic Ethanol: A Joint Research Agenda*, U.S. Department of Energy, DOE/SC-0095, U.S. Department of Energy Office of Science and Office of Energy Efficiency and Renewable Energy, USA, 2006.
4. K. Sheram, *The Environmental Data Book*. The World Bank, Washington DC, USA, 1993.
5. J. Jaffe, M. Ranson and R. N. Stavins, Linking tradable permit systems: A key element of emerging international climate policy architecture, *Ecol. Law Quarterly*, 2009, **36**, 789–808.
6. B. A. Babcock, The impact of US biofuel policies on agricultural price levels and volatility, *Center for Agricultural and Rural Development*, Iowa State University, Issue Paper No. 35, June 2011.
7. OECD, *Economic Assessment of Biofuel Support Policies*, OECD Directorate for Trade and Agriculture, Press Conference, Paris, 16 July 2008. http://www.oecd.org/dataoecd/54/15/40715381.pdf.
8. A. E. Farrell, R. J. Plevin, B. T. Turner, A. D. Jones, M. O'Hare and D. M. Kammen, Ethanol can contribute to energy and environmental goals, *Science*, 2006, **311**, 506–508.

9. J. Fargione, J. Hill, D. Tilman, S. Polasky and P. Hawthorne, Land clearing and the biofuel carbon debt, *Science*, 2008, **319**, 1235–1238.
10. Y. Zheng, Z. Pan and R. Zhang, Overview of biomass pretreatment for cellulosic ethanol production, *Int. J. Agric. Biol. Eng.*, 2009, **2**, 51–58.
11. Biobased industrial products: Research and commercialization priorities. *National Academy Press*, Washington, DC, 2000. Available at: http://www.nap.edu/books/0309053927/html4.
12. FAO, *Pulp and Paper Capacities: Survey 2007–2012*. Food and Agriculture Organisation of the United Nations. Rome, Italy, 2008.
13. D. Raymond and G. Closset, Forest products biorefinery: Technology for a new future, *Solutions*, 2004, September, 49–53.
14. L. Kang, W. Wang and Y.Y. Lee, Bioconversion of paper mill sludges to ethanol by SSF and SSCF, *Appl. Biochem. Biotechnol.*, 2010, **161**, 53–66.
15. Y. Lin and S. Tanaka, Ethanol fermentation from biomass resources: current state and prospects, *Appl. Microbiol. Biotechnol.*, 2006, **69**, 627–642.
16. J. R. Mielenz, Ethanol production from biomass: technology and commercialization status, *Curr. Opin. Microbiol.*, 2001, **4**, 324–329.
17. D. G. Olson, J. E. McBride, A. J. Shaw and L. R. Lynd, Recent progress in consolidated bioprocessing, *Curr. Opin. Biotechnol.*, 2011, **23**, 1–10.
18. E. Sjostrom, *Wood Chemistry. Fundamentals and Applications.* Academic Press, San Diego, USA, 1993.
19. B. Saha, Hemicellulose bioconversion, *J.Ind. Microbiol. Biotechnol.*, 2003, **30**, 279–291.
20. S. Kato, Ultrastructure of the plant cell wall: biochemical view point, *Encyclop. Plant Physiol.*, 1981, **30**, 409–425.
21. D. Fengel and G. Wegener, *Wood-Chemistry, Ultrastructure, Reactions.* Walter de Gruyter, Berlin, Germany, 1984.
22. R. L. Casebier and J. K. Hamilton, Alkaline degradation of glucomannans and galactoglucomannans, *Tappi J.*, 1965, **48**, 664–669.
23. C. A. S. Gustavsson and W. W. Al-Dajani, The influence of cooking conditions on the degradation of hexenuronic acid, xylan, glucomannan, and cellulose during kraft pulping of softwood, *Nord. Pulp. Pap. Res.*, 2000, **15**, 160–167.
24. A. J. Kerr and D. A. I. Goring, The role of hemicellulose in the delignification of wood, *Can. J. Chem.*, 1975, **53**, 952–959.
25. T. E. Amidon, C. Wood, A. Shupe, Y. Wang, M. Graves and S. Liu, Biorefinery: Conversion of woody biomass to chemicals, energy and materials, *J. Biobased Mater. Bioenergy*, 2008, **2**, 100–120.
26. W. J. Frederick, S. J. Lien, C. E. Courchene, N. A. DeMartini, A. J. Ragauskas and K. Iisa, Co-production of ethanol and cellulose fibers from southern pine: A technical and economic assessment, *Biomass Bioenergy*, 2008, **32**, 1293–1302.
27. A. van Heiningen, Converting a kraft pulp mill into an integrated forest biorefinery, *Pulp Pap. Can.*, 2006, **107**, 38–43.

28. H. Resalati, H. Kermanin, F. Fadavi and F. Feizmand, Effect of extraction on Soda-AQ pulping, *Lignocellulose*, 2012, **1**, 71–80.

29. H. Mao, J. Genco, A. van Heiningen and H. Pendse, Technical economic evaluation of a hardwood biorefinery using the near-neutral hemicellulose preextraction process, *J. Biobased Mater. Bioenergy*, 2008, **2**, 1–9.

30. T. E. Amidon, B. Bujanovic, S. Liu and J. R. Howard, Commercializing biorefinery technology: A case for the multi-product pathway to a viable biorefinery, *Forests*, 2011, **2**, 929–947.

31. T. E. Amidon, The biorefinery in New York: woody biomass into commercial ethanol, *Pulp Pap. Can.*, 2006, **107**, 47–50.

32. T. E. Amidon and S. Liu, Water-based woody biorefinery, *Biotechnol. Adv.*, 2009, **27**, 542–550.

33. M. P. Coughlan and G. P. Hazlewood, s-1,4-D-Xylan-degrading enzyme systems: biochemistry, molecular biology and applications, *Biotechnol. Appl. Biochem.*, 1993, **17**, 259–289.

34. L. Viikari, M. Tenkanen, J. Buchert, M. Rättö, M. Bailey, M. Siika-aho and M. Linko, Bioconversion of forest and agricultural plant residues. In: *Hemicellulases for Industrial Applications*, ed., J. N. Saddler, 131–182, CAB International, Wallingford, UK, 1993.

35. K. Y. W. Wong, L. U. L. Tan and J. N. Saddler, Multiplicity of ß-1,4-xylanase in microorganisms: functions and applications, *Microbiol. Rev.*, 1988, **52**, 305–317.

36. G. Szakacs, K. Urbanszki, and R. P. Tengerdy, In: *Glycosyl Hydrolases for Biomass Conversion*, eds., M. E. Himmel, J. O. Baker and J. N. Saddler, American Chemical Society, ACS Symposium Series 769, Washington DC, USA, 2001, 190–203.

37. J. Szendefy, G. Szakacs and L. Christopher, Potential of solid-state fermentation enzymes of *Aspergillus oryzae* in biobleaching of paper pulp, *Enzyme Microb. Technol.*, 2006, **24**, 1149–1156.

38. A. Pandey, P. Selvakumar, C. R. Soccol and P. Nigam, Solid state fermentation for the production of industrial enzymes, *Curr. Sci.*, 1999, **77**, 149–162.

39. J. C. Mueller and C. C. Walden, Microbiological utilisation of sulphite liquor, *Process Biochem.*, 1970, **6**, 35–42.

40. N. Kosaric, K. K. Ho and Z. Duvnjak, Effect of spent sulphite liquor on growth and ethanol fermentation efficiency of Saccharomyces ellipsoideus, *Water Poll. Res. J. Can.*, 1980, **16**, 91–98.

41. Z. A. Chipeta, J. C. du Preez, G. Szakacs and L. Christopher, Xylanase production by fungal strains on spent sulphite liquor, *Appl. Microbiol. Biotechnol.*, 2005, **69**, 71–78.

42. S. S. Helle, A. Murray, J. Lam, D. R. Cameron and S. J. Duff, Xylose fermentation by genetically modified *Saccharomyces cerevisiae* 259ST in spent sulfite liquor, *Bioresour. Technol.*, 2004, **92**, 163–71.

43. S. Reidy, Making green ethanol, *Biofuels Business*, 2010, May, 28–31.

44. J. C. Royer and J. P. Nakas, Xylanase production by *Trichoderma longibrachiatum*, *Enzyme Microb. Technol.*, 1989, **11**, 405–410.

45. K. A. Onysko, Biological bleaching of chemical pulps: A review, *Biotech. Adv.*, 1993, **11**, 179–198.

46. L. Viikari, A. Kantelinen, J. Sundquist and M. Linko, Xylanases in bleaching: From an idea to the industry, *FEMS Microbiol. Rev.*, 1994, **13**, 335–350.

47. L. Christopher, Xylanases: properties and applications. In: *Concise Encyclopedia of Bioresource Technology*, ed., A. Pandey, The Haworth Press, Binghamton, New York, USA, 2004, 601–609.

48. K. Y. W. Wong, E. De Jong, J. N. Saddler and R. W. Allison, Mechanism of xylanase aided bleaching of kraft pulp, *Appita J.*, 1997, **50**, 509–518.

49. A. Kantelinen, B. Hortling, J. Sundquist, M. Linko and L. Viikari, Proposed mechanism of the enzymatic bleaching of kraft pulp with xylanases, *Holzforschung*, 1993, **47**, 318–324.

50. S. H. Bhosale, B. R. Mala and V. V. Deshpande, Molecular and industrial aspects of glucose isomerase, *Microbiol. Rev.*, 1996, **60**, 280–300.

51. K. Parker, M. Salas and V. C. Nwosu, High fructose corn syrup: Production, uses and public health concerns, *Biotechnol. Mol. Biol. Rev.*, 2010, **5**, 71–78.

52. L. Hyvonen and P. Koivistoinen, Fructose in food systems. In: *Nutritive Sweeteners*, eds., G. G. Birch, and K. J. Parker, 133–144, Applied Science Publishers, London, UK, 1982.

53. G. A. Bray, S. J. Nielson and B. M. Popkin, Consumption of high-fructose corn syrup in beverages may play a role in the epidemic of obesity, *Am. J. Clin. Nutr.*, 2004, **79**, 537–543.

54. J. Puls and B. Saake, Industrially isolated hemicelluloses. In: *Hemicelluloses: Science and Technology*, eds., P. Gatenholm and M. Tenkanen, 24-37, American Chemical Society, Washington DC, USA, 2004.

55. A. Ebringerova and T. Heize, Xylan and xylan derivatives – biopolymers with valuable properties 1. Naturally occuring xylans, structures, isolation procedures and properties, *Macromol. Rap. Commun.*, 2000, **21**, 542–556.

56. M. Ishihara, M. Nojiri, N. Hayashi and K. Shimizu, Isolation of xylan from hardwood by alkali extraction and steam treatment, *Mokuzai Gakkaishi*, 1996, **42**, 1211–1220.

57. A. Embringerova and Z. Hromadkova, Der Einsatz von Amoniak-lösungen bei der Gewinnung von Hemicellulosen des D-Xylantyps aus Laubholz, *Holz als Roh- und Wekrstoff*, 1996, **54**, 127–129.

58. W. G. Glasser, W. E. Kaar, R. K. Jain and J. E. Sealey, Isolation options for non-celllulosic heteropolysaccharides (HetPS), *Cellulose*, 2000, **7**, 299–317.

59. M. Nilsson, L. Saulnier, R. Andersson and P. Aman, Water extractable polysaccharides from three milling fractions of rye grain, *Carbohydr. Polym.*, **30**, 229–237.

60. R. Sun, X. F. Sun and J. Tompkinson, Hemicelluloses and their derivatives. In: *Hemicelluloses: Science and Technology*. P. Gatenholm and

M. Tenkanen, ed., American Chemical Society, Washington, DC, USA, 2004, 2–22.

61. A. Matsushika, H. Inoue, T. Kodaki and S. Sawayama, Ethanol production from xylose in engineered Saccharomyces cerevisiae strains: current state and perspectives, *Appl. Microbiol Biotechnol.*, 2009, **84**, 37–53.

62. M. Zhang, C. Eddy, K. Deanda, M. Finkelstein and S. Picataggio, Metabolic engineering of a pentose metabolism pathway in ethanologenic *Zymomonas mobilis*, *Science*, 1995, **267**, 240–243.

63. A. Eliasson, C. Christensson, C. F. Wahlbom and B. Hahn-Hägerdal, Anaerobic xylose fermentation by recombinant *Saccharomyces cerevisiae* carrying XYL1, XYL2, and XKS1 in mineral medium chemostat cultures, *Appl. Environ. Microbiol.*, 2000, **66**, 3381–3386.

64. S. Watanabe, A. A. Saleh, S. P. Pack, N. Annaluru, T. Kodaki and K. Makino, Ethanol production from xylose by recombinant *Saccharomyces cerevisiae* expressing protein-engineered NADH-preferring xylose reductase from *Pichia stipitis*, *Microbiology*, 2007, **153**, 3044–3054.

65. S.-M. Lee, T. Jellison and H. S. Alper, Directed evolution of xylose isomerase for improved xylose catabolism and fermentation in the yeast *Saccharomyces cerevisiae, Appl. Environ. Microbiol.*, 2012, Published online in June 2012, doi: 10.1128/AEM.01419-12.

66. J. S. Tolan and R. K. Finn, Fermentation of D-xylose to ethanol by genetically modified *Klebsiella planticola*, *Appl. Environ. Microbiol.*, 1987, **53**, 2039–2044.

67. K. Ohta, D. S. Beall, J. P. Mejia, K. T. Shanmugam and L. O. Ingram, Metabolic engineering of *Klebsiella oxytoca* M5A1 for ethanol production from xylose and glucose, *Appl. Environ. Microbiol.*, 1991, **57**, 2810–2815.

68. B. E. Wood, L. P. Yomano, S. W. York and L. O. Ingram, Development of industrial-medium-required elimination of the 2,3-butanediol fermentation pathway to maintain ethanol yield in an ethanologenic strain of *Klebsiella oxytoca*, *Biotechnol. Progr.*, 2005, **21**, 1366–1372.

69. H. Yanase, K. Nozaki and K. Okamoto, Ethanol production from cellulosic materials by genetically engineered *Zymomonas mobilis*, *Biotechnol. Lett.*, 2005, **27**, 259–263.

70. S. B. Piriya, P. T. Vasan, V. S. Padma, U. Vidhyadevi, K. Archana and S. J. Vennison, Cellulolytic ethanol production by recombinant cellulolytic bacteria harbouring pdc and adh II genes from Zymomonas mobilis, *Biotechnol. Res. Intl.*, 2012. Published online doi: 10.1155/2012/817549.

71. H. Tao, R. Gonzalez, A. Martinez, M. Rodriguez, L. O. Ingram, J. F. Preston and K. T. Shanmugam, Engineering a homo-ethanol pathway in *Escherichia coli*: Increased glycolytic flux and levels of expression of glycolytic genes during xylose fermentation, *J. Bacteriol.*, 2001, **183**, 2979–2982.

72. J. Zaldivar, J. Nielsen and L. Olsson, Fuel ethanol production from lignocellulose: a challenge formetabolic engineering and process integration, *Appl. Microbiol. Biotechnol.*, 2001, **56**, 17–34.

73. E. Guedon, M. Desvaux and H. Petitdemange, Improvement of cellulolytic properties of *Clostridium cellulolyticum* by metabolic engineering,
Appl. Microbiol. Biotechnol., 2002, **68**, 53–58.

74. Y.-H. P. Zhang and L. R. Lynd, New generation biomass conversion:
Consolidated bioprocessing. In: *Biomass Recalcitrance*. ed., M. E. Himmel,
480–494, Blackwell Publishing, Oxford, UK, 2008.

75. M. Desvaux, *Clostridium cellulolyticum*: model organism of mesophilic
cellulolytic clostridia, FEMS, *Microbiol. Rev.*, 2005, **29**, 741–764.

76. K. C. Jennert, C. Tardif, D. I. Young and M. Young, Gene transfer
to *Clostridium cellulolyticum* ATCC 35319, *Microbiol.*, 2000, **146**,
3071–3080.

77. W. T. Islam, J. C. Combs, B. C. Lynn and H. J. Strobel, Proteomic profile
changes in membranes of ethanol-tolerant *Clostridium thermocellum*,
Appl. Microbiol. Biotechnol., 2007, **74**, 422–432.

78. H. Alper, J. Moxley, E. Nevoigt, G. R. Fink and G. Stephanopoulos,
Engineering yeast transcription machinery for improved ethanol tolerance
and production, *Science*, 2006, **314**, 1565–1568.

79. R. D. Haan, J. E. McBride, D. C. La Grange, L. R. Lynd and W. H. Van
Zyl, Functional expression of cellobiohydrolases in *Saccharomyces cere-
visiae* towards one-step conversion of cellulose to ethanol, *Enzyme
Microb. Technol.*, 2007, **40**, 1291–1299.

80. W. H. Van Zyl, L. R. Lynd, R. Den Haan and J. E. McBride, Consolidated bioprocessing for bioethanol production using *Saccharomyces
cerevisiae*, *Adv. Biochem. Eng. Biotechnol.*, 2007, **108**, 205–235.

81. T. Werpy and G. Pedersen, *Top Value Added Chemicals from Biomass –
Volume 1*. U.S. Department of Energy, 2005.

82. J. H. Clark, F. E. I. Deswarte and T. J. Farmer, The integration of green
chemistry into future biorefineries, *Biofuels Bioprod. Bioref.*, 2009, **3**,
72–90.

83. M. Mattey, The production of organic acids, *Crit. Rev. Biotechnol.*, 1992,
12, 87–132.

84. K. Okano, S. Yoshida, R. Yamada, T. Tanaka, C. Ogino, H. Fukuda and
A. Kondo, Improved production of homo-D-lactic acid via xylose fermentation by introduction of xylose assimilation genes and redirection of
the phosphoketolase pathway to the pentose phosphate pathway in lactate dehydrogenase gene-deficient *Lactobacillus plantarum*, *Appl. Environ.
Microbiol.*, 2009, **75**, 7858–7861.

85. J. K. Magnuson and L. L. Lasure, Organic acid production by fila-
mentous fungi. In: *Advances in Fungal Biotechnology for Industry, Agriculture, and Medicine*. eds., J. J. Tkacs and L. Lange, 307–340, Kluwer
Academic/Plenum Publishers, New York, USA, 2004.

86. J. J. Dibner and P. Butin, Use of organic acids as a model to study the
impact of gut microflora on nutrition and metabolism, *J. Appl. Poultry
Res.*, 2002, **11**, 453–463.

87. L. Jarvis, Prospects for lactic acid are healthy as demand for all end uses
grows, *Chemical Market Reporter*, 2003, **263**, 12.

88. Y.-Y. Wee, J.-N. Kim and H.-W. Ryu, Biotechnological production of lactic acid and its recent applications, *Food Technol. Biotechnol.*, 2006, **44**, 163–172.

89. D. J. Hayes, S. Fitzpatrick, M. H. B. Hayes and J. R. H. Ross, The Biofine process – production of levulinic acid, furfural and formic acid from lignocellulosic feedstocks, *Bioref. Ind. Proc. Prod.*, 2006, **1**, 139–164.

90. A. M. R. Galletti, C. Antoanetti, V. de Luise, D. Licursi and N. N. o di Nasso, Levulinic acid production from waste biomass, *BioResources*, 2012, **7**, 1824–1835.

91. G. Huber, S. Iborra and A. Corma, Synthesis of transportation fuels from biomass: chemistry, catalysts, and engineering, *Chem. Rev.*, 2006, **106**, 4044–4098.

92. J. Yanowitz, E. Christensen and R. L. McCormick, Utilization of renewable oxygenates as gasoline blending components, Technical Report NREL/TP-5400-50791, National Renewable Energy Laboratory, 2011.

93. C. P. Kubicek, O. Zehentgruber, H. El-Kalak and M. Röhr, Regulation of citric acid production by oxygen: Effect of dissolved oxygen tension on adenylate levels and respiration in *Aspergillus niger*, *Eur. J., Appl. Microbiol. Biotechnol.*, 1980, **9**, 101–115.

94. C. W. Yang, Z. J. Lu and G. T. Tsao, Lactic acid production by pellet-form *Rhizopus oryzae* in a submerged system, *Appl. Biochem. Biotechnol.*, 1995, **51**, 57–71.

95. H. Kautola, M. Vahvaselka, Y. Y. Linko and P. Linko, Itaconic acid production by immobilized *Aspergillus terreus* from xylose and glucose, *Biotechnol. Lett.*, 1985, **7**, 167–172.

96. M. S. Dunn and B. W. Smart, DL-Aspartic acid, *Org. Synth.*, 1950, **30**, 7.

97. S. A. Overman and A. H. Romano, Role of pyruvate carboxylase in fumaric acid accumulation by *Rhizopus nigricans*, *Bacteriol. Proc.*, 1969, **69**, 128.

98. Y. Peleg, A. Barak, M. C. Scrutton and I. Goldberg, Malic acid accumulation by *Aspergillus flavus*. 3. 13C-NMR and isoenzyme analyses, *Appl. Microbiol. Biotechnol.*, 1989, **30**, 176–183.

99. J. G. Zeikus, M. K. Jain and P. Elankovan, Biotechnology of succinic acid production and markets for derived industrial products, *Appl. Microbiol. Biotechnol.*, 1999, **51**, 545–552.

100. A. S. Mamman, J.-M. Lee, Y.-C. Kim, I. T. Hwang, N.-J. Park, Y. K. Hwang, J.-S. Chang and J.-S. Hwang, Furfural: hemicellulose/xylose derived biochemical, *Biofuel Bioprod. Bioref.*, 2008, **2**, 438–454.

101. K. J. Zeitsch, *The Chemistry and Technology of Furfural and its Many By-Products*. Elsevier, Köln, Germany, 2000.

102. J. R. Weil, B. Dien, R. Bothast, R. Hendrickson, N. S. Mosier and M. C. Ladisch, Removal of fermentation inhibitors formed during pretreatment of biomass by polymeric adsorbents, *Ind, Eng. Chem. Res.*, 2002, **41**, 6132–6138.

103. H. E. Hoydonckx, W. M. Van Rhijn, W. Van Rhijn, D. E. De Vos and P. A. Jacobs, Furfural and derivatives, *Ullmann's Encyclopedia of Industrial Chemistry*, 2007, Wiley-VCH, Weinheim, Germany.

104. R. Ozer, Vapor phase decarbonylation process, WIPO Patent Application WO/2011/026059, 2011.
105. D. T. Win, Furfural – gold from garbage, *AU. J. Technol.*, 2005, **58**, 185–190.
106. G. W. Huber, J. N. Chheda, C. J. Barrett and J. A. Dumesic, Production of liquid alkanes by aqueous-phase processing of biomass-derived carbohydrates, *Science*, 2005, **308**, 1446–1450.
107. L. C. Kemp, Jr., G. B. Hamilton and H. H. Gross, Furfural as a selective solvent in petroleum refining, *Ind. Eng. Chem.*, 1948, **40**, 220–227.
108. R. N. Selai, D. Irudayaraj, B. Mani, V. Phoobalan, R. B. Singh, R. B. Singh, B. A. Kumar, Extraction of aromatics from hydrocarbon oil using furfural-co-solvent extraction process, U.S. Patent 6,866,772, 2005.
109. M. A. Daous and Y. Yorulmaz, *Alternate Energy Sources*, 1989, **1**, 815.
110. R. Rodriguez-Kabana and G. Walters, Method for treatment of nematodes in soil using furfural, U.S. Patent 5084477, 1992.
111. C. Calvet, J. Pinochet, A. Camprubi, V. Estaun and R. Rodriguez-Kabana, Evaluation of natural chemical compounds against root-lesion and root-knot nematodes and side effects on the infectivity of Arbuscular mycorrhizal fungi, *Eur. J. Plant Pathol.*, 2001, **107**, 601–605.
112. U.S. EPA, Pesticide fact sheet, *U.S. Environmental Protection Agency*, Office of Prevention Pesticide and Toxic Substance, 2006.
113. Kirk-Othmer, *Encyclopedia of Chemical Technology*, 1966, **10**, 243.
114. A. Day, K. K. Sen and D. Sardar, Xylitol from jut stick, *Res. Indust.*, 1989, **34**, 202–204.
115. M. P. Karimkulova, Y. S. Khakimov and M. F. Abidova, Activity of modified catalysts in hydrogenation of xylose to xylitol, *Chem. Nat. Comp.*, 1989, **25**, 370–371.
116. J.-P. Mikkola, T. Salmi, R. Sjöholm, P. Mäki-Arvela and H. Vainio, Hydrogenation of xylose to xylitol: three-phase catalysis by promoted Raney nickel, catalyst deactivation and in-situ sonochemical catalyst rejuvenation, *Stud. Surf. Sci. Catal.*, 2000, **130**, 2027–2032.
117. T. D. Leathers, Bioconversion of maize residues to value-added coproducts using yeast-like fungi, *FEMS Yeast Res.*, 2003, **3**, 133–140.
118. R. S. Prakasham, R. R. Sreenivas and P. J. Hobbs, Current trends in biotechnological production of xylitol and future prospects, *Curr. Trends Biotech. Pharm.*, 2009, **3**, 8–36.
119. L. Preziosi-Belloy, V. Nolleau and J. M. Navarro, Xylitol production from aspen wood hemicellulose hydrolysate by *Candida guilliermondii*, *Biotech. Lett.*, 2000, **22**, 239–243.
120. J. M. Domínguez, N. J. Cao, M. S. Krishnan, C. S. Gong and G. T. Tsao, Xylitol production from hybrid poplar wood chips pretreated by the ammonia steeping process, *Biotechnol. Techn.*, 1997, **11**, 339–341.
121. S. S. Silva, G. A. Maria, B. A. Silva and M. R. Arnaldo, Acid hydrolysis of *Eucalyptus grandis* chips for microbial production of xylitol, *Process Biochem.*, 1998, **33**, 63–67.

122. C. Roberto, I. M. Mancilha and C. A. de Souza, Evaluation of rice straw hemicellulose hydrolysate in the production of xylitol by *Candida guilliermondii, Biotechnol. Lett.*, 1994, **16**, 1211–1216.
123. M. J. Cruz, J. M. Domínguez, H. Domínguez and J. C. Parajó, Xylitol production from barley bran hydrolysates by continuous fermentation with *Debaromyces hansenii, Biotechnol. Lett.*, 2000, **22**, 1895–1898.
124. B. Rivas, P. Torre, J. M. Domfínguez, P. Perego, A. Converti and J. C. Parajó, Carbon material and bioenergestic balances of xylitol production from corncob by *Debaryomyces hansenii, Biotechnol. Progr.*, 2003, **19**, 706–713.
125. E. Winkelhausen and S. Kuzmanova, Microbial conversion of D-xylose to xylitol, *J. Ferm. Bioeng.*, 1998, **86**, 1–14.
126. A. Pandey, C. R. Soccol, P. Nigam and V. T. Soccol, Biotechnological potential of agro-industrial residues. I. Sugarcane bagasse, *Bioresour. Technol.*, 2000, **74**, 69–80.
127. B. S. Dien, N. N. Nichols, P. J. O'Bryan and R. J. Bothast, Development of new ethanologenic Escherichia coli strains for fermentation of lignocellulosic biomass, *Appl. Biochem. Biotechnol.*, 2000, **84**, 181–196.
128. M. Mancilha and M. N. Karim, Evaluation of ion exchange resins for removal of inhibitory compounds from corn stover hydrolysate for xylitol fermentation, *Biotechnol. Progr.*, 2003, **19**, 1837–1841.
129. I. M. J. Cruz, J. M. Domínguez, H. Domínguez and J. C. Parajó, Solvent extraction of hemicellulosic wood hydrolysates: a procedure useful for obtaining both detoxified fermentation media and polyphenols with antioxidant activity, *Food Chem.*, 1999, **67**, 147–153.
130. E. M. Lohmeier-Vogel, C. R. Sopher and H. Lee, Intracellular acidification as a mechanism for the inhibition by acid hydrolysis-derived inhibitors of xylose fermentation by yeasts, *J. Ind. Microbiol. Biotechnol.*, 1998, **20**, 75–81.
131. C. Martin and L. J. Jönsson, Comparison of the resistance of industrial and laboratory strains of *Saccharomyces* and *Zygosaccharomyces* to lignocellulosic-derived fermentation inhibitors, *Enzyme Microb. Technol.*, 2003, **32**, 386–395.
132. P. Cassland and L. J. Jönsson, Characterization of a gene encoding trametes versicolor laccase A and improved heterologous expression in *Saccharomyces cerevisiae* by decreased cultivation temperature, *Appl. Microbiol. Biotechnol.*, 1999, **52**, 393–400.
133. B. S. Dien, R. B. Hespell, L. O. Ingram and R. J. Bothast, Conversion of corn milling fibrous co-products into ethanol by recombinant *Escherichia coli* strains K011 and SL40, *World J. Microbiol. Biotechnol.*, 1997, **13**, 619–625.
134. M. G. A. Felipe, Biotechnological production of xylitol from lignocellulosic materials. In: *Lignocellulose Biodegradation*, eds., B. C. Saha and K. Hayashi, American Chemical Society, Washington, DC, USA, 2004, 300–315.

135. L. Trahan, Xylitol: a review of its action on mutans streptococci and dental plaque-its clinical significance, *Int. Dent. J.*, 1995, **45**, 77–92.
136. P. Milgrom, K. A. Ly, M. C. Roberts, M. Rothen, G. Mueller and D. K. Yamaguchi, Mutans streptococci dose response to xylitol chewing gum, *J. Dent. Res.*, 2006, **85**, 177–181.
137. P. T. Mattila, M. J. Svanberg, T. Jämsä and M. L. Knuuttila, Improved bone biomechanical properties in xylitol-fed aged rats, *Metabolism*, 2002, **51**, 92–96.
138. M. Uhari, T. Kontiokari and M. Niemelä, A novel use of xylitol sugar in preventing acute otitis media, *Pediatrics*, 1998, **102**, 879–974.
139. J. Zabner, M. P. Seiler, J. L. Launspach, P. H. Karp, W. R. Kearny, D. C. Look, J. J. Smith and M. J. Welsh, The osmolyte xylitol reduces the salt concentration of airway surface liquid and may enhance bacterial killing, *Proc. Natl. Acad. Sci. U.S.A.*, 2000, **97**, 11614–11619.
140. Y.-H. P. Zhang, Reviving the carbohydrate economy via multi-product lignocellulose biorefineries, *J. Ind. Microbiol. Biotechnol.*, 2008, **35**, 367–375.
141. F. Carvalheiro, L. C. Duarte and F. M. Gírio, Hemicellulose biorefineries: a review on biomass pretreatments, *J. Sci. Ind. Res.*, 2008, **67**, 849–864.
142. V. Chambost and P. R. Stuart, Selecting the most appropriate products for the forest biorefinery, *Ind. Biotechnol.*, 2007, **3**, 112–119.
143. E. Adler, Lignin chemistry – past, present and future, *Wood Sci. Technol.*, 1977, **11**, 169–218.
144. G. Brunow, K. Lundquist and G. Gellerstedt, Lignin. In: *Analytical Methods in Wood Chemsitry, Pulping and Papermaking*, eds., E. Sjöström and R. Alén, Springer, Berlin, 1998, 77–124.
145. S. J. Duff and W. D. Murray, Bioconversion of forest products industry waste cellulosics to fuel ethanol: a review, *Bioresour. Technol.*, 1996, **55**, 1–33.
146. P. Singh, O. Sulaiman, R. Hashim, P. F. Rupani and L. C. Peng, Biopulping of lignocellulosic material using different fungal species: a review, *Rev. Environ. Sci. Biotechnol.*, 2010, **9**, 141–151.
147. J. Gierer, J. and S. Wannstrom, Formation of alkali-stable C-C bonds between lignin and carbohydrate fragments during kraft pulping, *Holzforschung*, 1984, **38**, 181–184.
148. P. Axegard and J. Wiken, Delignification studies - Factors affecting the amount of residual lignin, *Svensk Papperstidn.*, 1983, **86**, R178–R184.
149. G. A. Smook, *Handbook for Pulp and Paper Technologists*, Angus Wilde Publications, Vancouver, Canada, 1992.
150. J. D. Gargulak and S. E. Lebo, Commercial use of lignin-based materials. In: *Lignin: Historical, Biological, and Materials Perspectives*, eds., W. G. Glasser, R. A. Northey and T. P. Schultz, ACS Symposium Series, American Chemical Society, Washington, DC, 2000, 307.
151. R. J. A. Gosselink, E. de Jong, B. Guran and A. Abächerli, Co-ordination network for lignin – standardisation, production and applications

adapted to market requirements (EUROLIGNIN), *Ind. Crops Prod.*, 2004, **20**, 121–129.

152. G. Gellerstedt and G. Henriksson, Lignins: Major sources, structure and properties. In: *Monomers, Polymers and Composites from Renewable Resources*, eds., M. Bengacem and A. Gandini, Elsevier, Oxford, UK, 2008, 201–224.

153. X. Yan, Y. Zeng and X. Zhang, Influence of biotreatment on the character of corn stover lignin as shown by thermogravimetric and chemical structural analyses, *BioResources*, 2010, **5**, 488–498.

154. A. Vishtal and A. Kraslawski, Challenges in industrial applications of technical lignins, *BioResources*, 2011, **6**, 3547–3568.

155. J. E. Holladay, J. J. Bozell, D. Johnson and J. F. White, Top value added chemicals from biomass. Volume II: Results of screening for potential candidates from biorefinery lignin, U.S. Department of Energy, PNNL-16983, 2007.

156. E. K. Pye, Industrial lignin production and applications. In: *Biorefineries-Industrial Processes and Products*, eds., B. Kamm, P.R. Gruber and M. Kamm, Wiley-VCH Verlag GmbH, Weinheim, Germany, 2006, 165–200.

157. J. L. McCarthy and A. Islam, Lignin chemistry, technology, and utilization: a brief history. In: *Lignin: Historical, biological, and materials perspectives*, eds., W. G. Glasser, R. A. Northey and T. P. Schultz, 2–99, ACS Symposium Series 742, Washington, DC, 2000.

158. T. Q. Hu, ed., *Chemical Modification, Properties, and Usage of Lignin.* Kluwer Academic/Plenum Publishers, New York, 2002.

159. Y. Li and S. Sarkanen, Thermoplastics with very high lignin contents. In: *Lignin: Historical, Biological, and Materials Perspectives*, eds., W. G. Glasser, R. A. Northey and T. P. Schultz, ACS Symposium Series 742, Washington, DC, 2000, 351–366.

160. W. de Oliveira and W. G. Glasser, Multiphase materials with lignin, *J. Wood Chem. Technol.*, 1994, **14**, 119–126.

161. M. Lund and C. Felby, Wet strength improvement of unbleached kraft pulp through laccase catalyzed oxidation, *Enzyme Microb. Technol.*, 2001, **28**, 760–765.

162. R. P. Chandra, L. K. Lehtonen and A. J. Ragauskas, Modification of high lignin content kraft pulps with laccase to improve paper strength properties. 1. Laccase treatment in the presence of gallic acid, *Biotechnol. Progr.*, 2004, **20**, 255–261.

163. H. Dyer, ed., Lockwood-Post's Directory of the Pulp, Paper, and Allied Trades for 1998. Miller Freeman, New York, 1997.

164. K. Sudo and K. Shimizu, A new carbon-fiber from lignin, *J. Appl. Polym. Sci.*, 1992, **44**, 127–34.

165. S. Kubo and J. F. Kadla, Lignin-based carbon fibers: Effect of synthetic polymer blending on fiber properties, *J. Polym. Environ.*, 2005, **13**, 97–105.

166. J. Nakano, Y. Izuta, T. Orita, H. Hatakeyama, K. Kobashigawa, K. Teruya and S. Hirose, Thermal and mechanical properties of

polyurethanes derived from fractionated kraft lignin, *Sen'i Gakkaishi*, 1997, **53**, 416–422.

167. N.-E. El Mansouri, Q. Yuan and F. Huang, Characterization of alkaline lignins for use in phenol-formaldehyde and epoxy resins, *BioResources*, 2011, **6**, 2647–2662.

168. M. V. Alonso, M. Oliet, F. Rodríguez, G. Astarloa and J. M. Echeverría, Use of a methylolated softwood ammonium lignosulfonate as partial substitute of phenol in resol resins manufacture, *J. Appl. Polym. Sci.*, 2004, **94**, 643–650.

169. L. W. Smith, H. K. Goering and C. H. Gordon, Relationships of forage compositions with rates of cell wall digestion and indigestibility of cell-walls, *J. Dairy Sci.*, 1972, **55**, 1140–1147.

170. H. C. Trowell, D. A. Southgate, T. M. S. Wolever, A. R. Leeds, A. R., M. A. Gassull and D. J. A. Jenkins, Dietary fiber redefined, *Lancet*, 1976, **1**, 967.

171. K. Schedle, C. Plitzner, T. Ettle, L. Zhao, K. J. Domig and W. Windisch, Effects of insoluble dietary fibre differing in lignin on performance, gut microbiology, and digestibility in weanling piglets, *Arch. Anim. Nutr.*, 2008, **62**, 141–151.

172. W. D. Holloway, C. Tasman-Jones and S. P. Lee, Digestion of certain fractions of dietary fiber in humans, *Am. J. Clin. Nutr.*, 1978, **31**, 927–930.

173. X. Dong, M. Dong, Y. Lu, A. Turley, Z. T. Jin and C. Wu, Antimicrobial and antioxidant activities of lignin from residue of corn stover to ethanol production, *Ind. Crops Prod.*, 2011, **34**, 1629–1634.

174. B. Košiková and J. Lábaj, Lignin-stimulated protection of polypropylene films and DNA in cells of mice against oxidation damage, *BioResources*, 2009, **4**, 805–815.

175. X. Pan, J. F. Kadla, K. Ehara, N. Gilkes and J. N. Saddler, Organosolv ethanol lignin from hybrid poplar as a radical scavenger: Relationship between lignin structure, extraction conditions, and antioxidant activity, *J. Agric. Food Chem.*, 2006, **54**, 5806–5813.

176. H. Jameel, D. R. Keshwani, S. F. Carter and T. H. treasure, Thermo-chemical conversion of biomass to power and fuels. In: *Biomass to Renewable Energy Processes*, ed., J. Cheng, CRC Press, Boca Raton, FL, USA, 2010, 437–489.

177. A. G. Barneto, J. A. Carmona, A. Galvez and J. A. Conesa, Effects of the composting and the heating rate on biomass gasification, *Energ. Fuels*, 2009, **23**, 951–957.

178. C. Couhert, J. M. Commandre and S. Salvador, Is it possible to predict gas yields of any biomass after rapid pyrolysis at high temperature from its composition in cellulose, hemicellulose and lignin?, *Fuel*, 2009, **88**, 408–417.

179. G. Wang, W. Li, B. Q. Li and H. K. Chen, TG study on pyrolysis of biomass and its three components under syngas, *Fuel*, 2008, **87**, 552–558.

180. T. Qu, W. Guo, L. Shen, J. Xiao and K. Zhao, Experimental study of biomass pyrolysis based on three major components: Hemicellulose, cel-lulose, and lignin, *Ind. Eng. Chem. Res.*, 2011, **50**, 10424–10433.

181. M. J. A. Tijmensen, A. P. C. Faaij, C. N. Hamelinck and van M. R. M. Hardeveld, Exploration of the possibilities for production of FT liquids and power via biomass gasification, *Biomass Bioenerg.*, 2002, **23**, 129–152.

182. T. Furusawa, T. Sato, H. Sugito, Y. Miura, Y. Ishiyama, M. Sato, N. Itoh and N. Suzuki, Hydrogen production from the gasification of lignin with nickel catalysts in supercritical water, *Int. J. Hydrogen Energy*, 2007, **32**, 699–704.

183. M. E. Dry, High quality diesel via the Fischer-Tropsch process – a review, *J. Chem. Technol. Biotechnol.*, 2001, **77**, 42–50.

184. R. B. Anderson, *The Fischer-Tropsch Synthesis*. Academic Press, New York, USA, 1984.

185. S. Srinivas, R. K. Malik and S. M. Mahajani, Fischer-Tropsch synthesis using bio-syngas and CO_2, *Energ. Sustain. Develop.*, 2007, **11**, 66–71.

186. J. S. Kim, S. Lee, S. B. Lee, M. J. Choi and K. W. Lee, Performance of catalytic reactors for the hydrogenation of CO_2 to hydrocarbons, *Cat. Today*, 2006, **115**, 228–234.

187. K. W. Jun, H. S. Roh, K. S. Kim, J. S. Ryu and K. W. Lee, Catalytic investigation of FTS from biomass derived syngas, *App. Cat. A*, 2004, **259**, 221–226.

188. P. L. Spath and D. C. Dayton, Preliminary screening-Technical and economic assessment of synthesis gas to fuels and chemicals with emphasis on the potential for biomass-derived syngas, U.S. Department of Energy, Office of Scientific and Technical Information, National Renewable Energy Laboratory, NREL/TP-510-34929, 2003.

189. J. C. Erickson, Thermochemical biorefinery technologies, *Biofuels Business*, April 2007, 68–71.

190. R. Alen, E. Kuoppala and P. Oesch, Formation of the main degradation compound groups from wood and its components during pyrolysis, *J. Anal. Appl. Pyrol.*, 1996, **36**, 137–148.

191. F. Verheijen, S. Jeffery, A. C. Bastos, M. van der Velde and, I. Diafas, Biochar application to soils: A critical scientific review of effects on soil properties, processes and functions, European Commission, EUR 24099 EN, 2010.

192. J. L. Gaunt and J. Lehmann, Energy balance and emissions associated with biochar sequestration and pyrolysis energy production, *Environ. Sci. Technol.*, 2008, **42**, 4152–4158.

193. D. A. Laird, R. C. Brown, J. E. Amonette and J. Lehmann, Review of the pyrolysis platform for coproducing bio-oil and biochar, *Biofuels, Bioprod. Bioref.*, 2009, **3**, 547–562.

194. A. V. Bridgwater, *Fast Pyrolysis of Biomass: A Handbook*. CPL Scientific Publishing, Newbury, UK, 2003.

195. X.-F. Zhu and Q. Lu, Production of Chemicals from Selective Fast Pyrolysis of Biomass. In: *Biomass*, eds., M. Momba and F. Bux, Sciyo, Croatia, 2010, 147–164.

196. C. Mohan, C. U. Pittman and P. H. Steele, Pyrolysis of wood/biomass for bio-oil: A critical review, *Energ. Fuels*, 2006, **20**, 848–889.

197. M. Kleinert and T. Barth, Phenols from lignin, *Chem. Eng. Technol.*, 2008, **31**, 736–745.
198. A. V. Bridgwater, Production of high grade fuels and chemicals from catalytic pyrolysis of biomass, *Catal. Today*, 1996, **29**, 285–295.
199. B. V. Babu, Biomass pyrolysis: a state-of-the-art review, *Biofuels, Bioprod. Bioref.*, 2008, **2**, 393–414.
200. A. V. Bridgwater and G. V. C. Peacocke, Fast pyrolysis processes for biomass, *Renew. Sust. Energ. Rev.*, 2000, **4**, 1–73.
201. F. Davoudzadeh, B. Smith, E. Avni and R. W. Coughlin, Depolymerization of lignin at low-pressure using Lewis acid catalysts and under high-pressure using hydrogen donor solvents, *Holzforschung*, 1985, **39**, 159–66.
202. M. Garcia-Perez, S. Wang, J. Shen, M. Rhodes, W. J. and C. Z. Li, Effects of temperature on the formation of lignin-derived oligomers during the fast pyrolysis of Mallee woody biomass, *Energ. Fuel*, 2008, **22**, 2022–2032.
203. B. Scholtze and D. Meier, Characterization of the water-insoluble fraction from pyrolysis oil (pyrolytic lignin). Part I. PY-GC/MS, FTIR, and functional groups, *J. Anal. Appl. Pyrolysis*, 2001, **60**, 41–54.
204. R. Bayerbach and D. Meier, Characterization of the water-insoluble fraction from fast pyrolysis liquids (pyrolytic lignin). Part IV: Structure elucidation of oligomeric molecules, *J. Anal. Appl. Pyrol.*, 2009, **85**, 98–107.
205. L. Deng, Z. Yan, Y. Fu and Q. X. Guo, Green solvent for flash pyrolysis oil separation, *Energ. Fuel*, 2009, **23**, 3337–3338.
206. R. Giroux, B. Freel and R. Graham, Novel natural resin formulations, U.S. Patent 6326461, 2001.
207. Y. D. Yeboah, K. B. Bota and Z. Wang, Hydrogen from biomass for urban transportation, *Proceedings of the 2002 U.S. DOE Hydrogen Program Review*, NREL/CP-610-32405, 2002, 1–19.
208. C. Di Blasi, A. Galgano and C. Branca, Influences of the chemical state of alkaline compounds and the nature of alkali metal on wood pyrolysis, *Ind. Eng. Chem. Res.*, 2009, **48**, 3359–3369.
209. A. Vigneault, D. K. Johnson and E. Chornet, Base-catalyzed depolymerization of lignin: Separation of monomers, *Can. J. Chem. Eng.*, 2007, **85**, 906–916.
210. Q. Lu, Z. Tang, Y. Zhang and X. F. Zhu, Catalytic upgrading of biomass fast Pyrolysis vapors with Pd/SBA-15 catalysts, *Ind. Eng. Chem. Res.*, 2010, **49**, 2573–2580.
211. M. J. W. Dignum, J. Kerler and R. Verpoorte, Vanilla production: Technological, chemical, and biosynthetic aspects, *Food Rev. Int.*, 2001, **17**, 199–219.
212. H. Priefert, J. Rabenhorst and A. Steinbuchel, Biotechnological production of vanillin, *Appl. Microbiol. Biotechnol.*, 2001, **56**, 296–314.
213. T. D. Bugg, M. Ahmad, E. M. Hardiman and R. Singh., The emerging role for bacteria in lignin degradation and bio-product formation, *Curr. Opin. Biotechnol.*, 2011, **22**, 394–400.
214. T. Higuchi, Biodegradation of lignin – biochemistry and potential applications, *Experientia*, 1982, **38**, 159–166.

215. K.-E. L. Eriksson, R. A. Blanchette and P. Ander, *Microbial and Enzymatic Degradation of Wood Components*. Springer-Verlag, Berlin, Germany, 1990.
216. F. J. Ruiz-Dueñas and Á. T. Martínez, Microbial degradation of lignin: how a bulky recalcitrant polymer is efficiently recycled in nature and how we can take advantage of this, *Microb. Biotechnol.*, 2009, **2**, 164–177.
217. Y. H. Chen, L. Y. Chai, Y. H. Zhu, Z. H. Yang, Y. Zheng and H. Zhang, Biodegradation of kraft lignin by a bacterial strain *Comamonas* sp. B-9 isolated from eroded bamboo slips, *J. Appl. Microbiol.*, 2012, **112**, 900–906.
218. E. Masai, Y. Katayama and M. Fukuda, Genetic and biochemical investigations on bacterial catabolic pathways for lignin derived aromatic compounds, *Biosci. Biotechnol. Biochem.*, 2007, **71**, 1–15.
219. A. Raj, M. M. K. Reddy and R. Chandra, Identification of low molecular weight aromatic compounds by gas chromatography–mass spectrometry (GC–MS) from Kraft lignin degradation by three *Bacillus* sp., *Int. Biodeter. Biodeg.*, 2007, **59**, 292–296.
220. C. Trigo and A. S. Ball, Production of extracellular enzymes during the solubilisation of straw by *Thermomonospora* fusca BD25, *Appl. Microbiol. Biotechnol.*, 1994, **41**, 366–372.
221. D. L. Crawford, A. L. Pometto and R. L. Crawford, Lignin degradation by *Streptomyces viridosporus*: isolation and characterization of a new polymeric lignin degradation intermediate, *Appl. Environ. Microbiol.*, 1983, **45**, 898–904.
222. R. L. Nicholson and R. Hammerschmidt, Phenolic compounds and their role in disease resistance, *Ann. Rev. Phytopathol.*, 1992, **30**, 369–89.
223. H. Faustino, N. Gil, C. Baptista and A. P. Duarte, Antioxidant activity of lignin phenolic compounds extracted from kraft and sulphite black liquors, *Molecules*, 2010, **15**, 9308–9322.
224. S. Mitsuhashi, T. Kishimoto, Y. Uraki, T. Okamoto and M. Ubukata, Low molecular weight lignin suppresses activation of NF-[kappa]B and HIV-1 promoter, *Bioorg. Med. Chem.*, 2008, **16**, 2645–2650.
225. O. R. Gotlieb, Phytochemicals: Differentiation and function, *Phytochem.*, 1990, **29**, 1715–1724.
226. P. Sannigrahi, A. J. Ragauskas and G. A. Tuskan, Poplar as a feedstock for biofuels: A review of compositional characteristics, *Biofuels, Bioprod. Bioref.*, 2010, **4**, 209–226.
227. M. B. Jansson, F. Alvarado, A.-K. Berquist and O. Dahlman, Analysis of pulp extractives with off-line supercritical fluid extraction (SFE) and supercritical fluid chromatography (SFC), *Int. Symp. Wood Pulp. Chem. Proc.*, CTAPI, Beijing, 1993, Vol. II, 795.
228. M. B. Jansson and N.-O. Nilvebrant, Wood extractives. In: *Pulp and Paper Chemistry and Technology, Vol. 1, Wood Chemistry and wood Biotechnology*, eds., M. Elk, G. Gellerstedt and G. Henriksson, Walter de Gruyter, Berlin, Germany, 2009, 147–171.

229. H. Kohda, R. Kasal, K. Yamasaki, K. Murakami and O. Tanaka, New sweet diterpene glucosides from *Stevia reboudiana*, *Phytochem.*, 1976, **15**, 981–983.

230. B. R. Noller, *Chemistry of Organic Compounds*. Saunders, Philadelphia, PA, USA, 1961.

231. T. Robinson, *The Organic Constituents of Higher Plants*. Cordus Press, North Amherst, MA, USA, 1991.

232. F. W. Barlow, Rubber, gutta and chicle. In: *Natural Products of Woody Plants II*, ed., J. W. Rowe, Sringer-Verlag, Berlin, Germany, 1989, 1028–1050.

233. A. de Jong, J. Plat and R. P. Mensink, Metabolic effects of plant sterols and stanols, *J. Nutr. Biochem.*, 2003, **14**, 362–369.

234. A. Anderson and B. Richmond, The chemistry of wood durability and decay. Structure of fungicidal components in some cedars, *Symp. Phytochem., Univ. Hong Kong*, 1964, 101–116.

235. T. C. Scheffer and E. B. Cowling, Natural durability of wood to microbial deterioration, *Ann. Rev. Phytopathol.*, 1966, **4**, 57–58.

236. C. C. Winterbourn, Nutritional antioxidants: their role in disease prevention, *N.Z. Med. J.*, 1995, **108**, 447–448.

237. A. Bagchi, A. Garg, R. L. Krohn, M. Bagchi, M. X. Tran and S. J. Stohs, Oxygen free radical scavenging abilities of vitamins C and E, and a grape seed proanthocyanidin extract in vitro, *Res. Commun. Mol. Pathol. Pharmacol.*, 1997, **95**, 179–189.

238. J. E. Wood, S. T. Senthilmohan and A. V. Peskin, Antioxidant activity of procyanidin-containing plant extracts at different pHs, *Food Chem.*, 2002, **77**, 155–161.

239. A. Pizzi, *Wood Adhesives Chemistry and Technology*, Vol. 1, Marcel Dekker, New York, USA, 1983.

240. E. Kulvik, Chestnut wood tannin extract in plywood adhesives, *Adhesives Age*, 1976, **20**, 19–21.

241. U. C. Bhargava and B. A. Westfall, Antitumor activity of Juglas nigra (black walnut) extractives, *J. Pharm. Sci.*, 1968, **57**, 1674–1677.

242. A. Pizzi, Wood/bark extracts as adhesives and preservatives. In: *Forest Products Biotechnology*, eds., A. Bruce and J. F. Palfreyman, 151–182, Taylor & Francis, London, UK, 1998.

243. O. R. Gottlieb and M. Yoshida, Lignans. In: *Natural Products of Woody Plants I*. ed., J. W. Rowe, Springer-Verlag, Berlin, Germany, 1989, 349–511.

244. H. Hovelstad, I. Leirset, K. Oyaas and A. Fiksdahl, Screening analyses of pinosylvin stilbenes, resin acids and lignans in Norwegian conifers, *Molecules*, 2006, **11**, 103–114.

245. D. F. Zinkel, Fats and fatty acids. In: *Natural Products of Woody Plants II*. ed., J. W. Rowe, 369–399, Springer-Verlag, Berlin, Germany, 1989.

246. D. J. Durzan, Nitrogennous extractives. In: *Natural Products of Woody Plants I*. ed., J. W. Rowe, Springer-Verlag, Berlin, Germany, 1989, 179–200.

247. T. Aniszewski, *Alkaloids – Secrets of Life*. Elsevier, Amsterdam, The Netherlands, 2007.

248. A. G. Gilman, L. S. Goodman, T. W. Rall and F. Murad, eds., *The Pharmacological Basis of Theurapics*, Macmillan, New Your, USA, 1985.

249. A. A. Kadir, Drugs from plants. In: *Forest Products Biotechnology*, eds., A. Bruce and J. F. Palfreyman, Taylor & Francis, London, UK, 1998, 209–234.

250. M. C. Wani, H. C. Taylor, M. E. Wall, P. Coggon and A. T. McPhail, Plant antitumor agents. VI. The isolation and structure of taxol, a novel anti-leukemic and antitumor agent from *Taxus brevifolia*, *J. Am. Chem. Soc.*, 1971, **93**, 2325–2327.

251. E. M. Croom, Taxol: science and application. In: *Taxus for Taxol and Taxoids*, ed., M. Suffness, CRC Press, Boca Raton, FL, USA, 1995, 37–70.

252. M. Pandil, R. S. Kumaran, Y.-K. Choi, H. J. Kim and J. Muthumary, Isolation and detection of taxol, an anticancer drug produced from *Lasiodiplodia theobromae*, an endophytic fungus of the medicinal plant *Morinda citrifolia*, *Afr. J. Biotechnol.*, 2011, **10**, 1428–1435.

253. R. Alén, Basic chemistry of wood delignification. In: *Forest Products Chemistry*. ed., P. Stenius, Fapet Oy, Helsinkin, Finland, 2000, 58–104.

254. T. M. Grace, Overview of kraft recovery, In: *Pulp and Paper Manufacture. Vol. 5. Alkaline Pulping*, ed., M. J. Kocurek, Joint Textbook Committee of the Paper Industry, Atlanta, GA, USA, 1989, 473–477.

255. D. D. S. Liu, J. Monnier, G. Tourigny, J. Kriz, E. Hogan and A. Wong, Production of high quality cetane enhancer from depitched tall oil, *Petrol. Sci. Technol.*, 1998, **16**, 597–609.

256. H.-D. Belitz and W. Grosch, *Food Chemistry*. Springer, Heidelberg, Germany, 1987.

257. J. Bjorklund, P. Wormald and O. Dahlman, Reactions of wood extractives during ECF and TCF bleaching of kraft pulp, *Pulp & Pap. Can.*, 1995, **96**, 42–45.

258. C. Chirat and V. De la Chapelle, Heat- and light-induced brightness reversion of bleached chemical pulps, *J. Pulp Pap. Sci.*, 1999, **25**, 201–205.

259. I. Croon, S. Dillén and J.-E. Olsson, Brightness reversion of birch sulfate pulp, *J. Polym. Sci. Part C: Polym. Symp.*, 1965, **11**, 173–195.

260. M. Mahagaonkar and H. Suss, The effect of different alkaline treatments on extractive removal prior to TCF bleaching of *Radiata pine* magnesium bisulfite pulp, *Appita Ann. Gen. Conf. Proc.*, 1999, 433–439.

261. R. C. Sun, D. Salisbury and J. Tomkinson, Chemical composition of lipophilic extractives released during the hot water treatment of wheat straw, *Bioresour. Technol.*, 2003, **88**, 95–101.

262. J. Bouchard, L. H. Allen, C. Lapointe and M. Pitz, Improved deresination during oxygen delignification, *Pulp & Pap Can.*, 2003, **104**, 39–43.

263. P. Aarto and J. Jukka, Method for reducing the extractives content of high-yield pulps and method for producing bleached high-yield pulps, WO 2005/080672 A2, Helsinki, Finland, 2005.

264. R. L. Farrell, K. Hata and M. B. Wall, Solving pitch problems in pulp and paper processes by the use of enzymes or fungi. In: *Advances in Biochemical Engineering/Biotechnology*, ed., T. Sheper, Springer-Verlag, Berlin, Germany, 1997, 197–212.

265. K. Fischer, M. Akhtar, K. Messner, R. A. Blanchette and T. K. Kirk, Pitch reduction with the white-rot fungus *Ceriporiopsis subvermispora*. In: *Biotechnology in the Pulp and Paper Industry*, ed., E. Srebotnik and K. Messner, Facultas-Universitätsverlag, Vienna, Austria, 1996, 193–198.

266. H. Ghezzaz, L. Pelletier and P. R. Stuart, Biorefinery implementation for recovery debottlenecking at existing pulp mills – Part I: Potential for debottlenecking, *Tappi J.*, 2012, **11**, 17-.

267. E. D. Larson, S. Consonni, R. E. Katofski, K. Lisa and W. J. Frederick, An assessment of gasification-based biorefining at kraft pulp and paper mills in the United States, Part A: Background and assumptions, *Tappi J.*, 2008, **7**, 8–14.

268. E. D. Larson, S. Consonni, R. E. Katofski, K. Lisa and W. J. Frederick, An assessment of gasification-based biorefining at kraft pulp and paper mills in the United States, Part B: Results, *Tappi J.*, 2009, **8**, 27–35.

269. C. Bergeron, D. J. Carrier, S. Ramaswamy, *Biorefinery Co-Products: Phytochemicals, Primary Metabolites and Value-Added Biomass Processing*, John Wiley & Sons, Chichester, UK, 2012.

270. P. Söderholm and R. Lundmark, The development of forest-based biorefineries: Implications for market behaviour and policy, *Forest Prod. J.*, 2009, **59**, 6–16.

271. V. Chambost, J. McNutt and P. R. Stuart, Partnership for successful enterprise transformation of forest industry companies implementing the forest biorefinery, *Pulp. Pap. Can.*, 2009, **110**, 19–24.

272. H. Weyerhaeuser, Biotechnology in forestry: The promise and the economic reality, *Solutions*, 2003, October, 32–34.

273. T. W. Jeffries and L. Viikari, ed., *Enzymes for Pulp and Paper Processing*, Amercan Chemical Society, Washington, DC, USA, 1996.

274. R. A. Young and M. Akhtar, eds., *Environmentally Friendly Technologies for the Pulp and Paper Industry*, John Wiley & Sons, New York, NY, USA, 1997.

275. Viikari and R. Lantto, eds., *Biotechnology in the Pulp and Paper Industry: 8 ICBPPI Meeting*, Elsevier, Amsterdam, The Netherlands, 2002.

276. S. D. Mansfield and J. N. Saddler, eds., *Application of Enzymes to Lignocellulosics*, Amercan Chemical Society, Washington, DC, USA, 2003.

277. P. R. Stuart and M. M. El-Halwagi, eds., *Integrated Biorefineries: Design, Analysis, and Optimization*, CRC Press, 2012.

CHAPTER 2
Economic and Policy Aspects of Integrated Forest Biorefineries

JIANBANG GAN

Department of Ecosystem Science and Management, Texas A&M
University, College Station, TX 77843-2138, USA
Email: j-gan@tamu.edu; Tel.: + 1-979-862-4392; Fax: + 1-979-845-6049

2.1 Introduction

Integrated biorefineries convert biomass to biofuels, biopower, and bioproducts
to maximize the utilization efficiency of biomass resources. Because of their
technical and economic advantages, the prospect of integrated biorefineries is
very attractive in the development and deployment of a biobased industry.
Among future integrated biorefineries are integrated forest biorefineries
(IFBRs), which will primarily use forest biomass as feedstock and have potential
to be integrated with the production of traditional forest products.

The prospect of IFBRs is even more promising for several reasons. First,
wood has been and will continue to be a major energy source globally. Wood
fuels use almost 60% of total roundwood harvested in the world, with most
wood fuels being consumed in developing countries.[1] The amount of woody
biomass available for bioenergy purposes in developed countries is also significant.[2,3] Secondly, wood and woody biomass can be used for many purposes
including the production of energy, building materials, pulp and paper products, and other bioproducts. Many of these products can and should be jointly
produced to take advantage of technological advances and to improve the
utilization efficiency of wood resources. Thus, integrated production systems

RSC Green Chemistry No. 18
Integrated Forest Biorefineries
Edited by Lew Christopher
© The Royal Society of Chemistry 2013
Published by the Royal Society of Chemistry, www.rsc.org

such as IFBRs are a good fit. Thirdly, it is economically crucial to enhance the efficient utilization of wood resources in a systematic manner. IFBRs coupled with the supply chains of traditional forest products can reduce the costs of forest biomass harvesting and transporting.[4,5] Integrating biofuel production with pulp/paper milling can also reduce the cost of converting forest biomass to biofuels.[6] These economic advantages will ultimately translate into the improved cost-competitiveness of biofuels, biopower, and bioproducts as well as traditional forest products. Fourthly, the forest products industry has a long history of using wastes (*e.g.* harvest and mill residues) and byproducts of its production processes.[2] For example, many wood products and pulp/paper mills use wastes and residues to generate power and heat for their own use. These practices and experiences along with the available infrastructures of the forest products industry will be helpful for developing IFBRs, though IFBRs entail expanded partnerships between the forest products industry and other sectors.

This chapter focuses on the economic and policy aspects of IFBRs with an emphasis on IFBR supply chains. The next section will provide a brief introduction to the traditional forest products sector and its supply chains. Section 2.3 will describe the supply chains of IFBRs and their key economic aspects. A decision support model for determining the key economic aspects of IFBR supply chains will be presented in Section 2.4. Policy issues in the development and deployment of IFBRs will be discussed in Section 2.5. And the chapter will be closed with a summary.

2.2   Traditional Forest Products Sector

2.2.1   Conditions and Outlook of Forest Products Markets

Forestry and the forest products industry are important economic sectors in many developed and developing countries. The world's forests supplied 1.42×10^9 m^3 of industrial roundwood and 1.85×10^9 m^3 of fuelwood in 2009.[1] Industrial roundwood is primarily used as building materials for housing and as raw materials for making furniture and paper products. The traditional forest products industry and markets have witnessed considerable changes in the past few decades. The demand for lumber is cyclical, depending upon housing markets and ultimately overall economic conditions. The demand for pulpwood in many developed countries has weakened due to shifting global demand and competition, paper recycling, and electronic media, though the outlook for paper markets in emerging economies is strong. These changes coupled with technological advances and environmental pressure precipitate both challenges and opportunities for exploring alternative uses of wood resources.

At present, most fuelwood is produced and consumed in developing countries, primarily used for cooking and heating. The use of forest biomass for energy purposes in developed countries is likely to increase as biofuels are viewed as a means to mitigate greenhouse gas (GHG) emissions, reduce dependence on imported oil, and promote economic development, especially in rural areas. Yet the realization of this potential for forest biofuels will be dictated by many factors including energy prices, commitment to GHG emission reductions, breakthroughs in biofuel technologies, land availability and use constraints, and social acceptance, among other things.

Expanding the production and market scope of the traditional wood products industry by incorporating bioenergy and bioproducts will make the industry more resilient to changes in markets and policies. While leading to more efficient use of forest resources, this expansion can increase the profitability and robustness of the industry. Additionally, it will create more jobs, contributing to the sustainability and prosperity of forest-dependent communities.

2.2.2 Supply Chains of Traditional Forest Products

The supply chain of traditional forest products involves growing trees, harvesting timber, transporting wood, processing wood into various (intermediate and end) products, and distributing wood and paper products (as shown in the left part of Figure 2.1). Though natural forests still contribute to timber supply, forest plantations have played an increasingly important role in meeting the world's needs for wood and paper products. Timber-harvesting systems vary across regions, largely depending upon terrain and climate conditions, forest characteristics, and technologies that are available and affordable.

These supply chains often have one or very few major end-products such as lumber or wood pulp/paper products. Thus, they are relatively simple, without addressing the interactions in producing and distributing different end-products. For instance, a supply chain for a lumber mill or industry mainly concerns the growth, harvest and transport of sawlogs, conversion of sawlogs to lumber, and distribution of lumber.[7,8] Similarly, a supply chain for a pulp/paper mill or industry focuses on growing and procuring pulpwood, converting

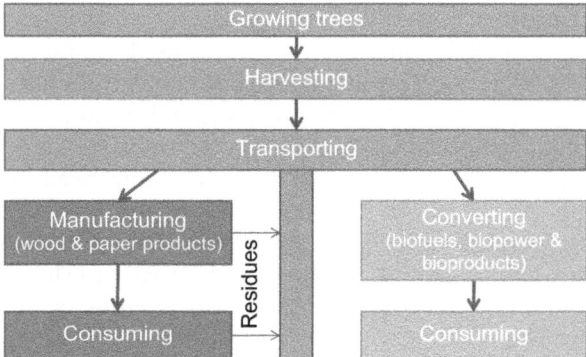

Figure 2.1 Integration of supply chains of both traditional forest products and integrated forest biorefineries.

pulpwood to pulp and paper, and distributing paper products.[9] Because the production highly focuses on a few products, byproducts and wastes/residues are not fully utilized. Additionally, as a mature industry, the wood and paper products industry uses quite standardized equipment and production systems. Thus, the technoeconomic aspects of supply chains of traditional forest products are not too complex.

2.3 Integrated Forest Biorefineries (IFBRs)

2.3.1 Supply Chains of IFBRs

IFBRs involve converting forest biomass into biofuels, biopower, and bio-products (biochemicals and biomaterials) (Figure 2.1). There are several forest biomass sources that can be potentially used as feedstocks for IFBRs. They include logging residues, thinning residues, mill residues, small-diameter trees including those traditionally used for pulping, and energy plantations, among other sources. Various conversion technologies are available or under development, including the biochemical, thermochemical, and hybrid processes.

A supply chain for IFBRs generally consists of growing trees; harvesting, transporting, and processing (including storage and pretreatments) biomass, converting biomass to biofuels and bioproducts, and distributing end-products. Given the variety of forest biomass sources, harvesting and transporting methods, conversion technologies, and end-products, IFBR supply chains can vary considerably. The variety and choice of IFBRs will be driven by market demands for end-products, feedstock availability and quality (including physical and chemical characteristics), conversion technologies, and government policy. These factors are often interrelated. For example, feedstock availability or end-products demanded by society could dictate the development or choice of conversion technologies; available conversion technologies in turn can influence what end-products to produce and what type of biomass to grow. Similarly, policies interact with the choice of end-products, feedstocks, and technologies. These interlinks call for systematical approaches to designing IFBR supply chains.

Future IFBRs can build on existing models used by the forest products industry. Some existing wood products and pulp/paper mills have incorporated power and heat generation fueled by forest and mill residues. The power and heat generated are primarily used by the mills themselves. This primitive model can be expanded by adopting new technologies and improved supply chains to produce multiple products. Mill residues and byproducts (*e.g.* lignin from pulping), for instance, can be used to produce biofuels or bioproducts as well as biopower and heat. Such expansions will enable IFBRs to share some existing infrastructures with the wood and paper products industry, reducing investment and production cost.

Though supply chains for IFBRs will be more complex than those for the traditional wood and paper industry, they share some common components such as growing, harvesting, and transporting wood/biomass. Additionally, residues and wastes resulting from manufacturing of wood and paper products can be used as feedstock for IFBRs, and power and heat generated from IFBRs in turn can be used in the production of wood and paper products. All these point to the potential and need for integration or at least coordination between the supply chains of traditional wood and paper production and IFBRs. Such an integration or coordination will enhance the efficiency of resource use and the overall production system. This has been confirmed by existing studies. Integrating harvesting systems can significantly reduce biomass harvest cost.[5] Bundling forest residues can take advantage of traditional logging trucks, while increasing the bulk density of biomass transported, thus reducing biomass transport costs.[10] Integrated production of traditional wood products and bioenergy will increase returns on capital and generate more jobs than standalone production of traditional wood products or bioenergy.[11]

Given the purposes of IFBRs, their supply chains will/should have some unique features. Unlike the supply chains of traditional forest products, IFBR supply chains should be able to coordinate the efficient utilization of multiple biomass materials to produce multiple end-products. Hence, IFBR supply chains are expected at least to:

- include the production and distribution of multiple end-products (major products and byproducts/coproducts);
- seek the optimization of the entire system rather than the production and distribution of a single product;
- allow for the use of multiple biomass feedstock sources, ideally including both woody and nonwoody biomass sources (Note: Using multiple feedstocks can take full advantages of not only biomass resources abundant in the region but also the economies of scale of IFBRs due to increased spatial distribution density of biomass.[12] Moreover, different feedstocks that are available at different seasons may be able to better meet the year-round production need for IFBRs.);
- allow for the use of a specific biomass feedstock to produce different and multiple products;
- allow for the use of wastes and byproducts to produce value-added products;
- allow for the shared use or integration of infrastructures and equipment for producing and distributing different products;
- allow for allocation of biomass materials and other inputs among different uses according to changes in market demands and policy goals; and
- encourage the development and adoption of efficient, integrative technologies.

2.3.2 Key Economic Aspects of IFBRs

IFBRs should be able to optimize the efficient and sustainable use of forest resources and other inputs, while helping meet society's needs for biofuels, biopower and bioproducts. There are several key economic aspects attributable to the efficiency of IFBRs, including the choices of end-products, feedstock sources, logistics and conversion/production technologies, and locations and scales of IFBRs. I describe each of these aspects below. Though these aspects are discussed separately here, they are interrelated and should be considered systematically in analyzing and designing IFBRs.

2.3.2.1   End-Product Portfolio

What end-products to produce is an important decision for IFBRs. Many factors can influence this decision. Among them are market demand, feedstock characteristics, available technologies, GHG credits/taxes, and energy security consideration. Market demands is a fundamental driving force for choosing an end-product portfolio. Some feedstock may be more appropriate for producing a specific product than others due to its chemical and physical properties. Available technologies for feedstock production (*e.g.* biotechnology to manipulate the properties of feedstock), logistics, and biofuel conversion or bioproduct production can enable the production of biofuel and bioproducts technically feasible and cost effective. Increases in GHG offset credits or emission taxes would favor the production of biopower over biofuels,[13] whereas the opposite would be true when national energy/oil security is a priority.

2.3.2.2   Feedstock Choices

There are several forest biomass sources that can be used as feedstocks for IFBRs. These biomass sources include harvest residues; thinning residues; mill residues; energy plantations; small-diameter trees; trees damaged by fires, pests or storms; trees with low value or productivity; invasive plant species; and urban wood wastes. Abundance and costs of these biomass sources may vary spatially and temporally. Several biomass sources may coexist at a given location or time. Production cost, environmental consequences, and social acceptance will determine the actual supply level of these biomass sources for IFBRs.[14]

Harvest residues are tied to timber harvest. They have been used for power and heat production to some extent,[2] yet they can be redirected to other uses and a lot of them are still available. The production cost of harvest residues is relatively low, particularly when integrative harvest systems are used in timber and residue procurements.[4,5] Under some circumstances, removals of harvest residues may negatively affect long-term soil productivity. Hence, balancing current residue harvest with future forest productivity and sustainability (including soil C, N and H_2O maintenance) is essential in determining the appropriate removal intensity of harvest residues.[15]

Thinning residues include those generated from traditional silvicultural thinnings for timber production and from forest fuel treatments for mitigating wildfire risk. The availability of thinning residues will depend upon the needs for and costs of thinning practices. In general, the production cost of this biomass is high. Yet the benefits of improved forest stand quality and productivity or reduced fire risk can be significant, and at least can partially offset the production cost of thinning residues.

Mill residues are, in general, uniform and good in quality. They are concentrated on specific locations, thus reducing the costs of collection. Most mill residues have been used for different purposes. They, however, could be reallocated to producing biofuels or bioproducts.

Short-rotation woody biomass plantations can allow for spatial concentration of feedstock production. Their relatively high biomass yield can increase the spatial distribution density of biomass, which is clearly an economic advantage for IFBRs.[12] However, these plantations take additional land areas and often require high inputs of labor, capital, fertilizers, and water, which will incur additional costs and may lead to undesired environmental consequences. Besides, monocultural practices associated with the plantations could have adverse impacts on biodiversity.

Small-diameter trees include pulpwood and those that are too small or too low quality for timber production. The quantity of this biomass source is significant, yet competing uses could limit its availability for producing biofuels and bioproducts. In some regions like the southern United States, current weak demand for wood pulp indicates the potential of this biomass source as feedstock for IFBRs. Existing pulpwood procurement and logistic systems can be modified and adopted to serve IFBRs.

Additionally, there are other sources of forest biomass, including trees damaged by storms, fires or pests and the low-quality or low-value trees removed by stand improvement practices. The availability of these biomass sources is hard to predict, which could create a challenge for designing a logistic system for IFBRs.

2.3.2.3 Logistics and Conversion Technologies

Logistics and conversion technologies play an essential role in determining the production cost of bioenergy and bioproducts. Tremendous research efforts are underway to develop effective and efficient logistic systems and conversion technologies. Among the major technologies being tested for converting cellulosic biomass to biofuels are the biochemical and thermochemical approaches and their combinations. Process efficiency, fossil fuel use, GHG emissions, water use, and economic profitability vary across different conversion technologies. In general, processes that produce both biofuels and power have higher efficiencies and displace more GHG emissions than those producing either biofuels or power. Integrating biological and thermochemical processing also lead to higher efficiencies than stand-alone processes.[16]

Gasification-based biorefineries appears particularly promising both technologically and economically for the pulp and paper industry.[17,18] The internal rate of return for this type of biorefineries is estimated to be between 14% and 18%. The return will be even higher when environmental benefits such as reduced pollution and GHG emissions are valued.

Biorefining technologies are still under development and advance rapidly. New improvements on these technologies are expected before they can be used for commercial production, and new technologies will also emerge. The uncertainty and evolution of conversion technology should be properly considered in designing IFBR supply chains.

2.3.2.4 Siting and Size of IFBRs

Selecting the location and size of IFBRs has important implications for the production cost of biofuels and bioproducts. The location of an IFBR should be carefully selected to minimize the overall transportation cost of biomass and end-products. Additionally, other factors like environmental conditions and access to roads may also influence the siting of IFBRs. A smaller IFBR will reduce the unit cost of feedstock transportation, but increase the unit capital cost because of the economies of scale. Hence, there exists an economically optimal size for an IFBR.[19] Decisions on IFBR siting and size can be complex and often involve the use of modeling tools such as mathematical programming, geographical information system, and other tools.[20]

2.4 A Decision Support Model for IFBRs

The key economic aspects of IFBRs as described above need to be considered systematically in designing IFBR supply chains. To this end, a mathematical (nonlinear) programming model is presented below. The model maximizes the net benefit of all IFBRs in a study area subject to a set of constraints.

Table 2.1 lists the decision variables (denoted by capital letters) that represent the key economic aspects of IFBRs discussed previously, and other symbols used to depict the model. This model is drawn on the one developed by Gan and Smith.[13]

(a) Objective function

$$Max\ \pi = \sum_f \left(p_f + b_f\right)Q_f - \sum_{il} cb_{il}X_{il} - \sum_{ijlm} ct1_{ijlm}Y1_{ijlm} - \sum_{iklm} ct2_{iklm}Y2_{iklm}$$
$$- \sum_{kjlm} ct3_{kjlm}Y3_{kjlm} - \sum_{iklm} cs_{kl}Y2_{iklm} - \sum_{jln} cr_{jln}U_{jln}$$

(2.1)

The objective function is to maximize the net benefit (π) of IFBRs in a given time period (*e.g.* one year or multiple years where the benefits or costs can be converted to equivalent annual benefits or costs) in a region that consists of multiple basic spatial units. The basic spatial unit can be considered as a geographic area that is small enough so that the difference in biomass transport

Table 2.1 Symbols for variables and parameters used in the model.

Symbol	Description
Subscripts	
f	Type of end-products
i	Basic spatial unit where biomass is grown
j	Location of refineries
k	Location of biomass storages
l	Type of biomass
m	Type of transportation modes
n	Type of conversion technologies
Parameters	
b_f	Unit cobenefit of end-product f (*e.g.* energy security premium, carbon offset credit, *etc.*)
cb_{il}	Unit cost of biomass *l* supplied at basic spatial unit *i* (including biomass growing, harvesting, and transporting to the centralized location within a basic spatial unit)
$ct1_{ijlm}$	Unit cost of transporting biomass *l* from basic spatial unit *i* to refinery *j via* transportation mode *m*
$ct2_{iklm}$	Unit cost of transporting biomass *l* from spatial unit *i* to storage *k via* transportation mode *m*
$ct3_{kjlm}$	Unit cost of transporting biomass *l* from storage *k* to refinery *j via* transportation mode *m*
cr_{jln}	Unit cost of converting biomass *l via* technology *n* at refinery *j*
cs_{kl}	Unit cost of storing biomass *l* at storage *k*
p_f	Price of end-product *f*
qr_j	Capacity of refinery *j*
qs_k	Capacity of storage *k*
$\bar{x}_{il}$	Maximum allowable amount of biomass *l* to be harvested in spatial unit *i*
θ	Proportion of biomass lost during storage
η_{lfn}	Conversion rate from biomass *l* to end-product f using technology *n*
Decision variables	
Q_f	Quantity of end-product *f* produced
U_{jln}	Amount of biomass *l* converted using conversion technology *n* at refinery *j*
X_{il}	Amount of biomass *l* produced/supplied from basic spatial unit *i*
$Y1_{ijlm}$	Amount of biomass *l* transported from spatial unit *i* to refinery *j via* transportation mode *m*
$Y2_{iklm}$	Amount of biomass *l* transported from spatial unit *i* to storage *k via* transportation mode *m*
$Y3_{kjlm}$	Amount of biomass *l* transported from storage *k* to refinery *j via* transportation mode *m*

cost within the unit is negligible. Thus, biomass supply from the unit can be approximated by a supply from the unit's center. Equation (2.1) is a nonlinear function because the unit storage and conversion costs are not constant. Instead, they are nonlinear functions of the production capacity of storage facilities and refineries (qs_k and qr_j), respectively. Benefits include (a) the revenues from the sales of end-products (including major products and byproducts/coproducts) and (b) cobenefits such as GHG offset credits, energy security premium, *etc.* Costs are incurred in growing, harvesting, and transporting biomass to a centralized location in the basic spatial unit; biomass transportation from a centralized site to a refinery or storage and from a storage facility to a refinery; biomass storage; and conversion of biomass to end-products. For simplicity, we assume that the cobenefits on a per-unit basis are constant for a specific end-product and that biomass will be supplied from a basic spatial unit at a constant price within its allowable supply/harvest level.

(b) Constraints

Several constraints are needed for the net-benefit maximization model. The first one is the biomass harvest constraint.

$$X_{il} \leq \bar{x}_{il} \tag{2.2}$$

This constraint ensures that the amount of biomass harvested from a basic spatial unit is below its maximum allowable level for each biomass source.

Biomass transportation constraints are also necessary. There are two biomass transportation constraints. One is to ensure that the amount of biomass shipped to all refineries and storages from each basic spatial unit does not exceed the amount of biomass harvested from the unit, *i.e.*

$$\sum_{jlm} Y1_{ijlm} + \sum_{klm} Y2_{iklm} \leq X_{il} \tag{2.3}$$

The other transportation constraint is intended to maintain the balance between the inflows and outflows of biomass to/from a storage facility. The total amount of biomass transported to a storage facility from all units, after adjusted for handling loss, should be equal to the amount of biomass shipped out to all refineries, *i.e.*

$$(1 - \theta) \sum_{im} Y2_{iklm} = \sum_{jm} Y3_{kjlm} \tag{2.4}$$

Next, the quantity of biomass transported to a storage or a refinery should not exceed the capacity of the storage or the refinery. That is,

$$\sum_{ilm} Y2_{iklm} \leq qs_k \tag{2.5}$$

$$\sum_{ilm} Y1_{ijlm} + \sum_{klm} Y3_{kjlm} \leq qr_j \tag{2.6}$$

Here we assume that the storage facilities and refineries can handle all types of biomass sources, an expected feature of advanced IFBRs. If this is not the case, constraints (2.5) and (2.6) can be modified accordingly.[13]

Furthermore, we need a biomass allocation/utilization constraint that takes the following form.

$$\sum_{n} U_{jln} = \sum_{im} Y1_{ijlm} + \sum_{km} Y3_{kjlm} \tag{2.7}$$

This constraint is to ensure that the total amount of a biomass source converted using various technologies at a refinery is equal to the amount of the biomass shipped to that refinery from all spatial units and storage facilities.

It also allows for the incorporation of conversion technology selection into the analysis.

Also, the quantity of each end-product produced can be expressed as

$$Q_f = \sum_{jln} \eta_{lfn} U_{jln} \qquad (2.8)$$

Obviously, eqn (2.8) can allow for the joint production of multiple end-products at a refinery, reflecting another key feature of IFBRs. Finally, all variables in eqns (2.1)–(2.8) are non-negative.

This model has several unique features that can guarantee the simultaneous optimization of key economic aspects of an IFBR supply chain. These key aspects include the type and amount of feedstocks to grow and harvest, feedstock transport and storage (it can also include pretreatments), siting and size of storage facilities and refineries, conversion-technology selection, and choices of end-products. It incorporates the economies of scale into cost computation as cs_{kl} and cr_{jln} are functions of the capacity of storages and refineries, respectively. The model allows for joint production of multiple end-products using various types of feedstocks and conversion technologies.

2.5 Policy Aspects of IFBRs

2.5.1 Major Barriers to IFBR Development and Deployment

There exist several barriers to the commercial development and deployment of IFBRs. Among these barriers, cost and uncertainty are perhaps the two most critical. First, the concept of IFBRs is relatively new, and many key technologies for IFBRs including conversion technology are still under development. As a result, the production cost of biofuels remains a major barrier to their market proliferation. Significant cost reductions will largely depend upon technology breakthroughs and are the premise for biofuels to play a significant role in energy markets and GHG emission mitigation. Secondly, given that IFBRs are new there exist uncertainties in feedstock supply, technology advancements, interactions with the markets of related products, environmental consequences, government policy, *etc.* Such uncertainties could impede the rapid deployment of IFBRs. Thirdly, energy, economies, and the environment are highly interrelated. Yet, market, institutional, and policy platforms for systematically addressing the interrelationships among energy, economies, and the environment are still lacking. This also represents a roadblock to the development of biofuels and bioproducts.

2.5.2 Policy for Enhancing IFBR Development and Deployment

There are several policy options that can promote the development and deployment of IFBRs. First, because key technologies for IFBRs are still under development, support for RD&D (Research, Development, and Deployment)

investments is essential. Such support should target the biofuels and bioproducts that are truly reflective of cost effectiveness, energy efficiency, environmental sustainability, and social acceptance. Secondly, among the technologies
needed for IFBRs, conversion technology is probably most critical to reduce
the overall production cost of biofuels.[19] Logistics is another important aspect
of IFBRs as it has a large influence on the overall production cost as well.
Hence, policy support for the advances of conversion technologies and logistic
systems should be a priority. Thirdly, experiences in OECD countries reveal
that market deployment policies have made a significant contribution to
increasing the market share of bioenergy.[21] These policies are particularly
important for new products like biofuels and bioproducts. Finally, the goals or
drivers of bioenergy are multifaceted. Unfortunately, these goals are often
reflected separately in energy, environmental or economic policies. Achieving
the overall effectiveness and synergy calls for the seamless coordination and
integration of energy, environmental, and economic policies.

Some of the above recommended polices are related to incentives and support for IFBRs. These incentives and support policies are necessary for
developing a new industry, though they could be a tough sale politically.
Certain social benefits of bioenergy such as GHG emission offsets and energy-
security enhancement are not reflected in the market price of bioenergy. Thus,
the policy support is essentially transfer payments for or endorsement to these
nonmarket benefits of bioenergy. Moreover, some of the incentives and policy
support can phase out as the industry is established.

2.6   Summary and Discussion

This chapter is intended to describe some economic and policy aspects of
IFBRs. IFBR supply chains are explained and analyzed. A mathematical
programming model for decision support on the optimization of IFBR supply
chains is presented. Policy options for overcoming barriers to IFBR development and deployment are discussed.

It is clear that IFBRs have great potential, especially when integrated with
the supply chains of traditional forest products. The development of a new
biofuels and bioproducts sector can complement the traditional forest products
industry. The complementarity between the two sectors will enhance the efficiency of resource use, overall economic return, and job creation, contributing
to the prosperity and sustainability of the industries and forest-based communities. On the other hand, the needed integration of IFBRs with the traditional forest products industry calls for systematical approaches to designing
IFBR supply chains.

Since there is no IFBR in commercial operation yet, there is a lack of realistic
data for economic analysis. Thus, many economic questions of IFBRs remain
to be answered. With the development of IFBRs, the economics of IFBRs will
evolve. Some new economic issues may emerge, though some existing ones will
be resolved, entailing new or improved policy measures and institutional
structures. For instance, how will or should IFBRs interact or cooperate with

the energy sector? To what extent, will IFBRs compete with the traditional forest products sector (most likely the pulp and paper sector) in terms of the demand for woody raw materials? And how will such competitions affect consumers and the sustainability of forest ecosystems? Hence, much work on the economics and policy of IFBRs remains to be done in order to better guide the development and deployment of IFBRs.

Acknowledgement

The work leading to this publication was supported by Texas AgriLife Research. However, the opinions are the author's.

References

1. United Nations Food and Agriculture Organization, *FAOSTAT*, Rome, Italy, 2011. http://faostat.fao.org/DesktopDefault.aspx?PageID=626& lang=en (last accessed May 2011).
2. U.S. Department of Energy, *U.S. Billion-ton Update: Biomass Supply for a Bioenergy and Bioproducts Industry,* Oak Ridge National Laboratory, Oak Ridge, TN, 2011. http://www.ascension-publishing.com/BIZ/Billion-Ton-2.pdf (last accessed August 2011).
3. M. de Wit and A. Faaij, *Biomass Bioenerg.*, 2010, **34**, 188.
4. J. Gan and C. T. Smith, *Biomass Bioenerg.*, 2006, **30**, 1011.
5. G. Puttock, *Biomass Bioenerg.*, 1995, **8**, 73.
6. M. Galbe and G. Zacchi, *Appl. Microbiol. Biotechnol.*, 2002, **59**, 618.
7. D. Beaudoin, L. LeBel and J. M. Frayret, *Can. J. For. Res.*, 2007, **37**, 128.
8. D. H. Burger and M. S. Jamnick, *For. Chron.*, 1995, **71**, 89.
9. D. Carlsson, S. D'Amours, A. Martel and M. Rönnqvist, *INFOR*, 2009, **47**, 167.
10. K. Kärhä and T. Vartiamäki, *Biomass Bioenerg.*, 2006, **30**, 1043.
11. Forest Products Association of Canada, *Transforming Canada's Forest Products Industry—Summary of Findings from the Future Bio-pathways Project,* http://www.fpac.ca/publications/Biopathways%20ENG.pdf (last accessed July 2011).
12. J. Gan, *Energ. Policy*, 2007, **35**, 6003.
13. J. Gan and C. T. Smith, *Int. J. For. Eng.*, in review.
14. J. Gan, A. Jarrett and C. Johnson, *J. Sustain. For.*, in press.
15. J. Gan and C. T. Smith, *Biofuels*, 2010, **1**, 539.
16. M. Laser, E. Larson, B. Dale, M. Wang, N. Greene and L. R. Lynd, *Biofuel. Bioprod. Bior.*, 2009, **3**(2), 247.
17. E. D. Larson, S. Consonni, R. E. Katofsky, K. Iisa and J. Frederick, *TAPPI J.*, 2008, **7**(11), 8.
18. E. D. Larson, S. Consonni, R. E. Katofsky, K. Iisa and J. Frederick, *TAPPI J.*, 2009, **7**(12), 4.
19. J. Gan and C. T. Smith, *Biomass Bioenerg.*, 2011, **35**, 3350.
20. H. An, W. E. Wilhelm and S. W. Searcy, *Biomass Bioenerg.*, 2011, **35**, 3763.
21. J. Gan and C. T. Smith, *Biomass Bioenerg.*, 2011, **35**, 4497.

CHAPTER 3

Integrated Forest Biorefineries: Sustainability Considerations for Forest Biomass Feedstocks

MARCIA PATTON-MALLORY,*[1] KENNETH E. SKOG[2] AND VIRGINIA H. DALE[3]

[1] USDA Forest Service, Pacific Northwest Research Station, 400 N. 34th Street, Suite 201, Seattle, WA, 98103, USA; [2] USDA Forest Service, Forest Products Laboratory, One Gifford Pinchot Drive, Madison, WI, 53726, USA; [3] Environmental Sciences Division, Oak Ridge National Laboratory, Bethel Valley Road, Building 1505, Room 200, Oak Ridge, TN, 37831, USA
*Email: mpattonmallory@fs.fed.us; Tel.: (206) 732-7846

3.1 Introduction

The concept of sustainability fundamentally considers three dimensions: economic, social and environmental. All three dimensions must be considered when discussing sustainability of specific biorefineries. Sustainability allows for levels of current use that do not compromise the capability to meet future needs. When considering sustainability over a region it is important to address the effect of multiple biorefineries, electric power plants that use wood, and traditional wood products production facilities. Forest bioenergy feedstocks for biofuels and biopower are potentially identical, and thus sustainability considerations discussed here apply to biofuels and biopower, in addition to production of traditional forest products. Biorefineries and biomass power

RSC Green Chemistry No. 18
Integrated Forest Biorefineries
Edited by Lew Christopher

plants will tend to initially be sited where there is low cost biomass and shorter hauling distances. This will favor locations where there are higher density forests and low-cost logging and mill residues from current forest products production.[1] As residues are used, plants will tend to be sited where pulpwood is available and competition is lowest.

For the purposes of this chapter, we use the term "biorefinery" to include all types of integrated biorefineries. Integrated biorefineries are similar to conventional refineries in that they produce a range of products to optimize both the use of the feedstock and production economics. Although most of the discussion will focus on practices and landscapes within the United States, some consideration is needed of wood used for energy and products in other countries because their level of use can influence our domestic use *via* global markets and changes in wood imports and exports.

Integrated biorefineries are expected to be large industrial facilities that produce 50–100 million gallons of biofuels per year.[2] Using an estimate of 70 gallons of biofuels per oven dry ton of forest biomass feedstock, this translates to using 0.714–1.430 million oven dry tons of forest biomass feedstock per year, which is about the amount used by a conventional sized pulp mill in the U.S.

Depending on the type of integrated biorefinery, and other coproducts (including energy), the feedstock demand could be higher. Furthermore, new biorefineries are more likely to be sited where there is currently a sufficient infrastructure in place that has historically or currently produces other forest products, such as lumber, pulp, wood fuel pellets and wood composite products, that may be complementary or compete for forest-based wood resources. Other biorefineries plan to add biofuels production to existing facilities, such as pulp mills, where the biofuel and traditional pulp products will both be produced from the same feedstocks.[3]

3.2 Background

The forest products industry in the United States has evolved with voluntary sustainability standards and best management practices (BMPs). State programs vary from voluntary BMPs to mandatory practices prescribed by statute.[4] A working definition of sustainable forestry is:

"The practice of meeting the forest resource needs and values of the present without compromising the similar capability of future generations. It involves practicing a land stewardship ethic that integrates reforestation, managing, growing, nurturing, and harvesting trees for useful products with conservation of soil, air and water quality, wildlife and fish habitat, and aesthetics."[5]

In addition to the guidelines for sustainable forestry, biorefineries are also expected to contribute to renewable energy goals linked with energy security, environmental improvement, rural wealth, and climate change mitigation goals. Sustainability guidelines for using agricultural or forest feedstocks for energy have been proposed by various groups. Examples of topics areas covered include:

- feedstock type (thinnings, harvesting residues, purpose-grown trees, pulp wood, *etc.*) and management;
- land use (type of land used to produce woody feedstocks and associated changes in land use such as conversion from naturally regenerating forests to plantations, or from farms to forests, public *vs.* private land ownership);
- harvest, collection, and processing of feedstock;
- transport of feedstock, fuel and location of refinery;
- fuel type, conversion process and blending;
- coproducts (heat, power, traditional forest products, biobased chemicals, *etc.*);
- net energy and energy efficiency (improvements over gasoline or diesel, energy used to produce the biofuels as compared to biofuel energy content);
- relation to existing forest industry (location, size, competition or integrated);
- social (participation by landowners, risk management, incentives, human health);
- economic (feedstock competition, transport, storage efficiency, risk, uncertainty, *etc.*).

Detailed discussions of sustainability for bioenergy as a system show how the economic, social, and environmental dimensions of sustainability are interconnected.[6] Economic aspects include coping with limited land resources; existing and emerging feedstocks; technical advances; and different feedstocks having different biological characteristics, resource requirements, and costs of production and transport. Social aspects include following applicable laws and international treaties, using open and transparent participatory processes that actively engage relevant stakeholders and establish rights and obligations, ensuring decent wages and working conditions and the safety of workers, and acknowledging worker rights to organize and collectively bargain. Major aspects of environmental sustainability for bioenergy systems are shown in Figure 3.1 and discussed by McBride and others, who identified 19 measurable indicators for soil quality, water quality and quantity, greenhouse gases, biodiversity, air quality, and productivity.[6] These indicators were selected to be both measureable and useful to decision makers in characterizing and assessing sustainable bioenergy.

3.3   U.S. Sustainability Frameworks and Policy

A comparison of existing U.S. forest certification schemes that provide management guidelines for sustainability in the forestry sector was compiled by

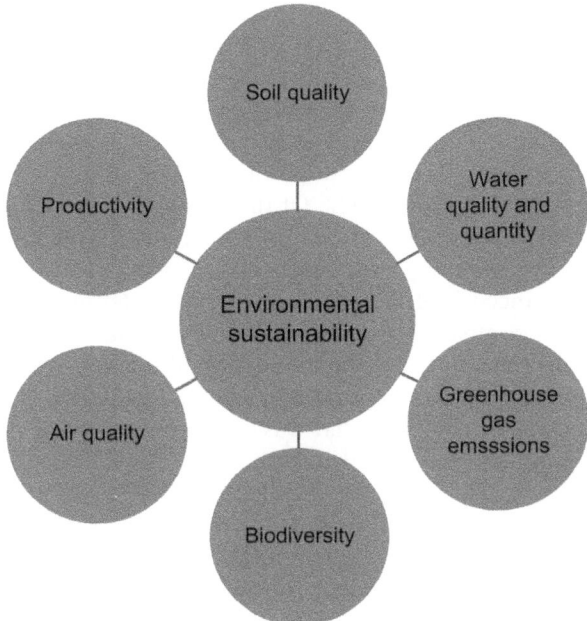

Figure 3.1 Aspects of environmental sustainability.

Rickenback with the Sustainable Forestry Partnership at Oregon State University.[7] The general features of the Sustainable Forest Initiative (SFI), Forest Stewardship Council (FSC) and American Tree Farm (ATF), Green Tag Forestry, and International Organization for Standards (ISO) are described, along with the management and ecology focal areas, operational considerations, and social and economic factors. In general, these programs are voluntary third-party certification systems that forest landowners or managers use to identify forestland that is managed to meet agreed-upon standards and sometimes to label products that originate from these forests.[8] Certification is a market-based, nonregulatory forest conservation tool designed to recognize and promote environmentally responsible forestry, and sustainability of forest resources.[9]

States with significant forestry activities have adopted Best Management Practices (BMPs) to ensure water quality. However, BMPs were not intended to directly address practices such as increased removal of logging residues. To address biomass removal levels more specifically, a number of states have developed voluntary state harvesting or retention guidelines that indicate the extent to which biomass may be removed from harvest sites for biofuels or biopower. States that currently have adopted harvesting and/or retention guidelines include Maine, New Hampshire, Vermont, New York, Rhode Island, Connecticut, Massachusetts, Michigan, Minnesota, Missouri, Pennsylvania, Maryland, Wisconsin, and California.[4]

The Energy Independence and Security Act (EISA 2007) contains a restrictive definition of "renewable biomass" as a way to address some of the sustainability concerns that had been raised in developing the Renewable Fuels Standard 2 (RFS2); however, this definition raised other concerns regarding its impact on public land management and private forest economics.[10] The RFS2 mandates U.S. production of biofuels with a goal of producing 21 billion gallons of advanced biofuels by 2022. Of this total, 16 billion gallons are to be made from cellulosic feedstocks. The definition of "advanced biofuels" includes thresholds for reductions in greenhouse-gas emission profiles as compared to gasoline, with advanced biofuels required to meet a 60 per cent reduction in GHG emissions as compared to gasoline. Woody biomass feedstocks that can be counted toward the cellulosic ethanol portion of the RFS2 under EISA 2007 only include forest residues from state and private forest plantations or woody energy crops harvested from land that was inactively managed tree plantations in 2007. This definition is designed to discourage land clearing for bioenergy feedstock production. However, the detailed biomass definition in EISA 2007 excludes all federal forests and private lands not managed as a tree plantation in 2007. It thereby restricts the use of a market that could facilitate thinning or residue removal and thus could reduce the high fire risk of many federal forests due to overstocked stands and other forest health concerns.[11] Table 3.1 summarizes the restrictions on feedstocks from forest biomass as the law is currently written, and contrasts the EISA 2007 definition with the broader definition of "renewable biomass" included in the Food Conservation and Energy Act of (2008).[12] The Farm Bill also has programs that support increased production of biofuels. Tracking the current status of biomass definitions in various laws is critical because producing biofuels from qualified feedstocks that meet the "renewable biomass" definition in the EISA 2007 allows biofuels producers to receive the $1.01/gallon tax incentive payment.

State Forestry agencies and the U.S. Forest Service provide guidelines for private family forest landowners that include sustainable forestry principles.[13] These agencies also develop Stewardship Plans for private forest landowners to apply these principles and achieve landowner objectives. Having a Stewardship Plan or Certification may become a requirement for qualifying under some federal incentives and payments for producing biofuels feedstocks.

A group called the Council on Sustainable Biomass Production (CSPB) is developing a voluntary certification system for bioenergy feedstocks.[14] Table 3.2 contrasts the CSBP certification system with existing voluntary forestry certification systems and state forestry stewardship guidelines.[13,15,16] A new challenge for CSBP and other groups trying to develop standards for certification is determining appropriate ways to assess greenhouse-gas emissions.

An example of local efforts to address sustainability of forest biomass feedstocks is the city of Gainesville, Florida. They developed a Forest Stewardship Incentive Plan to help assure "that the manner in which biomass was procured (for biopower) would not only minimize any environmental harm, but transform the forestry industry to improve biodiversity in the region and assure sustainable supplies of renewable biomass energy."[17]

Table 3.1 Different "Renewable Biomass" definitions in the EISA Renewable Fuels Standard and Farm Bill.

Energy Independence and Security Act of 2007	Food, Conservation, and Energy Act of 2008
Renewable biomass means each of the following (including any incidental, *de minimis* contaminants that are impractical to remove and are related to customary feedstock production and transport):	RENEWABLE BIOMASS- The term 'renewable biomass' means:
(1) *Planted crops and crop residue harvested from existing agricultural land cleared or cultivated prior to December 19, 2007 and that was nonforested and either actively managed or fallow on December 19, 2007.*	(A) *materials, precommercial thinnings, or removed exotic species that—*
**(2) *Planted trees and tree residue from a tree plantation located on nonfederal land (including land belonging to an Indian tribe or an Indian individual that is held in trust by the U.S. or subject to a restriction against alienation imposed by the U.S.) that was cleared at any time prior to December 19, 2007 and actively managed on December 19, 2007.*	(i) *are byproducts of preventive treatments (such as trees, wood, brush, thinnings, chips, and slash), that are removed—*
(3) *Animal waste material and animal byproducts.*	(I) *to reduce hazardous fuels;*
**(4) *Slash and precommercial thinnings from nonfederal forestland (including forestland belonging to an Indian tribe or an Indian individual, that are held in trust by the United States or subject to a restriction against alienation imposed by the United States) that is not ecologically sensitive forestland.*	(II) *to reduce or contain disease or insect infestation; or*
	(III) *to restore ecosystem health;*
	(ii) *would not otherwise be used for higher-value products; and*
	(iii) *are harvested from National Forest System land or public lands (as defined in Section 103 of the Federal Land Policy and Management Act of 1976 (43 U.S.C. 1702), in accordance with—*
	(I) *Federal and State law;*
	(II) *applicable land management plans; and*
	(III) *the requirements for old-growth maintenance, restoration, and management direction of paragraphs (2), (3), and (4) of subsection (e) of section 102 of the Healthy Forests Restoration Act of 2003 (16 U.S.C. 6512) and the requirements for large-tree retention of subsection (f) of that section; or*
**(5) *Biomass (organic matter that is available on a renewable or recurring basis) obtained from the immediate vicinity of buildings and other areas regularly occupied by people, or of public infrastructure, in an area at risk of wildfire.*	

Table 3.1 (Continued)

Energy Independence and Security Act of 2007	Food, Conservation, and Energy Act of 2008

(6) Algae.
(7) Separated yard waste or food waste, including recycled cooking and trap grease, and materials described in §80.1426(f)(5)(i).
	(B) any organic matter that is available on a renewable or recurring basis from non-Federal land or land belonging to an Indian or Indian tribe that is held in trust by the United States or subject to a restriction against alienation imposed by the United States, including:
	(i) renewable plant material, including
	  (I) feed grains;
	  (II) other agricultural commodities;
	  (III) other plants and trees; and
	  (IV) algae; and
	(ii) waste material, including—
	  (I) crop residue;
	  (II) other vegetative waste material (including wood waste and wood residues);
	  (III) animal waste and byproducts (including fats, oils, greases, and manure);
	  (IV) construction waste; and
	  (V) food waste and yard waste.

Table 3.2 Comparison of various sustainable forestry certification systems used in the United States.

Sustainability Principle or Criteria	Council on Sustainable Biomass Production (draft)	Sustainable Forestry Initiative (SFI) Certification	Forest Stewardship Council (FSC) Certification	National Association of State Foresters Stewardship Principles
Soil Productivity- Forest Productivity and Health	**3.1 Soil** This principle recognizes that soil stability is vital, and that soil fertility and organic matter are critical to the sustainable production of food, feed, fiber, and fuel. PRINCIPLE: Biomass production shall maintain or improve soil quality by minimizing erosion, enhancing carbon sequestration, and promoting healthy biological systems and chemical and physical properties.	**2. Forest Productivity and Health** To provide for regeneration after harvest and maintain the productive capacity of the forest land base, and to protect and maintain *long-term* forest and soil *productivity*. In addition, to protect forests from economically or environmentally undesirable levels of wildfire, pests, diseases, *invasive exotic plants and animals* and other damaging agents and thus maintain and improve *long-term forest health and productivity*.	**Principle 6. Environmental Impact** Forest management shall conserve biological diversity and its associated values, water resources, soils, and unique and fragile ecosystems and landscapes, and, by so doing, maintain the ecological functions and the integrity of the forest.	**Principle 2.** Maintain and improve productive capacity. **Principle 4.** Protect soil and water resources. **Principle 3.** Maintain and improve the health and vigor of the forest and itslandscape/watershed.
Biological Diversity	**3.2 Biological Diversity** The conservation of biological diversity is a critical component of sustainability at the field/ stand level as well as at the landscape level. This principle articulates the expectation that growers will deploy management	**4. Protection of Biological Diversity** To manage forests in ways that protect and promote *biological diversity*, including animal and plant species, *wildlife habitats*, and ecological or natural community types.	**Principle 6. Environmental Impact** Forest management shall conserve biological diversity and its associated values, water resources, soils, and unique and fragile ecosystems and landscapes and, by so doing, maintain the ecological	**Principle 1.** Contribute to the conservation and biological diversity of the forest and the surrounding landscape.

Table 3.2 (*Continued*)

Sustainability Principle or Criteria	Council on Sustainable Biomass Production (draft)	Sustainable Forestry Initiative (SFI) Certification	Forest Stewardship Council (FSC) Certification	National Association of State Foresters Stewardship Principles
	systems in their operations that maintain or enhance biodiversity. PRINCIPLE: Biomass production shall contribute to the conservation or enhancement of biological diversity, in particular native plants and wildlife.		functions and the integrity of the forest. **Principle 9. Maintenance Of High Conservation Value Forests** Management activities in high conservation value forests shall maintain or enhance the attributes that define such forests. Decisions regarding high conservation value forests shall always be considered in the context of a precautionary approach.	
Water	3.3 Water This principle recognizes the vulnerability of both the available water supply and the quality of available water. Biomass production should not contribute to the depletion of ground or surface water supplies. When irrigation is necessary, the most efficient irrigation technology appropriate to the circumstance should be used.	3. Protection of Water Resources To protect water bodies and *riparian* zones, and to conform with *best management practices* to protect water quality.	**Principle 6. Environmental Impact** Forest management shall conserve biological diversity and its associated values, water resources, soils, and unique and fragile ecosystems and landscapes; and, by so doing, maintain the ecological functions and the integrity of the forest.	

Climate Change	PRINCIPLE: Biomass production shall maintain or improve the quality and quantity of surface water, groundwater, and aquatic ecosystems. **3.4 Climate Change** One fundamental objective of biomass-based bioenergy systems is to mitigate GHG emissions, providing a low-carbon energy alternative to fossil fuels. This principle embraces full lifecycle assessment (LCA) as the primary tool for ensuring substantive reduction in GHG emissions. PRINCIPLE: Biomass production shall reduce GHG emissions as compared to fossil fuels. Emissions shall be estimated *via* a consistent approach to lifecycle assessment.		**Principle 5.** Pursue carbon-friendly management and promote biomass as a renewable energy source
Social and Economic	**3.5 Socioeconomic Well-Being** CSBP embraces a tripartite vision of sustainability, focusing on practices and products that are environmentally, socially and economically sound. This principle	**6. Protection of Special Sites** To manage forests and lands of special significance (ecologically, geologically or *culturally important*) in a manner that *protects* their integrity and takes into account their unique qualities.	**Principle 3. Indigenous People's Rights** The legal and customary rights of indigenous peoples to own, use and manage their lands, territories, and resources shall be recognized and respected.	**Principle 6.** Consider socioeconomic benefits

Table 3.2 (*Continued*)

Sustainability Principle or Criteria	Council on Sustainable Biomass Production (draft)	Sustainable Forestry Initiative (SFI) Certification	Forest Stewardship Council (FSC) Certification	National Association of State Foresters Stewardship Principles
	speaks to the need for sustainable distribution of socioeconomic benefit to the various participants in biomass and bioenergy production systems. A sustainable commercial model benefits from the support of wealth creation in local communities. PRINCIPLE: Biomass production shall take place within a framework that sustainably distributes overall socioeconomic opportunity for and among all stakeholders (including land owners, farm workers, suppliers, biorefiners, and local community), and ensures compliance with labor laws and human rights.	**7. Responsible Fiber Sourcing Practices in North America** To use and promote among other forest land-owners *sustainable forestry* practices that are both scientifically credible and economically, environmentally and socially responsible.	**Principle 4. Community Relations And Worker's Rights** Forest management operations shall maintain or enhance the long-term social and economic well being of forest workers and local communities.	
Legality	**3.6 Legality** Compliance with all legal requirements by a grower is a minimum expectation for the standard.	**8. Avoidance of Controversial Sources Including Illegal Logging in Offshore Fiber Sourcing** To avoid wood fiber from	**Principle 1. Compliance With Laws And FSC Principles** Forest management shall respect all applicable laws	**Principle 7.** Comply with laws, rules and guidelines.

	PRINCIPLE: Biomass production shall comply with applicable federal, provincial, state, and local laws, ordinances, and regulations.	illegally logged forests when procuring fiber outside of North America, and to avoid *sourcing fiber from countries without effective social laws*. **9. Legal Compliance** To comply with applicable federal, provincial, state, and local *forestry* and related environmental laws, statutes, and regulations.	of the country in which they occur, and international treaties and agreements to which the country is a signatory, and comply with all FSC Principles and Criteria. **Principle 2. Tenure And Use Rights And Responsibilities** Long-term tenure and use rights to the land and forest resources shall be clearly defined, documented and legally established.
Transparency Public Involvement	**3.7 Transparency** The interactions of a participant with stakeholders must be conducted in a transparent manner while protecting commercially sensitive information and maintaining intellectual property. PRINCIPLE: Production of certified biomass shall be transparent.	**12. Public Involvement** To broaden the practice of *sustainable forestry* on *public lands* through community involvement. **13. Transparency** To broaden the understanding of forest certification to the *SFI 2010–2014 Standard* by documenting certification audits and making the findings publicly available.	
Use of Best Available Science Continuous Improvement Training and Education	**3.8 Continuous Improvement** CSBP is committed to a process of continued assessment of the usefulness of the standard's practices to ensuring the	**10. Research** To support advances in sustainable forest management through *forestry* research, science and technology.	**Principle 8. Monitoring And Assessment** Monitoring shall be conducted appropriate to the scale and intensity of forest management to assess

Table 3.2 (Continued)

Sustainability Principle or Criteria	Council on Sustainable Biomass Production (draft)	Sustainable Forestry Initiative (SFI) Certification	Forest Stewardship Council (FSC) Certification	National Association of State Foresters Stewardship Principles
	desired sustainability outcomes. The standard will be updated periodically, incorporating scientific results that reveal better practices that are commercially viable. Growers are also expected to continuously improve performance as guided by annual certification audits. PRINCIPLE: Biomass production practices and outcomes shall continuously improve based on the best available science.	**11. Training and Education** To improve the practice of *sustainable forestry* through training and education *programs*. **14. Continual Improvement** To continually improve the practice of forest management, and to monitor, measure and report performance in achieving the commitment to *sustainable forestry*.	the condition of the forest, yields of forest products, chain of custody, management activities and their social and environmental impacts.	
Multiple Use/ Integrated Resource Planning/Sustained Yield	**3.9 Integrated Resource Management Planning** The preparation of and adherence to a complete management plan is considered essential to ensure that a grower can deliver on the multiple requirements for sustainable production.	**1. Sustainable Forestry** To practice *sustainable forestry* to meet the needs of the present without compromising the ability of future generations to meet their own needs by practicing a land stewardship ethic that integrates *reforestation* and the	**Principle 5. Benefits From The Forest** Forest management operations shall encourage the efficient use of the forest's multiple products and services to ensure economic viability and a wide range of environmental and social benefits.	**Principle 3.** Maintain and improve the health and vigor of the forest and its landscape/watershed.

Principle 7. Management Plan

A management plan appropriate to the scale and intensity of the operations shall be written, implemented, and kept up to date. The long-term objectives of management, and the means of achieving them, shall be clearly stated.

Principle 10. Plantations

Plantations shall be planned and managed in accordance with Principles and Criteria 1–9, and Principle 10 and its Criteria. While plantations can provide an array of social and economic benefits, and can contribute to satisfying the world's needs for forest products, they should complement the management of, reduce pressures on, and promote the restoration and conservation of natural forests.

PRINCIPLE: Biomass production shall be based on an integrated resource management plan that shall be completed, monitored and updated to address objectives of the CSBP standard, appropriate to the scale and intensity of the operation.

managing, growing, nurturing and harvesting of trees for useful products and ecosystem services such as the *conservation* of soil, air and water quality, carbon, *biological diversity, wildlife* and *aquatic habitats,* recreation, and aesthetics.

5. Aesthetics and Recreation

To manage the visual impacts of forest operations, and to provide recreational opportunities for the public.

3.4   International Sustainability Frameworks and Policy

Biorefineries operating in the U.S. may be producing biofuels that are traded in international markets. This will require an understanding of international standards and certification schemes for sustainable biofuels production. The Roundtable on Sustainable Biofuels (RSB) has released "Principles & Criteria for Sustainable Biofuels Production" as a basis for stakeholder discussion around requirements for sustainable biofuels.[18] Their intent is to implement the standards through certification systems for biofuels operations throughout the world, issuing certificates to recognize operations that meet their criteria.

Recently, the Global Bioenergy Partnership (GBEP) released their sustainability indicators for bioenergy, aiming to help countries assess and develop sustainable production and use of bioenergy.[19] The 24 indicators are divided into Environmental, Social and Economic Pillars.

The European Union recently released the "EU Renewable Energy Directive" that established a certification scheme for biofuels that include rising targets over time for greenhouse-gas saving over gasoline and diesel, and rigorous external auditing throughout the supply chain.[20] The focus to date has primarily been on agricultural rather than forest-based feedstocks.

3.5   Sustainability Topics to Watch

The policy discussion around accounting for indirect land-use changes associated with biofuels production, and their potential greenhouse gas effects has been a major point of public debate in the U.S and internationally. This concept is based on the premise that use of land to grow bioenergy feedstocks in one location results in other farmers responding to market signals (higher commodity prices) and producing the displaced crop, which can cause greenhouse emissions, depending on prior land use. Currently, global economic models (such as FAPRI) are used to estimate this effect by examining land-use change worldwide with more ethanol production in various countries, but those models do not have general land-use change in them.[21] The FASOM model for the U.S. does have the ability to consider tree planting on agricultural land for either biofuels feedstocks or timber production.[22] Furthermore, various disciplines explain land changes *via* distinct theories and have different spatial resolutions and interpretations of indirect land-use change and its implications. There is some discussion of this topic in the development of the U.S. Renewable Fuels Standard (RFS2) Summary and Analysis of Comments and a significant part of the EU standard's protocol.[23] The extent to which biofuel production induces indirect land-use change and the associated quantity of greenhouse-gas emission is not resolved.

A second topic relates to assessing the lifecycle emissions associated with biofuels, where it is often assumed that emissions from wood biomass are carbon neutral. This assumption is being examined by the U.S. EPA in development of guidelines to restrict greenhouse-gas emissions from certain stationary sources, such as electric power plants. The uncertainty about the carbon-offset benefits of wood biomass burning for energy was considered in a draft rule released by EPA that identifies stationary GHG emissions sources that need to have permits to emit GHGs.[24] US EPA requested information on the issue of wood biomass carbon neutrality and as a result of those responses has suspended for three years any rule requiring such biomass emissions to be limited in the same way as fossil emissions (Environmental Protection Agency 2011). Research is needed to clarify the carbon-offset benefits of using wood for energy, and how scale of the analysis, in both time and space, influences the outcome. It seems likely that findings about the GHG offset benefits of using wood for electric power could influence how GHG offset benefits are estimated for wood-based biofuels.

The definition of "renewable biomass" and what qualifies for various bio-fuels incentives is also likely to be a contentious point of debate as various federal and state incentives are adopted and revised. Current definitions are summarized in a recent Congressional Research Service report (Table 3.1).[25]

Acknowledgements

A portion of this research was supported by the U.S. Department of Energy (DOE) under the Office of the Biomass Program. Oak Ridge National Laboratory is managed by the UT-Battelle, LLC, for DOE under contract DE-AC05-00OR22725. Additional support was provided by the USDA Forest Service. Matt Langholtz, Marilyn Buford and Frederick Deneke provided useful comments in reviews of an earlier draft of the manuscript.

References

1. K. E. Skog, R. Rummer, B. Jenkins, N. Parker, P. Tittman, Q. Hart, R. Nelson, E. Gray, A. Schmidt, M. Patton-Mallory, and G. Gordon. 2009. A strategic assessment of biofuels development in the Western States. In: W. McWilliams, G. Moisen, R. Czaplewski, comps. Forest Inventory and Analysis (FIA) Symposium 2008; October 21–23, 2008; Park City, UT. Proc. RMRS-P-56CD. Fort Collins, CO: USDA, Forest Service, Rocky Mountain Research Station. 13 p. http://www.treesearch.fs.fed.us/pubs/33372.
2. Western Governors' Association. 2008. Strategic Assessment of Bioenergy Development in the West: Bioenergy conversion technology characteristics. Report by the Antares Group, Inc. September 1, 2008. Western Governors' Association, Denver, CO. http://www.westgov.org/component/joomdoc/doc_download/214-wga-bioenergy-assessment-conversion-tech.
3. D. Dodgeon, 2010. Pulp mills as modern biorefineries: Positioned for fiber, fuels and chemicals, Outlook: North America, 2010, Paper 360, 8–11.

4. A. M. Evans, R. T. Pershel, and B. K Kittler. 2010. Revised assessment of biomass harvesting and retention guidelines. Forest Guild, Sante Fe, NM, 33 p.
5. Society of American Foresters. The Dictionary of Forestry. J. Helms, ed. Bethesda, MD, 1998, 210 p.
6. A. McBride, V. H. Dale, L. Baskaran, M. Downing, L. Eaton, R. A. Efroymson, C. Garten, K. L. Kline, H. Jager, P. Mulholland, E. Parish, P. Schweizer and J. Storey., Indicators to support environmental sustainability of bioenergy systems, *Ecological Indicators*, 2011, **11**(5), 1277–1289.
7. M. Rickenback. Comparison of Forest Certification Schemes of Interest to USA Forest Owners. Sustainable Forestry Partnership at Oregon State University, 1999. (http://sfp.cas.psu.edu/pdfs/Certification_matrix.pdf).
8. Oregon State University Extension. Forest Certification in North America. 2006. EC 1518.
9. Pinchot Institute for Conservation. Certification: Definition and Background. 2011. http://www.pinchot.org/project/59.
10. EISA. 2007. Energy Independence and Security Act of 2007. http://energy.senate.gov/public/_files/getdoc1.pdf.
11. D. R. Becker, D. Larson and E. C. Lowell, Financial considerations of policy options to enhance biomass utilization for reducing wildfire hazards, *Forest Policy and Economics*, 2009, **11**(8), 628–635.
12. Food Conservation and Energy Act of 2008. http://www.usda.gov/documents/Bill_6124.pdf.
13. Forest Stewardship Council. Principles and Criteria for Forest Stewardship. Washington, DC. 1996. 10p. http://www.fscus.org/images/documents/FSC_Principles_Criteria.pdf.
14. Council on Sustainable Biomass Production. 2009. October 2009. http://www.csbp.org/files/survey/CSBP_Draft_Standard.pdf.
15. Sustainable Forestry Initiative. 2010. Sustainable Forestry Initiative 2010–1014 Standard. 15p. http://www.sfiprogram.org/files/pdf/Section2_sfi_requirements_2010–2014.pdf.
16. National Association of State Foresters. Stewardship handbook for family forest owners. NASF, Washington, DC. 2009. 28p. http://sfp.cas.psu.edu/pdfs/NASFStwdshphandbook.pdf.
17. City of Gainesville. Stewardship Incentive Plan for Biomass Fuel Procurement. City of Gainesville Regional Utility Committee and the Ad Hoc Sustainable Biomass Procurement Committee. April 2009. 128p.
18. Roundtable on Sustainable Biofuels. Principles & Criteria for Sustainable Biofuels Production, RSB-STD-01-001 version 1.0. 2009. http://www.csbp.org/.
19. Global Bioenergy Partnership. 2011. GBEP Sustainability Indicators. May 2011. http://www.globalbioenergy.org/news0/detail/en/news/79357/icode/.
20. European Union. EU Renewable Energy Directive. 2010. http://ec.europa.eu/energy/renewables/biofuels/sustainability_criteria_en.htm.
21. J. F. Fabiosa, J. C. Beghin, F. Dong, A. Eliobeid, S. Tokgoz and T.-H. Yu, Land allocation effects of the global ethanol surge: predictions

from the international FAPRI model, *Land Economics*, 2010, **86**(4), 687–706.

22. D. M. Adams, R. J. Alig, J. M. Callaway, B. A. McCarl, and S. W. Winnett. The forest and agriculture sector optimization model (FASOM): model structure and policy implications. Res. Pap. PNW-RP-495. Portland, OR. U.S. Department of Agriculture, Forest Service, Pacific Northwest Research Station. 1996. 60p.

23. Environmental Protection Agency. Renewable Fuels Standard Program (RFS2) Summary and analysis of comments. 2010. http://www.epa.gov/oms/renewablefuels/420r10003.pdf.

24. Environmental Protection Agency. Deferral for CO_2 Emissions From Bioenergy and Other Biogenic Sources Under the Prevention of Significant Deterioration (PSD) and Title V Programs: Proposed Rule. Federal Register Vol 76 (54). March 21, 2011.

25. K. Bracmort and R. W. Gorte. Comparison of definitions in legislation. Congressional Research Service, CRS7-5700, R40529, 2010. 30 p.

CHAPTER 4

Integrated Forest Biorefineries: Product-Based Economic Factors

BRIAN L. COOPER,* JEFFREY R. LONDON,
ROBERT J. MELLON AND MICHAEL A. BEHRENS

Hazen Research, Inc., 4601 Indiana St., Golden, CO 80403, USA
*Email: cooperb@hazenresearch.com; (303) 279-4501

4.1 Introduction

The integrated forest biorefinery has been described as a factory built to exploit the chemistry of natural resources for the manufacture of a wide variety of chemicals, fuels, and other intermediates. Taking advantage of the complex polymeric structures evident in naturally occurring biomass feedstock is a logical goal and one with enticing potential, given the reactivity (and digestibility) of carbohydrates as well as the similarity of the variety of phenylpropenyl units extant in lignin residues to valuable and complex chemical commodities. In principle, this makes good sense; however, recalcitrant lignocellulosic biomass processing encounters many challenges when moving from plant structures to salable products.

Most cost models put a definitive mark on yields at each step, below which the economics of the process are unfavorable. The greatest return for extra effort (in biorefining and just about any other process) lies in the early steps; as a direct cost driver, the feedstock and associated transport costs can seemingly

RSC Green Chemistry No. 18
Integrated Forest Biorefineries
Edited by Lew Christopher

make or break a venture. Increasing the amount per acre harvested has plant biologists across the world attempting to select or engineer biomass that allows for reduction in the cost of feedstock by any means (ignoring for now the complications regarding harvesting and drying procedures, the incremental cost of transporting biomass, and the cost of sizing the biomass). As in most research projects, the last bit of recovery proves to be the most difficult in processing lignocellulosic feed material into monomeric carbohydrate slurries appropriate for fermentation. Given that the resultant sugars will begin to degrade under the same pretreatment conditions in which they are liberated from their polymeric forms, those sugars released early in the process are difficult to collect because the reaction is still underway. The mature pretreatment/ saccharification process is therefore a delicate balance of closely controlled aggressive chemistry that will yield the desired product and remove it from the system before the majority of the carbohydrates begin to react into furans and furfurals. All of this is performed in a vessel that can handle high temperatures, pressures, and aggressive chemistry in a reducing environment, normally under constant duty.

As one would expect, this requires very specialized equipment, particularly when dilute acid hydrolysis is the primary pretreatment mechanism. Alternative chemistries have shown promise and may well be superior in terms of carbohydrate liberation; however, each still results in a difficult process to control. Even minor changes in feedstock particle size, not to mention structural chemistry, can render yields and reagent usage unacceptable. Contrast this with the goals of the bulk feedstock supplier. It is in the interest of the forest harvester to transport feedstock to the refinery in bulk form; it is therefore in the interest of the refinery to be able to accept this bulk form without excessive handling. The incremental costs of progressive size reduction eat at the margin of specialized processes and are of particular interest with biomass because regional climate variations can produce very different feeds in a forest, not to mention similar feeds that respond to sizing techniques differently due to their age or relative water content. Can a biorefining process be developed for a particular feedstock that will result in low costs and high recoveries? Absolutely. The issue is when the feedstocks that are evaluated in the design basis of the plant either change or become too expensive; the impact of either development on the process economics is identical in that the cost of goods increases.

Consideration of whole-process design close to the beginning of the experimental stage can yield surprising and potentially advantageous results. While a particular unit operation may initially appear costly and inefficient, the downstream benefit can be shown to offset these perceived drawbacks. This same paradigm can be applied to bulk feedstocks, those that are deemed too variable and unacceptable for one mode of operation can be utilized more readily in a plant designed to accept the greatest variation possible. A small number of experimental test series can be carried out to look at the relationships between processing parameters and give the process engineer hope that managing a number of sensitive unit operations in the plant will yield a profitable, integrated process.

For the business manager of such an enterprise, there is also potential for a partial respite from the hair-raising volatility of energy and fuels markets in supporting the development of a product mix from the biorefinery that can be altered without extensive plant retrofits to suit economic realities. The key lies in a commitment to serve multiple markets with biorefinery products, along with the requisite capability and planning to develop a flexible biorefinery that can modify product output to some extent. This type of design flies in the face of those of us schooled in the science of unit operation optimization, but we should take solace in the knowledge that purposely accepting a low yield in one operation need not impact the overall process efficiency and may be the key to competing in mature energy and chemical markets, such as they are.

4.2   Biorefinery Operational Parameters

The biorefinery typically operates according to some variation of the diagram shown in Figure 4.1.

As the variations on this basic flowsheet are many, it behooves one to examine those processes that are shared by all: feedstock supply and pretreatment. In many cases this would be defined as harvesting and sizing the feedstock, while in others it may involve extensive processing in order to supply satisfactory woody materials to the more advanced processes of the biorefinery. In general, if the process being studied is a regional forest-supplied biorefinery, the most important parameters are tonnage fees paid to suppliers and operational expenses at the refinery involved with sizing and chemical treatment of the lignocellulosic materials. Because the chemical makeup of the available

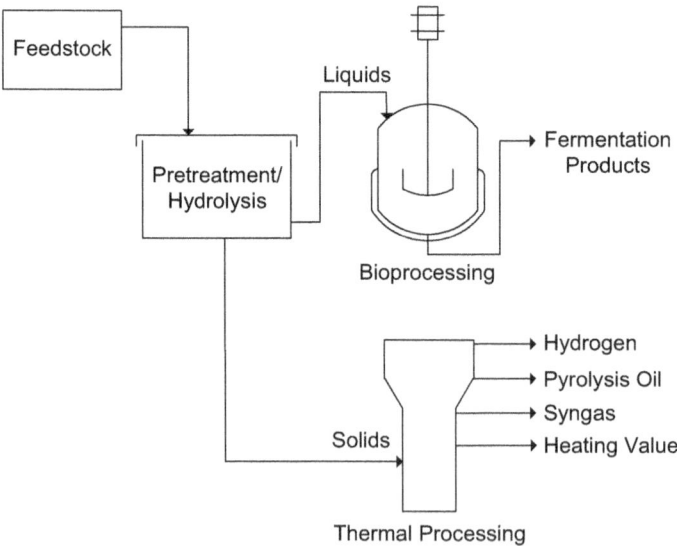

Figure 4.1   A common biorefinery flowsheet.

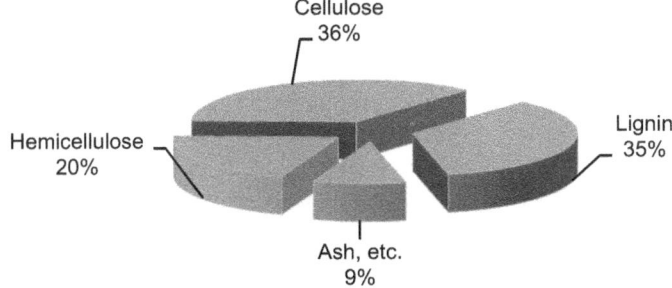

Figure 4.2   The chemical makeup of a generic softwood feedstock.

feedstocks for biorefineries has been covered in detail elsewhere in this publication, this section will assume the use of softwood for discussion purposes. Though there are slight variations, the basic makeup of the softwoods studied in detail at Hazen Research, Inc. are as shown in Figure 4.2 (from *Pinus ponderosa* and *P. contorta*).

The crucial parameters for the engineer attempting to realize value from these potential sources assume a maximal yield of carbohydrate in the pretreatment/hydrolysis step. Normally a foregone conclusion as a worthy operational goal, this target bears scrutiny.

4.3   Hydrolysis Yield Impact on Economic Models

The United States Department of Energy National Renewable Energy Laboratory (NREL) has been developing an integrated biorefinery Aspen[1] model for general distribution and in hopes of generating interest and providing process knowledge to the biorefining industry in the United States. The model assumes the target product is primarily cellulosic ethanol, but hydrolysis residues contribute to the cost model by way of heating value. This model, in addition to the Laboratory Analytical Procedures (LAP)[2] that are released and updated for the public's use, provides the researcher with some valuable tools in the science of feedstock characterization. Unless otherwise noted, all the analytical results in this chapter as they relate to feedstocks and specific carbohydrates were generated following the NREL LAPs. The most recent distribution of the associated Aspen model was announced and released in May 2011.[3] It should be noted that the NREL model utilizes both an acid pretreatment and an enzyme hydrolysis; we have focused on the pretreatment here so as to limit the scope of the discussion.

Besides providing a great deal of data surrounding actual yields in biorefining operations, the model provides sufficient information about the cost drivers of the products, so that one can change some of the assumptions made and calculate what the expected impact of the change will be. Using this type of data manipulation, one can then plot the various outcomes of different yield scenarios. While not entirely linear, the theoretical selling price of ethanol from the model NREL biorefinery increases with a decrease in hemicellulose hydrolysis. This is shown in Figure 4.3.

This is a completely predictable result, given the assumptions of the model, which again posit that the residues are contributing to product revenue only in regard to the potential power generation realized by combustion. Given that carbohydrates with their associated oxygen will yield a less energy-intensive residue, one would naturally assume that the best use of feedstock is in removing the sugars prior to power generation. It is apparent that increasing acid concentration in pretreatment (to a point) will yield higher carbohydrate recoveries as well as higher heating value residues, as shown in the proximate and ultimate analysis. This is shown in Table 4.1.

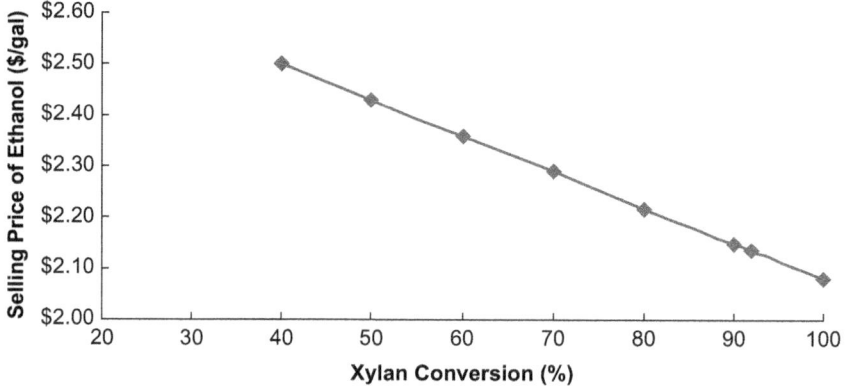

Figure 4.3 Selling price of ethanol generated utilizing the process modeled by NREL as impacted by carbohydrate yields in acid pretreatment.

Table 4.1 Biomass residues after acid hydrolysis subjected to proximate and ultimate analysis.

First-stage temperature (°C)			175			140		
Acid (wt%)	0.05	0.10	0.20	0.40	0.05	0.10	0.20	0.40
Sugar (% yield)	36.8	54.7	39.7	74.9	37.5	55.3	58.6	67.7
Proximate								
Volatile	84.2	83.8	81.5	76.2	83.2	84.6	79.8	77.1
Fixed Carbon	15.0	15.5	17.8	23.4	14.4	11.8	18.4	20.9
Btu/lb (HHV)	9023	9149	9295	9542	8913	8972	9157	9484
Ultimate								
Carbon	52.3	52.1	54.4	56.3	50.9	51.3	53.0	54.2
Hydrogen	5.72	5.62	5.81	5.77	5.61	5.48	5.61	5.42
Oxygen	41.0	41.4	38.0	36.1	40.9	39.4	39.3	37.9
Sulfur	0.05	0.03	0.05	0.12	0.04	0.06	0.09	0.19
F-Factor (DSCF/million Btu)	9091	8861	9369	9505	8931	8960	9123	9004

The combustion exhaust in each case is roughly equivalent when looking at the F-factors, but in this case the sulfur content of the exhaust (and the associated SO_x) increases with more aggressive application of sulfuric acid. One would assume the same correlation for NO_x and those processes that perform pretreatment with nitric acid. It is clear that looking at the residues as strictly a combustion feedstock, the data steer the process toward complete carbohydrate removal prior to any thermal processing. As such, the sunk costs associated with this removal are assumed to be a given in the cost model of many processes.

These assumptions are sound for the set of conditions specified in most cost models, yet have a fundamental weakness due to the variability of the chemistry of feedstocks. While general assumptions can be made about the makeup of various feeds, the yield assumptions in the models assume a particular chemistry that can be understood and incorporated into process economics, and will remain essentially unchanged. This can be a grievous error, and one that can impact processing costs to a large degree. Just as the agricultural residues available for processes will change year on year depending on many different environmental factors and fertilization regimes, so trees harvested from different regions within the same forest can and do present measurable variation, both within the same harvest year and in year to year comparisons.

The answer to this issue has been to paint feedstock with the widest brush possible, developing processes looking at what are assumed to be the largest deviations from the norm and setting process parameters that can accommodate these. This is in direct conflict with the requirements of the process that need to recover the maximum amount of carbohydrate from these same feedstocks. While some techniques are more robust than others, it is generally understood that pretreatment reagent usage and equipment sizing must be calibrated to a feed that is fairly well understood, unchanged, and available in bulk year after year.

It may well be that this feedstock does not exist. As an example, the Colorado forests of the Rocky Mountains comprise lodgepole pine, ponderosa pine, spruce, limber pine, and Douglas fir (*P. contorta, P. ponderosa, Picea pungens, Pinus flexilis, Pseudotsuga menziesii subsp. Glauca,* respectively) as well as many other associated woody trees and shrubs. Bulk harvesting of these forests will include some mix of these, with the primary constituent understood and marketed as such. Therefore a load of "ponderosa pine" mulch will likely contain many different species in some unknown and changing concentration. This disparity of feedstock consistency would of course be greater when harvesting forests still in the Rockies but a thousand miles to the North; one assumes that a bulk harvest from any forested region would present similar diversity. In order to examine what effects this changing mix would have on a biorefinery, Hazen has looked at a number of feedstocks in isolation. While the argument could be made that product effects seen in isolated feeds may not manifest in the same manner in a mixed feed, the discussion must start somewhere.

Our examination included two pines, lodgepole and ponderosa, that are both evergreen American hard pines (Section *Trifoliae*), which are normally harvested for various construction and landscaping purposes on both sides of the North American continental divide. In addition to these two, a subset of lodgepole pine that represents something of a special case was evaluated. In the last few years, large portions of the Rocky Mountain forests in North America have been beset by an infestation of the mountain pine beetle (*Dendroctonus ponderosae*) that has left large swaths of forest literally standing dead. While the causes for this infestation are beyond the scope of the current text, the end result has been that wood harvested from any of these forests will likely contain both desiccated dead wood as well as fresh trees. The appearance of these forests is striking, because the needles of the dead trees turn red and are noticeable even to the casual observer. Given the mood of the population that resides in this region, there has been considerable political and therefore research interest in attempting to make use of this wood. We began our evaluation of biorefinery feedstock using harvests of "beetle kill" trees because the material was readily available. As we engaged in our study, we included the fresh-harvested lodgepole, as well as ponderosa, for comparative purposes. Our initial findings were surprising and became the first few data points in what would evolve into this study of the biorefinery from a product-driven perspective.

Each of the feedstocks was utilized in an as-received state from a lumberyard; higher carbohydrate recovery could certainly have been attained with a finer particle size distribution. It was our intent to examine the possibilities of pretreating/hydrolyzing biomass in bulk harvest to get an idea of how much sugar material could be recovered without optimizing the feed. As such, samples included in our analysis regularly contained rocks, dirt, and other detritus commonly mixed in with large wood piles. This feed was slurried with a dilute acid solution in a titanium pressure vessel, then brought to hydrolysis temperatures utilizing direct steam injection. Figure 4.4 shows the carbohydrate

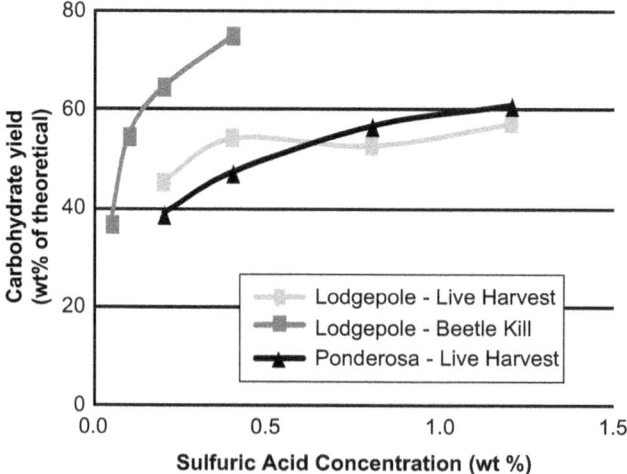

Figure 4.4 Total sugar recovery from softwood feedstocks.

liberation from live harvest lodgepole, dead standing lodgepole (beetle kill), and live harvest ponderosa feed as the acid concentration in pretreatment is increased, all other conditions being equal.

The results were somewhat unexpected, particularly with regard to the beetle kill when compared with the fresh harvest lodgepole. The variables that could explain the observed differences are many, as are the questions that are suggested. Did the beetle kill (standing dead) lodgepole have a greater carbohydrate content because the passive dessication and subsequent loss of any volatile organics? Perhaps this drier feedstock is more friable in a wood chipper, resulting in more fines and a subsequent surface area increase. Is 0.4% acid really more effective on lodgepole than it is on ponderosa? Would it benefit the biorefiner to implement a passive drying of feedstock prior to chipping, and if so, how does this add to the incremental cost of the feedstock? Finally, because the data set was culled from a single year's harvest, will the results be identical as different regions of the forest are harvested in the coming year, or will a year-on-year change in local weather patterns cause the hydrolysis data to show a different trend if the experiments are repeated?

The data are stark in demonstrating the issues, for example – how much time does a bioprocessor have to store biomass before it begins to break down and affect the chemistry of the process? Which conditions should be selected for the commercial plant? How aggressive should the chemistry be, and what equipment (size and materials of construction) will be specified to run the process at scale? These questions have answers, but in general they are qualified and thus not comprehensive. The early goal of the lignocellulosic process to be feedstock-agnostic was discarded along with the processes that would present consistent challenges in order to fit this definition.

4.4   Benefits of Product-Driven Operational Parameters

Understanding the point at which many North American biorefinery companies find themselves is aided by a (very) brief summary of recent history. The market factors that have influenced the development of biorefineries in the United States are complex and varied, and they are still sending mixed signals to all interested parties. In the last 24–30 months, the United States has seen a shift in priorities of biofuel startup companies; the passage of the Energy Independence and Security Act of 2007 encouraged many of these companies to focus on cellulosic ethanol production under the requirements of the Renewable Fuel Standard. The early focus was justifiably set on maximizing the ethanol production out of a biorefinery process, a somewhat challenging task when utilizing a bioagent that respires carbon dioxide as a part of fermentation.

The publication of Policy Research Working Paper 4682 (WPS4682) by the World Bank in 2008 is generally accepted as the first shot in the food-versus-fuel debate, asserting that increasing biofuel production in the United States was a major factor in the worldwide increase in food commodity prices. Despite the subsequent release of WPS5371, which backed away from the initial charge against biofuels, the political winds in the United States had changed, calling into question the continued existence of mandated tax rebates for ethanol producers. While lobbying extensively for their tax-advantaged status, startup companies also surveyed their options, in most cases finding that the potential for alternative products from their processes represented a way to add certainty to their business plans by taking government mandates out of the equation, if only in part. Therefore, a company that had completed its original fundraising as XYZ Biofuels now changed its focus, and changed its name to XYZ Bio-products, or simply XYZ Technologies, or some similar variation.

The product offerings also changed; where ethanol was the initial primary product with the opportunity to use process side streams for other purposes, chemicals – particularly replacements for petrochemical-derived intermediates – became the target. Many of the new process flowsheets included not only a fermentation, but a catalytic reaction that could produce these chemicals. Often the reaction was a hydrogenation, which is not surprising given the aromatic content of most proposed lignin structures.

The production of hydrogen in synthesis gas (syngas) then becomes much more important to the biorefiner than its role as a way of producing process heat. The value of residual carbohydrate in thermal feedstocks then goes up, as the more favorable H/C ratio of carbohydrates (1.4–1.7 as opposed to less than 1 in typical aromatic lignin units) changes the economics of a reduced carbo-hydrate yield in the pretreatment/hydrolysis stage. As a demonstration of how this can be exploited in the biorefinery, the syngas produced from feedstock subjected to different pretreatment procedures can be evaluated as it is pro-duced by the gasifier.

By way of a brief review, in typical dilute acid hydrolysis experiments the biomass charge is slurried with reagents in a pressure vessel, brought to tem-perature, and then the contents are flashed into a catch vessel. This allows fairly precise control of the time that the feed is subjected to hydrolysis conditions because the expanding slurry drops in temperature very quickly. A common adjunct to dilute acid hydrolysis is the steam explosion, where the same process is performed without the acid, with the theory that the rapid expansion of water inside the cellular systems in the biomass will rupture recalcitrant structures and facilitate hydrolysis, by reagents or by enzymes.

Figure 4.5 shows a plot of the syngas generated from hydrolysis residues that had been subjected to a steam explosion overlain with a plot of those that hadn't.

The difference appears slight; however, the trend was observed in multiple experiments utilizing different biomasses and different hydrolysis conditions. While the hydrogen effusion in all these reactions tends to tail off at a similar concentration in the gas (one assumes this is the baseline Gibbs free energy to liberate the hydrogen from the heart of the aromatic lignin), the relative amount of hydrogen can be estimated by the curve maxima as well as the incidence of hydrogen in the early stages of the experiment (essentially the duration). The addition of water in the steam explosion appears to be hydrogenating the residues to some degree, and this effect manifests in a higher

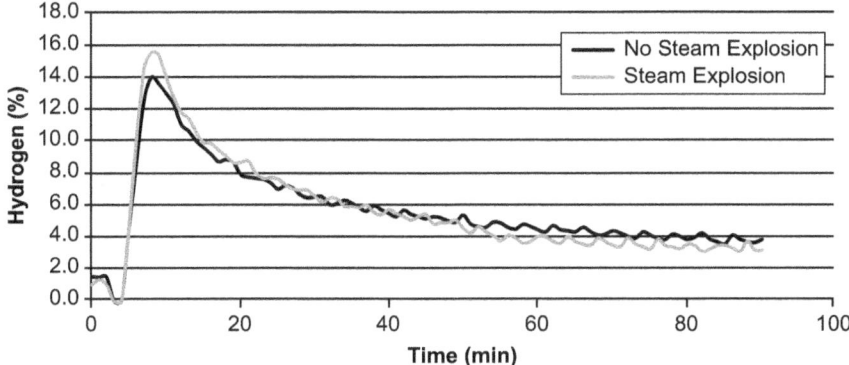

Figure 4.5 Hydrogen in syngas comparison between steam explosion biomass and nonsteam explosion biomass.

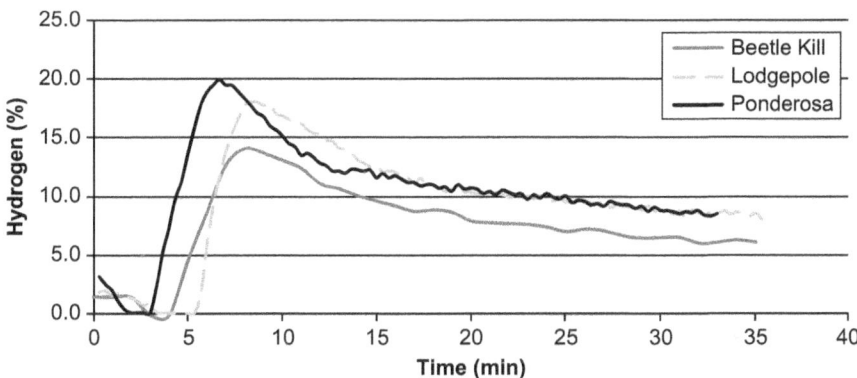

Figure 4.6 Hydrogen percentage in syngas from various feedstock hydrolysis residues.

hydrogen yield in the gasification. While the additional steam explosion did not appear to affect the carbohydrate yield for the better or for the worse, it may seem a waste of energy and time at first glance. However, the hydrogen yields imparted by the extra pretreatment step likely justify the extra work, provided that gasification is to be performed on the residues generated by the pretreatment.

As with the hydrolysis data, the gasification of residues that have very similar proximate and ultimate analyses yields results that would not be expected from the fuel values. Each of the experiments shown in Figure 4.6 activated the gas analysis equipment when the kiln heating cycle was started.

Note the difference in the times that hydrogen concentrations began building. It would appear that ponderosa's structural hydrocarbons are more amenable to gasification and will yield more hydrogen as well. With hydrogen generation in the product mix, one can see the advantage of processing this feedstock when it is available, or at least when it represents the majority of the various woody constituents of an available feed.

4.5   Value of Residues

The issue that complicates assigning relative values to the various product streams of a biorefinery is that one is compelled to assume a liquid fuel Btu is equivalent to a boiler fuel Btu. Each energy output from a biorefinery is competing in markets whose variations complicate a comparison to some degree. As a cursory analysis of the liquid fuel market (even in a single geopolitical region) and its contributing factors would likely require a volume as large as this entire textbook, heating value may present a simpler case when evaluating a process stream. In our analysis, it is taken to represent a worst-case valuation of side streams for budgeting purposes. Though an examination of the heating values does not account for all the various factors that may go into a stream's relative worth, it is a sufficient place to begin to build a comparative study.

Table 4.2 shows a simple study of the equivalent dollar value of a short ton of feedstock depending on the product ratio from the biorefinery as defined in the NREL models. In this example, it is assumed that the two product options are ethanol and heating value assigned to the resultant pyrolysis oil. Comparing a couple of test cases with a theoretical maximum, and the yield defined by NREL, and assuming that pyrolysis oil could be marketed as a heating oil with a reduced BTU value, one can see that substantial changes in product output yield revenues at roughly equivalent levels.

There is a slight revenue advantage to the conversion of carbohydrates in this model; however, one can imagine that the capital and operational expenditures needed to reach high levels of carbohydrate conversion would likely mitigate this advantage. These data show the diminishing returns in targeting the last bit of yield from a biomass: a \$2/st revenue advantage to increasing the carbohydrate yield from 90 to 100%. Knowing what we do about the difficulty of obtaining that last 10%, we are given confidence in the model because it captures the tradeoff very well.

As in the nature of the feedstock itself, it is folly to assume that all pyrolysis oil is equivalent. Take the example of the pyrolysis oil produced from the three feedstocks discussed in the previous section. Examining the resultant oil from these different residues from identical pretreatment experiments, one can see

Table 4.2   Product revenue comparison per dry ton of feedstock.

Assumptions	Carbohydrate Yield, %	Ethanol, lb	Pyrolysis Oil, lb	Total Revenues[a], $USD
Theoretical Maximum	100	783	382	269
NREL Model	90	688	549	267
Inefficient Hydrolysis	50	391	1070	262
No Hydrolysis	0	0	1759	256

[a]Assumes \$1.79/gallon Ethanol, \$1.05/gallon pyrolysis oil, based on proportional Btu content of heating oil.

Table 4.3 Fuel analysis of pyrolysis oil generated from hydrolysis residues.

	Single Phase	Heavy Oil		Light Oil	
Residue Type	Beetle Kill Ponderosa	Lodgepole	Ponderosa	Lodgepole	
	Lodgepole Fresh Harvest	Fresh Harvest	Fresh Harvest	Fresh Harvest	
Ultimate (weight %)					
Water	55.6	20.8	14.7	74.7	72.4
Ash	0.31	0.27	0.24	0.02	0.05
Sulfur	0.07	0.32	0.38	0.24	0.13
Carbon	32.1	59.7	60.8	11.1	12.8
Hydrogen	1.75	5.46	6.02	0.71	1.04
Nitrogen	0.02	0.03	<0.01	<0.01	<0.01
Oxygen	10.1	13.4	17.9	13.2	13.6
Proximate (weight %)					
Water	55.6	20.8	14.7	74.7	72.4
Ash	0.31	0.273	0.24	0.02	0.05
Volatile Matter	42.5	72.12	80.2	25.3	27.6
Fixed Carbon	1.58	6.81	4.88	<0.01	<0.01
Calorific Value					
BTU/lb	5238	9904	10147	1100	1934

variations that have significant impacts on the process design, Table 4.3 shows the fuel analysis of the oil.

Note that while the aggregate fuel values are roughly equivalent, there is a phase separation in the product of fresh harvest residue that is not present in the pyrolyzed beetle kill. This difference could certainly be exploited in the commercial operation, because the water partitions into the light-oil phase in large part, increasing the calorific value of the heavy oil by its subtraction. It is also apparent that whatever volatile organic compounds are lost during the desiccation that occurs after tree death, are likely oxygen-heavy, because the beetle kill lodgepole-derived product produces a more stable oil. An oil that is to be developed for commercial use benefits from the lower oxygen content because the polymerization of the organic acid content would be delayed to some degree. Alternatively, if the oil is to be used as a hog fuel boiler feed, additional oxygen works in favor of the application, imparting a higher heating value. In either application of course, emissions can be an issue and so the F-factors must be evaluated, and the resultant comprehensive value of the stream could be augmented or discounted appropriately.

One can imagine the impact of the appearance of products with these characteristics in a commercial operation that was not designed to handle a two-phase oil or was designed in expectation of the phase separation that does not occur due to the age of the feedstock. Nearly every measure that is used to put a value on the pyrolysis oil is impacted by the appearance (or lack of) the phase separation. Again, these disparate products are the result of identical experimentation in which the only variable is the condition of the wood at harvest in the field.

In this case, it is clear that a commercial biorefinery utilizing a bulk feedstock that is primarily beetle kill would likely benefit from an increased carbohydrate recovery in pretreatment/hydrolysis because the resultant pyrolysis oil is of lower value due to the water content/lack of phase separation. If process metrics are to require a given carbohydrate recovery with no allowance for process changes based on feedstock, this flexibility would have to come in the form of revenue generated from the pyrolyzed residue, which of course in the example would vary wildly with mixtures of feedstocks that produce either of the two products. It would not be unreasonable to look at this revenue picture in some instances and forego pyrolysis entirely, shunting the revenues right to incineration for their heat value alone.

These data represent a couple of series of experiments performed on some relatively pure feedstocks; for the commercial enterprise, a much broader study would obviously be warranted to substantiate the general findings. Given the number of correlated variables discovered in our small study, it could be expected that mixtures of the feeds would introduce even more variability to the product mix. As biorefineries struggle to enter extremely competitive and well-established energy markets with their products, any variability that can be understood and controlled in the commercial setting eliminates a potential barrier to entry.

4.6 Thermochemical Options

Utilizing the thermochemical product offerings to drive harvesting and pre-treatment process decisions is a legitimate pursuit. Pyrolysis and gasification are the initial process options to consider and thermochemical catalysis, long a staple of fine chemical production from petrochemicals (and upgrading of some of those same chemicals), is currently in development and used in commercial stages in many biorefineries. The product options are vast when one considers the value of carbohydrates, and even their degradation products, in a synthesis operation.

As a brief aside, the majority of the examples presented in this comparison assume that the thermochemical processing will take place after the carbo-hydrates are removed. Processes do exist whereby feedstocks are gasified as the initial process step in the biorefinery; and the resultant carbon monoxide, carbon dioxide, and hydrogen are used as nutrients in fermentation. One can imagine that feedstock variations would have less of an affect on this type of operation; one also assumes that these processes would have to be carried forward with supplemental hydrogen, which carries its own costs and hazards.

As with pyrolysis oil, syngas has a relative worth based on the chemical makeup, which is more complex than that suggested by the calorific value and that can be influenced by the pretreatment technology. Table 4.4 shows the makeup of syngas generated from a 0.4% acid hydrolysis residue from the three feedstocks.

The calorific values of these three gases are essentially the same, as are the hydrogen values and carbon monoxide/carbon dioxide ratios, suggesting that

Table 4.4   Syngas constituents from gasification of hydrolysis residue.

	Beetle Kill Lodgepole	Fresh Harvest Lodgepole	Fresh Harvest Ponderosa
Hydrogen (%)	14.1	13.1	13
Carbon Monoxide (%)	41.8	43.5	42.7
Carbon Dioxide (%)	16	10.9	14.3
Calorific Value (kJ/kg)	38 869	40 475	38 782

Table 4.5   Tedlar bag sample analysis of gasification products.

	Beetlekill	Lodgepole 0.40% Acid	Ponderosa
Hydrogen (%)	14.14	13.1	12.96
Carbon Monoxide (%)	41.77	43.46	42.7
Carbon Dioxide (%)	15.97	10.86	14.3
Calorific Value (kJ/kg)	38 868.59	40 475.07	38 781.62

	Lodgepole		Ponderosa	
Constituent (mole %)	No Steam	Steam Explosion	No Steam	Steam Explosion
Helium	0.0	0.0	0.0	0.0
Hydrogen	10.9	12.0	6.40	15.3
Oxygen	1.32	0.68	9.42	5.67
Carbon monoxide	47.0	42.6	39.6	38.0
Carbon dioxide	9.32	13.31	17.20	13.50
Methane	28.2	27.7	22.7	24.5
Ethane	1.32	1.30	1.33	1.11
Ethylene	0.65	0.74	0.86	0.58
Propane	0.38	0.50	0.67	0.47
N-butane	0.03	0.06	0.08	0.06
C4 Olefins	0.11	0.12	0.16	0.12
N-pentane	0.03	0.09	0.04	0.00
C5 Olefins	0.03	0.03	0.16	0.06
Hexanes	0.73	0.91	1.41	0.64
Calorific Value (kJ/kg)	37 484	38 538	28 034	40 605

gasification is not affected by the feedstock type to the degree that pyrolysis appears to be. When different pretreatment conditions are introduced to the same biomass, a change is seen in the residue gasification, as can be seen in Table 4.5; the hydrolysis experiments that generated these residues were conducted with 1.2% acid, and the samples were collected in Tedlar bags.

It is apparent that the effect of the steam explosion in the pretreatment extends beyond the hydrogen content for these two feedstocks. The trends are intriguing in that the addition of the steam explosion step is beneficial for both types of residue from the standpoint of the calorific value. In the lodgepole experiments, it appears that the higher alkanes are generated at the expense of carbon monoxide, while the ponderosa shows a general reduction in the higher alkane incidence, carbon monoxide is unaffected, and the carbon dioxide is significantly reduced.

This type of data set would suggest a repeat of the experiment, which was performed and yielded similar results. The mole fractions for some of the alkanes are low enough that general conclusions can be made about their presence; the overriding point is that thermochemical processes will produce different products given small changes in the treatment of the feed.

The generation of these products does not always work in favor of the biorefiner. A number of the hydrolysis experiments were conducted using more aggressive chemistry (essentially identical concentrations of acid but higher hydrolysis temperatures in the pressure vessel) in an attempt to yield more carbohydrates. While the carbohydrate yield did increase somewhat, the formation of hydroxymethylfuran and other degradation products also increased. These residues were then gasified, resulting in the data shown in Figure 4.7.

The delayed appearance of carbon monoxide in the syngas was also seen in the plots of data for hydrogen and for carbon dioxide. After verifying that this phenomenon was not an artifact of the experimental technique, it was surmised that the higher temperatures utilized in the hydrolysis began a gasification in the pressure vessel and that the syngas generated was lost in the flash at the end of the experiment. This results in a lower syngas yield, as well as in a shift in the time that syngas generation began. In a commercial operation, this should be avoided because it has the negative effect of degrading more carbohydrates and introducing fermentation inhibitors, as well as reducing yields in gasification. Such a change was seen in all experiments that attempted to utilize the higher temperatures and was consistently observed in plots of all the gasification products that were measured.

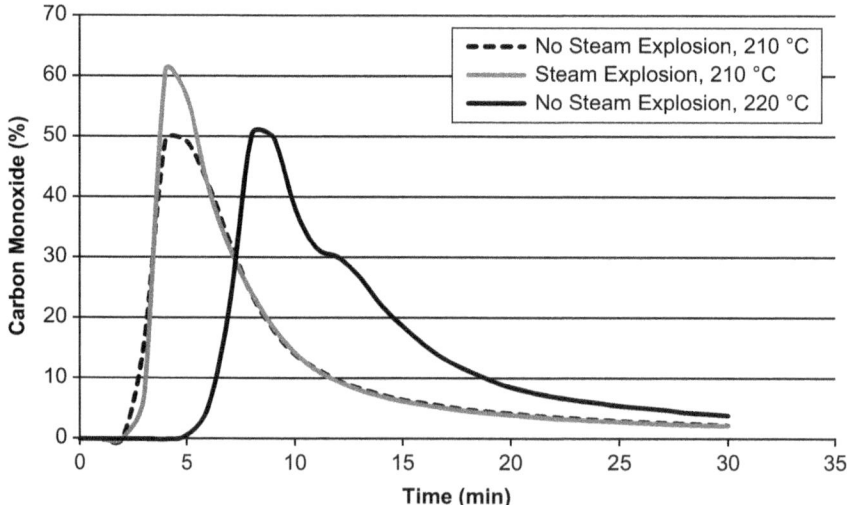

Figure 4.7 Carbon monoxide generation from the gasification of residues subjected to differing temperatures in hydrolysis.

4.7　Integrated Processing

These data point the process engineer to an integrated biomass refinery where pretreatment can be altered to fit the product demands of the business, which one hopes will be driven by the potential for revenues through the sale of energy-producing products. The biorefiner has control over the process design, but he must be cognizant of the need for operational flexibility; this control is largely surrendered once bulk shipments of biomass feedstock begin. The best laid plans will not prevent the situation arriving when material secured from forests differs somewhat in makeup from the feedstock that was used to develop the process. As one of the larger components of the cost of goods, accepting feedstock at preferential price points should always be a goal. In our experience, this results in a heterogeneity that is difficult to predict and therefore must be accommodated to the fullest extent possible.

The biorefiner utilizing the technologies examined in this work has the advantage of building flexibility into the unit operations in order to meet this need. Dilute acid hydrolysis can be run in batch or semibatch fashion, allowing for the addition of a steam explosion step if the overall process calls for it. Likewise, increasing or decreasing acid concentrations can be obtained without major changes to the installed capacity, assuming the materials of construction are such that the most aggressive conditions are acceptable. Additionally, thermal processing units are often installed with the capability of running in a gasification or a pyrolysis mode depending on the desired product generation. It should also be noted that competing pretreatment technologies that utilize different chemistries would likely see similar benefits with relatively small changes in operations. Just as the conclusions herein regarding dilute acid were drawn from a reasonable set of experiments, the study could be repeated for additional technologies in order to verify the assumption.

The basic flowsheet for a biorefinery capable of the variable processing suggested is less linear and more complex than the basic version presented at the start of the chapter, as seen in Figure 4.8.

An arrangement of this sort allows a spectrum of options for the feedstocks, from simple cofiring of those that are either of such low quality that further processing is not economically viable to an integrated catalysis that can make use of multiple processing technologies through hydrogenation or other synthetic reactions.

The design of such a process gives the biorefiner an option to divert feed to process heat recovery at just about every step in the process. Alternatively, high-grade or easily processed feed can be taken through as many operations as margins allow. In this way, the impact of both feedstock variation as well as market fluctuations can be absorbed by the process to a point; the understanding here is that not running the plant at all is not a viable long-term option. Of course there will certainly be imbalances brought about by the changing mode of operation; our desire is to present a different way of thinking about process development that will avoid too narrow a focus. Often the biotransformation step will utilize proprietary technology that changes the

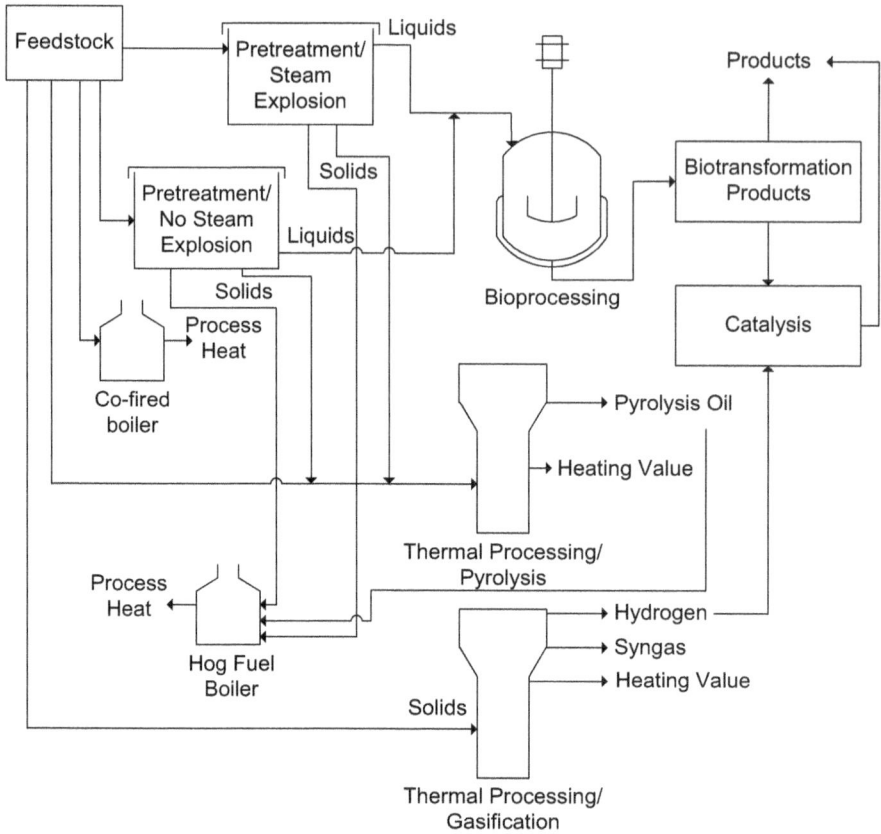

Figure 4.8   A basic integrated biorefinery flow sheet.

revenue picture and truly forms the backbone of many venture business plans. The goal of the integrated biorefinery model presented is the same, build in flexibility early in order to mitigate risk once you are to market.

4.8   Conclusion

It is clear from the limited data presented in this chapter that variations in the feed influence the downstream processing in an integrated biorefinery and that the incidence of this variation can take the form of something as simple as the age of the feed. Because others have presented a great deal of information on the heterogeneity of biomass harvested in bulk, we can take it as a given that the integrated biorefinery will have to accept process inefficiencies, unless it is designed with a certain amount of process variability that can mitigate the effect of early processing on the product mix. The studies conducted in support of this chapter were able to identify demonstrable correlations early on; one can imagine that a larger research effort on mixed feeds could contribute to this knowledge base, identifying additional cause and effect relationships that could be exploited in the name of stabilizing income from product sales.

The data also present a cautionary tale, in that unit operations that are optimized in development to a maximal recovery may not function this well in commercial practice. The research engineer is then encouraged to weigh the benefit of the effort it takes to squeeze the last bit of product out of early process steps. It may well be that the next year's harvest will not respond to the process in an identical manner, requiring a return to the development cycle. No process designer would want to work on such a process that will continually push the return on research investment back by continually eating up precious development resources.

As the integrated biorefinery is intended (at some level) to replace the traditional petroleum refinery, it stands to reason that a similar market-based manufacturing approach should be adopted. Just as the petroleum refiner initially set out to recover kerosene, the biorefiner likely began with the idea that they would be manufacturing fuel ethanol from lignocellulosic sources. The evolution of the modern petroleum refinery has carved the path that the biorefiner can follow, designing truly integrated operations with built-in flexibility to accommodate feedstocks that may or may not be well understood. One should not forget that the petroleum refinery has gone through many iterations in its long history, adapting to not only changing feeds, but changing market demands, changing environmental regulation, and changing geopolitical importance; it only stands to reason that the biorefinery is in for the same treatment. The larger concern is whether the biorefinery will be given sufficient time to adjust to changes in market realities; expecting these new plants to spring up fully integrated and optimized is folly. The researcher and the design engineer know how much effort must go into each unit operation, and are now learning that the race has just started. This same effort must be expended again to ensure that one part of the process is not degrading the performance of another.

A key factor that has not yet been addressed is the upstream push of the market forces of commodity products. The entry of biorefined petroleum replacements into the larger energy and chemical markets will certainly be a lengthy and messy affair. The notion that United States refiners used to dump gasoline in rivers as industrial waste is laughable to the modern engineer. Not quite so amusing is the history of international oil markets with their propensity to encourage collusion amongst multinational corporations (and countries for that matter), politically influenced tariffs and protectionism, and wild boom and bust periods. While some may see the biorefinery as the entity that can end this destructive cycle by returning at least some control over energy production to the local level, it is more likely that biorefineries will find the same market forces in a microcosm as fluctuations will hit regional areas as opposed to the global impact of movements in petroleum manufacturing. It would not be a great stretch to imagine two local energy/chemical producers in conflict over a local biomass supply or inadvertently driving up the cost of agricultural products in their demand for a particular feed. Such effects have only been speculated upon to date, but even hypothetical effects have been enough to influence policy.

A common criticism (often self-directed) of biorefining is that the collective research community is evaluating technologies that were used in earlier phases of industrialization and abandoned for the tempting energy density and availability of petroleum. For all the staggering success in biotechnology and microbiology in the last 50 years, there has been no great breakthrough that suddenly changes the game, allowing the energy/chemical markets the easy choice of product replacement with sustainable alternatives. It may well be that after 150 years of process development and market demands, some of the hardest problems the researcher faces may yield some of the greatest opportunities. As we are forced to deal with the here and now, steadfast design principles must be applied: integrate the process so the energy loss in one portion can be recovered in another, minimize waste, and strive to accept the most diverse feedstock available. Lacking success in those three areas, it is hard to see how a biorefinery can compete in the long term.

Of course advances in technology are not limited to biorefining, and a very real market mover will be the incremental cost of fossil fuels. One would assume that the general consensus among the readers of this book is that the need to find cost-equivalent replacements for fossil sources of energy and chemicals is urgent and real. We would posit that innovation is not limited to any particular branch of science, that there may be a breakthrough in energy production that changes the economic picture permanently, and it may put forward a new favored source: coal, natural gas, petroleum, or nuclear. The point is that one should not assume that expected market phenomena will gradually remove fossil fuels from the picture. Rather, it is best to assume that biorefinery products will be expected to compete on a level field with their intrinsic benefits laid bare. The investment in the research that will allow this has just begun.

References

1. Aspen Plus. Burlington, MA: Aspen Technology. http://www.aspentech.com/products/aspen-plus.cfm.
2. http://www.nrel.gov/biomass/analytical_procedures.html.
3. D. Humbird, R. Davis, L. Tao, C. Kinchin, D. Hsu, A. Aden, P. Schoen, J. Lukas, B. Olthof, M. Worley, D. Sexton and D. Dudgeon. "Process Design and Economics for Biochemical Conversion of Lignocellulosic Biomass to Ethanol: Dilute Acid Pretreatement and Enzymatic Hydrolysis of Corn Stover", US Department of Energy, Office of Efficiency and Renewable Energy, National Renewable Energy Laboratory, Golden Colorado Technical Report NREL/TP-5100-47764 May 2011. http://www.nrel.gov/biomass/pdfs/47764.pdf.

CHAPTER 5

Integrated Forest Biorefineries: Industrial Sustainability

EMMANUEL KOFI ACKOM

Global Network on Energy for Sustainable Development (GNESD),
UNEP RISOE Centre, Department of Management Engineering,
Denmark Technical University, Denmark
Email: emac@dtu.dk; emmackom@yahoo.com

5.1 Introduction

Biorefineries offer good potential for producing environmentally benign substitute products otherwise derived from fossil fuels. Analogous to crude oil refinery where several hydrocarbon products are obtained from petroleum, a suite of products are derived from wood in the forest biorefinery. Forest products, unlike a petroleum refinery, often come from distinct mills that operate independently from each other, compared to the latter where all products are often processed in the same facility (Table 5.1).

A result of integrated forest biorefinery, however, is the production within the same facility of traditional products in addition to new products that are typically derived from fossil fuels (Table 5.1). These products, for example plastics are currently used in applications such as packaging, automobiles and consumer electronic goods that must comply with stringent regulations on a lifecycle perspective. This in itself presents a challenge and will require considerable learning for the integrated forest biorefinery from the competing industry.

RSC Green Chemistry No. 18
Integrated Forest Biorefineries
Edited by Lew Christopher
© The Royal Society of Chemistry 2013
Published by the Royal Society of Chemistry, www.rsc.org

Table 5.1   Analogical comparison of crude oil refinery and forest biorefinery.

Refinery type	*Petroleum Industry*	*Forest Industry*		
Fractionation Process	Linear, highly integrated	Traditionally nonlinear, rather discrete activities but with some level of integration. The level of integration increases from solid wood products to retrofitted pulp and paper (having the highest level of integration in a forest biorefinery).		
Level of difficulty in terms of product fractionation	Relatively easy	Ranges from quite simple for solid wood products (lumber, plywood, oriented strand board) to intermediate difficulty (*e.g.* pulp and paper) to challenging integrated biorefinery (biomaterials, biochemicals and bioenergy).		
Variability in raw material input	Homogeneous	High heterogeneity		
Examples of products	Liquefied petroleum gas, Gasoline, Nephtha, Kerosene, Petroluem Diesel, Fuel oils, Lubricating oils, Parafin wax, Asphalt, Petroleum coke	Lumber, Engineered Wood Products (*i.e.* LVL, GLT,CLT,LSL,I-joists, Finger-jointed lumber), Plywood, Oriented Strand board, Particle Board, Medium Density Fiber Board, Pulp and Paper (traditional), Pulp and Paper Mills (retrofitted into integrated forest bio-refineries), Greenfield standalone bio-refineries (producing biochemicals, biomaterials and bioenergy).		
Facility type(s)	Petroleum Refinery Types: Similar	Forest Products Mill Types: Diverse		
		Lumber		Maximization of throughputs (volume) for cost reductions
		Engineered Lumber Products		
		Plywood		
		Oriented Strand board (OSB)		
		Particle Board		
		Medium Density Fiber Board		
		Traditional Pulp and Paper		
		Integrated Forest Biorefinery Example: Retrofitted Pulp and Paper into biorefinery (to produce pulp, paper, biomaterials, biochemicals and bioenergy)		Increasing profit margins through product diversification and value addition

Source: author's own compilation.

There is growing interest in forest biorefinery by several stakeholders including, the forest industry, the academic research community and policy makers.[1-4] It provides opportunities to improve rural economies, generate additional revenue streams to the forest industry and a hedge against price volatilities in oil prices by providing alternative energy sources and products that otherwise would have been derived from fossil fuels. It has been recommended that the adoption of the biorefinery concept could rejuvenate the forest industry through additional revenue stream and optimum utilization of raw materials and energy.[5,6] Biorefinery unlike petroleum refinery is characterized by heterogeneity in feedstock inputs and processing technologies. This has led to a number of varying definitions for a biorefinery and the need for terminology harmonization for the field.[7] For example, a biorefinery has been defined in some scholarly literature as the complete utilization of biomass feedstock and other input such as energy for the concurrent production of materials for paper products, chemicals and energy.[8,9] This seemingly excludes the solid wood industry such as lumber from being a biorefinery. However, future effort to improve competitiveness and technological advancements, could, for example, lead to lignin and ethanol manufacture with traditional lumber production where value-added biomaterials and chemicals are derived from biochemical conversion of residual white wood at an integrated forest industry facility. Excess heat from the facility could be utilized internally or sold to neighboring industries. This development will for example, enable solid wood product mills to balance the production of wood products, biomaterials, chemicals and energy depending on the market prices similar to the practice by the sugar-cane-ethanol facilities in Brazil, where for sustained viability of sugar mills, the production levels of sugar, ethanol and electricity are undertaken in response to prevailing market prices. The IEA Bioenergy, however, defines biorefinery as the environmentally benign processing of biomass into diverse products (including food, feed, materials and chemicals) and energy (such as heat, fuels and power).[7] The difference between the earlier definition[8,9] and the latter[7] is the inclusion of food and feed in biorefinery. This implies that the feedstock for biorefinery could include diverse biomass materials including; algae, municipal waste, agricultural produce and residues as well as forest materials. There seem to be significant market opportunities for the sector. For example, the global markets for biochemicals increased 23% from $1.59 billion (US) in 2007 to an estimated $1.96 billion (US) (Figure 5.1). It has been forecasted that the market for biochemicals will reach $3.54 billion (US) by 2014, representing a 123% increase from 2007 value (Figure 5.1).

An integrated forest biorefinery could comprise of greenfield standalone biorefineries, either through thermochemical or biochemical conversion technologies, as well as retrofitting traditional pulp and paper mills that incorporates additional process and product (biomaterials, biochemical and bioenergy streams).

The work reported here focuses on the latter, which is an integrated forest biorefinery in a pulp and paper mill and how industrial sustainability concepts could be applied.

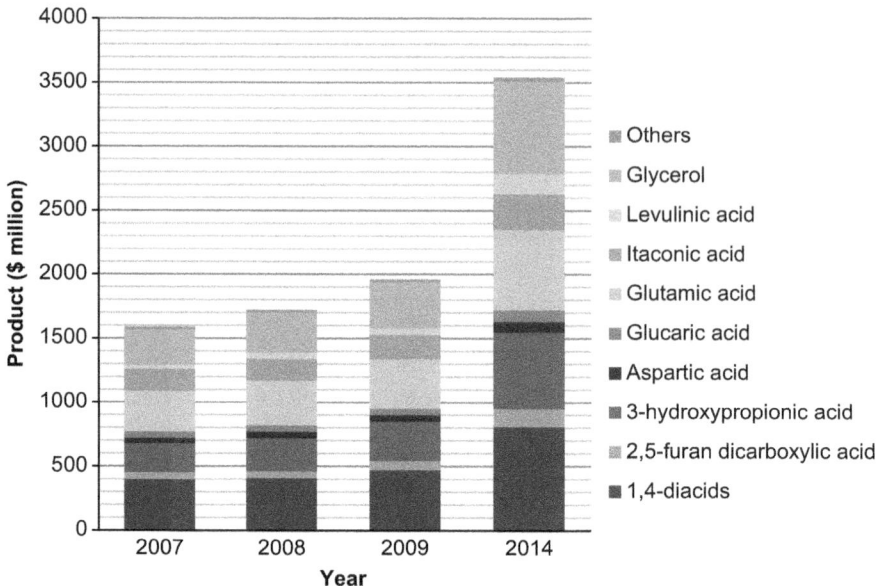

Figure 5.1   Global chemical markets from biomass (historic and projections). Author's graphical representation based on information.[41]

5.2   Industrial Sustainability: An Overview

The concept emanated from the United Nations Commission on Environment and Development (UNCED) meeting in 1987, was the industrial sustainability concept.[10] It was derived from the formal definition for sustainable development, stipulating that the use of natural resources to satisfy our present needs should not in any way jeopardize the ability of posterity to meet their own needs from the same resources.[11] Industrial activities and economic growth have often been closely linked to overexploitation of natural resources and lack of environmental stewardship. However, public pressure from informed groups coupled with efforts by business associations have led to significant due diligence by industries in being better stewards to our environment and natural resources over the last few decades.[12] Although there are still improvements that need to be made, a number of globally leading industries are presently able to achieve environmentally benign economic growth in a socially responsible manner. Several efforts had been made since UNCED (1987) to arrive at the current state of progress. Global effort, historically, includes, establishing the Business Council for Sustainable Development (BCSD) in 1990 with the support of UNCED; instituting a Business Charter for Sustainable Development in 1991; establishing the World Industry Council for Environment (WICE) in 1993; the creation of the World Business Council for Sustainable Development (WBCSD) in 1995 that comprised both the Business Council for Sustainable Development (BCSD) and the World Industry Council for

Environment (WICE). The formal definition of industrial sustainability came from the WBCSD meeting in 1995 indicates that production of goods and services to address current requirements should not be at the expense of future generations for the same goods and services. The definition of industrial sustainability was improved by scientist at the University of Cambridge to include conceptualization and design phases of products and services.[13,14] Although not specifically implied in the definitions, recycling, reuse and end-of-life handling of products are equally essential components in industrial sustainability. It is not enough for industries these days to focus solely on mitigating environmental impacts of their products and services. Rather, a focus wider than products and services to additionally, including, entire industrial operations, interaction with their physical surroundings (*i.e.* soil, water and air) and social environment or community in a responsible manner. Current industrial sustainability practice in companies relates very well with the field of industrial ecology and it is actually quite difficult to differentiate between the two.

5.3 Integrated Forest Biorefinery: An Overview

Integrated forest biorefineries utilize three main processes namely: mechanical/physical, biochemical and thermochemical, in converting raw forest biomass feedstocks into finished products. The physical/mechanical treatments are for pretreatment of the woody feedstock while thermochemical and biochemical processes are utilized in the bioconversion into the final products. Retrofitting pulp and paper mills to deliver a range of value-added products represents a good example of an integrated forest biorefinery.

5.3.1 Retrofitting Pulp and Paper Mills into Integrated Forest Biorefineries

Retrofitting pulp and paper mills into integrated forest biorefineries, however, requires important considerations that should be based on existing process configurations, modifications required as well as profit margins to be made from new value-added product streams. Essential factors to be considered in the decision making process for retrofitting pulp and paper mills into integrated biorefinery includes: location of the mill; availability of inexpensive feedstock; local market supply chains; current mill configuration; size and age, *etc.*[15] Based on the size of pulp and paper mills, it is advantageous, for example, to produce ethanol from hemicellulose in a larger mill (2000 tons of kraft pulp per day), compared to a smaller mill (300 tons of kraft pulp per day).[15] This is due to economy of scale resulting from the fact that commodity chemicals such as ethanol require high production levels. Regarding mill configuration, kraft mills for example are designed specifically to either produce market pulp or both market kraft pulp and paper. It should be expected in an integrated forest biorefinery that the latter consumes more energy and heat compared to the former. Much of the attention on integrated forest biorefinery has focused on development of process technologies, *i.e.* biochemical or thermochemical. However, a focus also on application of industrial sustainability concepts has good potential to sustain the viability of integrated forest biorefinery. Using publicly available information, this study provides a case study on the Tembec Temiscaming biorefinery. It investigates the company's operations in order to explore potential applications of industrial sustainability concepts.[16]

5.3.2 Integrated Forest Biorefinery with Industrial Sustainability Applications: A Case Study on Tembec Temiscaming

Tembec Temiscaming integrated forest biorefinery is located at Quebec, Canada. The products from Tembec Temiscaming integrated forest biorefinery are high value market pulp, ethanol, specialty cellulose, lignosulfonates, and thermal and electrical power (Figure 5.2). The company started with pulp production in 1973. Eight (8) years later it started utilizing waste ligno-sulfonates to offset energy that was otherwise obtained from fossil fuels. In 1983, Tembec Temiscaming begun sale of lignosulfonates on the market. The company was retrofitted in 1984 to produce resin production, followed by ethanol in 1991 and biogas in 2006, respectively.

The company produces 315 000 metric tons annually of high yield market pulp with certified wood from sustainably managed forests. It is reported that significant reduction in energy consumption, environmental impact, air emissions and effluent are associated with the company's high-yield pulp compared to other competing pulp types. Applications for high-yield market pulp applications include coated and uncoated papers, paperboard, linerboard, tissue and toweling.

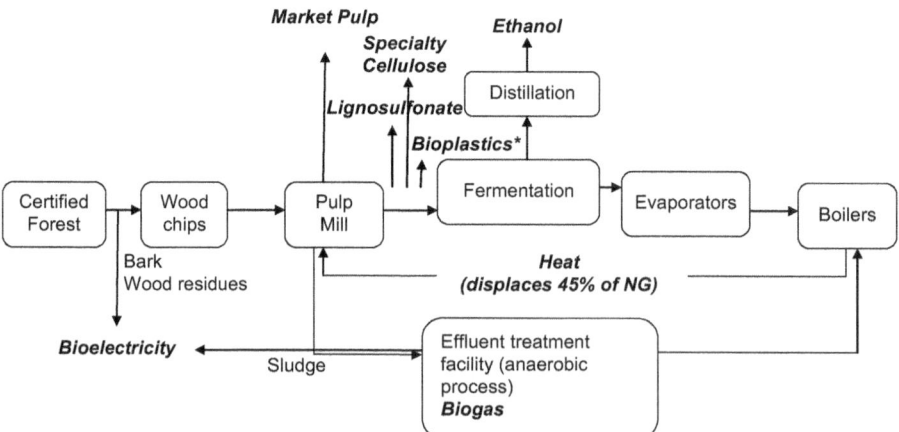

Figure 5.2 Simplified diagram of the Tembec Temiscaming Integrated Forest Biorefinery (Canada) indicating the product streams.
Research and Development phase.
Source: author's representation based on information.[16]

In sulfite pulping process, lignosulfonates are produced through sulfonation where lignin molecules are broken down and dissolved.[2] Current annual lignosulfonate production at Tembec Temiscaming is 90 000 metric tons. It is used as textile colorant, concrete additive, animal feed, binder for carbon black, wax and asphalt emulsion.

Fermentation by *Saccharomyces cerevisiae on* the spent sulfite liquor (wood hydrolysates) in the bioreactor produces alcohol that is then distilled to yield a high-purity ethanol (95%). Annual bioethanol production of 12 million liters makes Tembec Temiscaming a leading producer of industrial alcohol in Canada. Bioethanol is used in the production of vinegar, mouth washes and pharmaceutical products.

The company produces 160 000 metric tons of specialty cellulose annually. Applications for the specialty cellulose can be found in pharmaceutical products (as binders, as well as film-coating pills), food products (as texturizer and stabilizers), electronics (optical film for LCD technology), cosmetics (nail polish) and in paints (for quick drying lacquers and vanishes).

One of the product streams from the integrated forest biorefinery is resin. Total annual production of resin in Tembec Temiscaming is 47 000 metric tons. Resins are utilized in glueing panel wood products (*i.e.* plywood, oriented strand board and medium density fiberboard).

The company has the largest anaerobic wastewater treatment facility in North America. The production and use of biogas generated from the anaerobic treatment of wasterwater offsets 45% of natural gas required for pulp drying in 2010 compared to 2003. The treatment of effluent and utilization to displace fossil fuel can be showcased as a practical example of industrial sustainability in Tembec Temiscaming. Additionally, the company processes waste from the local township, thus further demonstrating corporate social responsibilities.

5.4 Opportunities in Industrial Sustainability for Integrated Forest Biorefinery: A Case Study

For sustained viability of any integrated forest biorefinery, it is imperative that the industrial sustainability concept be applied on a lifecycle perspective of operations. This includes:

- clear vision on environment management systems and the strong commitment to environmental due diligence throughout the company's operations;
- raw-material sourcing;
- manufacturing (*i.e.* material and energy efficiency, including reuse and recycling in production process, health and safety of workers, *etc.*);
- environmentally benign management of waste effluent and reutilization;
- end of product life management and reuse, recycling to provide secondary raw materials;
- socioeconomic aspects (*i.e.* job creation and improvements to rural economy, corporate social responsibilities, employee satisfaction, *etc.*).

5.4.1 Environmental Policy

The Tembec Temiscaming integrated forest biorefinery seem to apply industrial sustainability concepts in its operations to a large extent. It provides a good practical example for learning, towards establishing integrated forest bio-refineries. An environmental management system is applied throughout the company and the operations are ISO-14001 certified. The company maintain its competitiveness through sound economic, environmental and social practices dictated in part by its environmental policy. Demonstrating its strong commitment to sustainability, Tembec Temiscaming environmental policy states that "... converts forest resources into the quality products to satisfy our customers needs while protecting the environment and creating long-term economic and social benefits for our communities, employees, stakeholders and shareholders".[16] There seem to be striking similarities in the company's environmental policy with the WBCSD industrial sustainability definition.

5.4.2 Raw Material Sourcing

All woody feedstock for Tembec Temiscaming operations are sourced from sustainably managed forests that are either Programme for the Endorsement of Forest Certification (PEFC) or Forest Stewardship Council (FSC) certified.

5.4.3 Manufacturing

The company achieves significant reductions in energy utilization (and consequent greenhouse-gas emissions offsets) due in part to; retrofitting old equipments with modern technologies, modification to processes, and efficient utilization of residual material in a cogeneration facility. For example, a portion of the heat that is derived from combusting bark, wood shavings, sawdust, sludge and pulping liquor is utilized for steam production in pulping and paper manufacture, while the rest are utilized for bioelectricity generation. The company self-generates 10 MW of bioelectricity with a potential to upscale to over 30 MW. Tembec Temiscaming plans to become carbon neutral in the next 3 years by 2015.*

5.4.4 Environmentally Benign Management of Waste Effluent and Reutilization

Waste effluent from the company's operations are handled in an environmentally benign energy-efficient manner. Anaerobic treatment of Tembec Temiscaming waste water results in methane production that is utilized as fuel in the pulping process thus offsetting the need for natural gas. Biogas generated at the waste water treatment facility currently offsets 45% of natural gas consumption over an eight-year period.

---

* This is a commitment also shared by other members of the Forest Product Association of Canada (FPAC).

5.4.5 End-of-Life Management of Products

Nontraditional forest products such as plastics that typically are derived from fossil fuels could be produced in integrated forest biorefineries. There are, however, stringent regulations regarding end-of-life management of such materials in some jurisdictions. This in itself presents a challenge that requires considerable learning by the integrated forest biorefinery. This is an aspect of industrial sustainability that Tembec Temiscaming does not seem to cover in its operations. This could possibly be due to the fact that bioplastics are presently at the research and development stage in the company. In practice, however, end-of-life plastics are typically collected by municipalities through drop-collection depots and curbside collection points. The recycling process usually involves collection, separation, packaging, warehousing, and transportation. Biomaterials from an integrated forest biorefinery will be submitted to legislations similar to the fossil-fuel products that they substitute. For example, the German Packaging Ordinance is a recycling-oriented legislation that holds manufacturers accountable for the recycling or disposal of their packaging materials. This led to the industry in Germany to develop the Duales System Deutschland that allows participating companies to display a "green dot" logo on materials thereby ensuring free collection of such materials from consumers.[17] It has been reported that the amount of energy used in the collecting lightweight, low-value end-of-life products such as plastics, negatively affects the economics of its recycling and which varies for different plastic types.[17] Using examples from fossil-fuel-derived plastics, the total cost of manufacturing high-density plastic such as polyamide (PA) from virgin sources is estimated at 6 euro/kg compared to 1–1.50 euro/kg if it was obtained from recycled sources. It could therefore be inferred from the cost information that it makes economic sense for an industry to recycle its high-density plastics than to obtain from virgin sources.[17] Regarding low-density plastic such as polyvinylchloride (PVC) the total cost of manufacturing, however, from virgin sources is estimated at 0.50 euro/kg, which is lower than the most optimistic cost estimate associated with plastic material recycling (of 1 euro/kg).[17] Clearly, there is no financial incentive for manufacturing industries (such as integrated forest biorefineries) to engage in the business of recycling low-density plastic materials in the absence of any legislation. Therefore, legislative support and policy incentives by governments are required, especially when it comes to the recycling of low-density plastics.

5.4.6 Socioeconomic Aspects

Tembec Temiscaming operations result in considerable rural economy improvements in terms of job creation. Additionally, the company strives to ensure stakeholders, employees and customers satisfaction. Additionally, the recognition of rights and values of native communities within the area that the company operates is of paramount importance to sustained viability of the integrated forest biorefinery. The company seems to have gained the respect and trust from aboriginal communities within which it operates.

It renders essential services to the local community that includes treating the waste water from the local township at its facility further demonstrating corporate social responsibilities.

5.5 Challenges: Industrial Sustainability of Integrated Forest Biorefinery

In spite of the benefits associated with integrated forest biorefinery there exist considerable challenges that need to be resolved in order to ensure industrial sustainability. The nascent industry faces considerable environmental, economic, technical and social challenges that need to be addressed.

Reliable access to cheap feedstock at desired quality coupled with efficient supply chain and storage systems are essential for feedstock security and sustained viability of integrated forest biorefinery.

5.5.1 Environmental Sustainability Issues Related to Feedstock

Feedstock for integrated forest biorefinery should be sourced from certified forests that have been sustainably managed. The manner of deriving feedstock as raw material input for the integrated forest biorefinery is very essential for a sustained industry. Sustainability standards would be required for feedstock derived from forest plantations. Integrated forest biorefineries that derive raw materials from plantations are likely to be confronted with environmental sustainability issues similar to those associated with first-generation biofuels (from food sources).

Though first-generation biofuel does not fall under integrated forest biorefineries, the similarities in cultivation practices implies that forest plantations should comply with sustainability criteria such as:

- direct and indirect greenhouse-gas emissions;
- net energy balances;
- land for food, fuel and fiber conflicts and avoiding the use of culturally sensitive areas;
- water use (quality and quantity);
- biodiversity; and
- socioeconomic aspects (including the avoidance of underpaid labor in the forest plantation).[†]

Further studies are needed to establish minimum environmental sustainability requirements for the listed key indicators regarding plantation cultivation for integrated forest biorefineries. Native species should be used as much as possible in the forest plantation.

[†] The use of underpaid labor or children should also be avoided in manufacturing operations.

5.5.1.1 Greenhouse-Gas Emissions (Direct and Indirect)

A number of studies have provided estimations for direct and indirect GHG emissions.[18–25] The indirect land-use change (ILUC) GHG emission estimations seem to be very contentious in the overall sustainability discussion.[26] This is primarily due to the fact that the science and methodological issues that ILUC estimations are based on are still being developed and are inconclusive at this stage.

Forest plantations supplying feedstock to integrated forest biorefinery could potentially reduce their direct GHG emissions through the application of organic manure and biomass residues rather than using synthetic fertilizers from petrochemicals. This is because nitrous oxide emissions released in the field from synthetic fertilizers are 310 times more potent than carbon dioxide in terms of global-warming potential.[27]

Additionally, sustainable agricultural practices should be widely encouraged in forest plantations including "no till with biomass cover".[28–30] This results in significant improvements in soil-carbon sequestration.

A previous study suggests that a level playing field should be created whereby similar land use change GHG emissions requirements are applied to comparable products derived either from biorefineries or fossil fuels.[26] For example, the extraction of certain fossil-fuel types such as tar sands has been shown to being responsible for significant loss of forests and peatlands.[31]

While certain jurisdictions such as the European Union (EU) do not take into account ILUC in its GHG emission reduction mandate, others such as the United States of America consider ILUC.[‡]

5.5.1.2 Energy

Fossil-fuel-based energy consumption should be kept to a minimum in plantation operations that supply integrated forest biorefineries. Substituting synthetic fertilizers with organic manure and forest residues significantly improves the energy balance. This is due to the fact that a significant amount of energy is utilized in synthetic manure manufacture.

5.5.1.3 Land for Food, Fuel and Fiber

Land for forest plantations should be carefully selected in order to avoid potential conflicts regarding competing demand of the resource for food, fuel and wood fiber.

5.5.1.4 Water

Water requirements (quality and quantity) are essential criteria for forest plantation cultivation. Appropriate measures should be adopted to reduce

---

[‡] The United States Environmental Protection Agency (EFA) in its Renewable Fuel Standard 2 (RFS2) includes ILUC emissions in the overall GHG reductions mandate for ethanol production, which took effect from 1st July 2010.

water requirements both quality and quantity for forest plantations. For example, high water requirements have been reported for bioethanol production from first-generation sources.[32,33] Water requirements ranges from 2.3 to 8.7 million l/MWh for corn-derived bioethanol compared with 90–190 l/MWh for petroleum extraction/oil refinery.[34,35] Regarding practice, water use from direct precipitation is recommended.

5.5.1.5 Biodiversity

A negative impact on terrestrial and aquatic fauna should not be linked to forest plantations that supply feedstock for the integrated forest biorefinery. For example, aquatic fauna have been significantly impacted from the increased use of fertilization in corn ethanol operations that led to hypoxia in water bodies.[36,37] The use of pristine ecosystems such as forests and peatlands should be avoided for plantations. Marginal lands have been suggested for use in forest plantations.[38,39]

5.5.1.6 Socioeconomic Aspects

In the effort to provide adequate feedstock to supply integrated forest biorefinerry, it is very important that the rights of indigenous communities are respected and socioculturally sensitive areas avoided for forest plantation cultivation. Rather, local communities should benefit from considerable job creation and revenue from such plantations. Due to the similarities in cultivation operations between forest plantations and first-generation biofuels, the experience in East Africa could be adopted for forest plantation communities. For example in Mauritius, a fraction of the revenue from the operations is utilized by the industry,[§] to provide social amenities including health posts, schools and clean water.[40] Another example is improvements to road networks in rural communities in Kenya by a bioenergy plantation.** Though not specifically labeled as corporate social responsibility (CSR) by the companies themselves, these are, however, two clear examples of CSR that are very essential for sustained business activities nowadays.

5.5.2 Research and Development

There is the need for more research and development across the value chain, such as:

- Genetic manipulation to develop fast-growing timber species that for example require less nutrients and irrigation water (but rather depend on direct precipitation).

[§]This example is from sugar cane plantation and sale of cogenerated electricity from bagasse. It, however, provides some useful lessons that could be applied in integrated forest biorefinery.
**This example is just to showcase how experiences from other industries such as first-generation biofuel could be useful for adoption in feedstock operations in integrated forest biorefinery.

- Development and training of micro-organisms for highly efficient performance.
- To improve our understanding of enzymes and their interactions with substrate. Consequently, this would result in development of highly efficient enzymes and possibly lead to significant reductions in enzyme costs.

5.5.3 Logistics

It is particularly challenging for greenfield standalone biorefineries to develop logistical systems that ensures efficient collection, delivery and biomass storage.[41] Retrofitted pulp and paper biorefineries typically seem to have efficient developed systems already in place.

5.5.4 Investment

There is a challenge in several jurisdictions regarding access to capital investments to either retrofit existing mills or to establish new integrated forest biorefineries.[41]

5.5.5 Competition with Other Industries for Feedstock

A previous study shows that competition for the feedstock among forest industries is a major challenge that could result in higher demand for feedstock, leading to increased cost and consequently influencing the industrial sustainability.[12]

5.5.6 Processing

There is the challenge of completely substituting the use of fossil fuel for heat and power applications with alternative cleaner energy sources. Increased utilization of biomass residues, and waste products for heat and power applications in integrated forest biorefineries seem to offer considerable GHG emission reductions.[42] This also improves the economics of the operations. An additional challenge is gas clean-up and conditioning usually associated with thermochemical conversion platforms in biorefineries.

5.5.7 End-of-Life Legislations for New Products

The addition of new product streams presents a challenge for retrofitted mills as the industry need to comply with legislations that are new to traditional pulp and paper mills such as the "take-back" law and the Packaging Ordinance (Europe).

5.6 Policy Intervention: Improving Competitiveness of Integrated Forest Biorefinery Through Industrial Sustainability

Government-led policy intervention that supports access to cheaper wood feedstock is important to maintaining the economic viability of integrated forest biorefinery. Competition for the feedstock by other forest industries, in the absence of policy intervention, would lead to increased demand for the raw materials and higher cost, thereby affecting the industrial sustainability.[12]

It is recommended that governments develop clear sustainability guidelines for investors interested in the forest plantation business to supply the integrated biorefinery. It has been reported that the ongoing sustainability discussions associated with biomass feedstock for bioenergy (and biomaterials) provides preconditions for acceptability and long-term development, and should therefore be seen as an opportunity.[40]

A recommendation of this paper is that existing examples such as Tembec Temiscaming need to be carefully studied and the experience gained could be utilized by interested governments and corporations to support policy development in establishing successful integrated forest biorefineries.

An additional policy recommendation is that a careful comparative analysis that quantifies the economic, social and environmental benefits associated with competing product stream options in a particular biorefinery need to be undertaken to understand the industrial sustainability linked with respective options. This therefore requires mill-specific solutions regarding product streams rather than a "single tailored" solution for all retrofitted mills. It should possibly be the practice to establish future integrated forest biorefinery that has increased industrial sustainability applications. This will help ensure sustained viability of operations in integrated forest biorefineries along the value chain.

It is recommended that governments employ a range of policy tools to further the developments of integrated forest biorefineries in their jurisdictions. Targeted incentives, including infrastructure investment, tax credits and feed-in tariffs (for the excess bioelectricity produced) are support mechanisms that governments could utilize to facilitate the uptake of integrated forest biorefineries.

Government-led policy intervention is essential to facilitating recycling, especially of low density bioplastic materials. For example, governments can formulate policies that provide purchasing preference to products derived from recycled materials. It is expected that such policies would stimulate market developments through increased sales, resulting in possible production expansion and a consequent lower cost and/or profitability as a result of economy of scale.[17]

Applications of the industrial sustainability concept could be applied to the integrated forest biorefinery including value-chain operations, corporate social responsibility and public disclosure of activities.

5.7 Conclusions

From an industrial sustainability perspective, integrated forest biorefinery seem to offer a number of benefits compared to traditional forest industry. Careful analysis is required that takes into consideration existing process configurations, availability of inexpensive feedstock, markets for new value-added product streams and the local market supply chain. Equally important to the sustained viability of an integrated forest biorefinery is to cultivate a pragmatic environmentally responsible stewardship culture that ranges from sourcing of raw materials, through manufacture and processing, management of waste and "take back" policies of end-of-life products. A proactive policy that incorporates all aspects of industrial sustainability in integrated forest biorefineries is essential for achieving economic competitiveness at the same time meeting stringent environmental requirements and being socially responsible. A vibrant integrated forest biorefinery has far-reaching economic (new revenue streams), environmental (offsetting fossil-fuel-derived products, reduced energy and greenhouse-gas emissions) and social (job creation, waste water treatment, social responsibility) benefits to society. Clearly, application of industrial sustainability concepts along the value chain and good policy intervention will play key roles in maintaining a vibrant integrated forest biorefinery.

References

1. A. J. Ragauskas, C. K. Williams, B. H. Davison, G. Britovsek, J. Cairney, C. A. Eckert, W. J. Frederick Jr., J. P. Hallett, D. J. Leak, C. L. Liotta, J. R. Mielenz, R. Murphy, R. Templer and T. Tschaplinski, *Science*, 2006, **311**, 484.
2. L. Magdzinski, *Bioenergy and bioproducts at Tembec: synergies in integrated processing of biomaterials*, PAPTAC 92nd Annual Meeting, Pulp and Paper Technical Association, Montreal, 2006.
3. M. J. Realff and C. Abbas, *J. Ind. Ecol.*, 2004, **7**, 5.
4. M. Towers, T. Browne, R. Kerekes, J. Paris and H. Tran, *Pulp and Paper Canada*, 2007, **108**, 26.
5. B. Thorp, *Pulp and Paper*, 2005, **79**, 35.
6. A. Van Heiningen, *Converting a kraft pulp mill into an integrated forest products biorefinery*, PAPTAC 92nd Annual Meeting, Pulp and Paper Technical Association, Montreal, 2006.
7. International Energy Agency (IEA) Bioenergy Task 42, 2012, Accessed on 3rd March 2012: http://www.biorefinery.nl/fileadmin/biorefinery/docs/Brochure_Totaal_definitief_HR_opt.pdf.
8. P. Axegård, *The future pulp mill – a biorefinery*. Presentation at the 1st International Biorefinery Workshop. July 20–21. Washington DC, 2005.
9. V. Chambost, R. Eamer and P. R. Staurt, *Pulp and Paper*, 2007, **108**, 30.
10. S. Paramanathan, C. Farrukh, R. Phaal and D. R. Probert, *R&D Management*, 2004, **34**, 527.

11. G. Brundtland, ed., *Our Common Future*, The World Commission on Environment and Development, Oxford University Press, Oxford, 1987.
12. E. K. Ackom, W. E. Mabee and J. N. Saddler, *Appl. Biochem. Biotechnol.*, 2010, **162**, 2259.
13. S. X. Zeng, H. C. Liu, C. M. Tam and Y. K. Shao, *J. Clean. Prod.*, 2008, **16**, 1090.
14. P. M. Jansson, M. J. Gregory, C. Barlow, R. Phaal, C. J. P. Farrukh and D. R. Probert, *Industrial Sustainability—A Review of UK and International Research and Capabilities*. Cambridge, University of Cambridge, 2000.
15. U. Wising and P. Stuart, *Pulp and Paper*, 2006, **108**(6), 25.
16. Tembec, 2012, Accessed on 6th March 2012: http://tembec.com/en/products.
17. E. K. Ackom and J. Ertel in *Green Business*, ed. N. Cohen and P. Robbins, Sage, Thousand Oaks, California, 2010, p. 616.
18. J. E. Campbell, D. B. Lobell, R. C. Genova and C. B. Field, *Environ. Sci. Technol.*, 2008, **42**, 5791.
19. D. Tilman, J. Hill and C. Lehman, *Science*, 2006, **314**, 1598.
20. J. Fargione, J. Hill, D. Tillman, S. Polasky and P. Hawthorne., *Science*, 2008, **219**, 1235.
21. P. R. Adler, S. J. Del Grosso and W. J. Parton, *Ecol. Appl.*, 2007, **17**, 675.
22. H. K. Gibbs, M. Johnston, J. A. Foley, T. Holloway, C. Monfreda, N. Ramankutty and D. Zaks, *Environ. Res. Lett.*, 2008, **3**, 034001.
23. T. Searchinger, R. Heimlich, R. A. Houghton, F. Dong, A. Elobeid, J. Fabiosa, S. Tokgoz, D. Hayes and T.-H. Yu, *Science*, 2008, **319**, 1238.
24. G. Pineiro, E. G. Jobbagy, J. Baker, B. C. Murray and R. B. Jackson, *Ecol. Appl.*, 2009, **19**, 277.
25. US EPA, *Lifecycle Greenhouse Gas Emissions due to Increased Biofuel Production*. Model Linkage, Peer Review Report, Prepared by ICF International, United States Environmental Protection Agency (US EPA) 2009. Accessed on 29th February, 2012: http://www.epa.gov/otaq/renewablefuels/rfs2-peer-review-model.pdf.
26. E. K. Ackom, *Biofuels*, 2010, **8**, 237. Accessed on 1st March, 2012 at: http://www.future-science.com/toc/bfs/1/2.
27. UNFCCC, *Land use, land-use change and forestry*. Decision -/CMP.7. Advance unedited version. United Nations Framework Convention on Climate Change (UNFCCC). 2012. Accessed on February 12, 2012 at: http://unfccc.int/files/meetings/durban_nov_2011/decisions/application/pdf/awgkp_lulucf.pdf.
28. H. Kim, S. Kim and B. E. Dale, *Environ. Sci. Technol.*, 2009, **43**, 961.
29. K. Paustian, J. Six, E. T. Elliott and H. W. Hunt, *Biogeochemistry*, 2000, **48**, 147.
30. P. Smith, D. Martino, Z. Cai, D. Gwary, H. Janzen, P. Kumar, B. McCarl, S. Ogle, F. O'Mara, C. Rice, B. Scholes, O. Sirotenko, M. Howden, T. McAllister, G. Pan, V. Romanenkov, U. Schneider, S. Towprayoon, M. Wattenbach and J. Smith, *Philos. Trans. R. Soc.*, 2008, **363**, 789.

31. GFWC, Global Forest Watch Canada, 2012. Accessed on 29th February, 2012: http://www.globalforestwatch.ca/
32. S. E. Powers, R. Dominguez-Faus and P. J. J. Alvarez, *Biofuels*, 2010, **1**, 255.
33. R. Dominguez-Faus, S. E. Powers, J. G. Burken and P. J. Alvarez, *Environ. Sci. Technol.*, 2009, **43**, 3005.
34. Y.-W. Chiu, S. E. Powers, B. Walseth and S. Suh, *Environ. Sci. Technol.*, 2009, **43**, 2688.
35. US Department of Energy, *Energy demands on water resources*, report to Congress on the Interdependency of Energy and Water, US Department of Energy, Washington, DC. 2006, p. 80. Accessed on 12th March, 2010: www.leslieconsulting.com/lc/resources/121-RptToCongress-EWwEIA-comments-FINAL.pdf.
36. A. L. Mascarelli, *Environ. Sci. Technol.*, 2009, **43**, 7597.
37. S. C. Williams, *Science News*, 2007, **172**, 395.
38. M. J. Groom, E. M. Gray and P. A. Townsend, *Conserv. Biol.*, 2008, **22**, 602.
39. F. Danielsen, H. Beukema, N. D. Burgess, F. Parish, C. A. Bruhl, P. F. Donald, D. Murdiyarso, B. Phalan, L. Reijnders, M. Streuebig and E. B. Fitzherbert, *Conserv. Biol.*, 2009, **22**, 602.
40. GNESD, *Bioenergy: The potential for Rural Development and Poverty Alleviation*. Global Network on Energy for Sustainable Development (GNESD). Summary for policy-makers. 2011. Accessed on 1st March 2012: http://www.gnesd.org/Downloadables/Bioenergy_PotentialForDevelopment_SPM.pdf.
41. Parratt & Associates, *Scoping Biorefineries: Temperate Biomass Value Chains*, Prepared for Biotechnology Innovation Policy Section, Pharmaceuticals, Health Industries & Enabling Technologies Branch, Innovation Division, Department of Innovation, Industry, Science and Research, Canberra, Australia, 2010. Accessed on 6th March 2012: http://www.innovation.gov.au/Industry/Biotechnology/IndustrialBiotechnology/Documents/TemperateBiomassStudyExecutiveSummary.pdf.
42. E. K. Ackom, W. E. Mabee and J. N. Saddler, *Backgrounder: Major environmental criteria of biofuel sustainability*, International Energy Agency (IEA) Bioenergy Task 39 Report, 2010. Accessed on 29th February, 2012: http://www.task39.org/LinkClick.aspx?fileticket=wKf0TFLjXu0%3d&tabid=4426&language=en-US.

CHAPTER 6

Prehydrolysis Pulping with Fermentation Coproducts

T. H. WEGNER, C. J. HOUTMAN, A. W. RUDIE,
B. L. ILLMAN, P. J. INCE,* E. M. BILEK AND
T. W. JEFFRIES

USDA Forest Service, Forest Products Laboratory, Madison, Wisconsin
53726-2398, USA
*Email: pince@fs.fed.us; Tel.: +01-608-231-9364

6.1 Introduction and Background

Although the term "integrated biorefinery" is new, the concept has long been familiar to the pulp and paper industry, where processes include biomass boilers providing combined heat and power, and byproducts of pulping include turpentine, fatty acids and resin acids. In the dominant kraft (or sulfate) pulping process, dissolved lignin and chemicals from the pulp digester are concentrated by evaporation and burned in a recovery boiler to generate steam and to recover the inorganic chemicals that are recausticized and reused as fresh pulping chemicals. In addition, prior to pulping, bark is removed from pulpwood and fed to waste fuel boilers that raise additional steam. High-pressure steam is used in a cogeneration process to generate electricity for the plant, while lower-pressure steam is used for pulping process heat and paper drying. In recent years, many paper companies have added wood-fired boiler capacity and use slash from logging or wood residues from lumber mills to replace fossil

RSC Green Chemistry No. 18
Integrated Forest Biorefineries
Edited by Lew Christopher
© The Royal Society of Chemistry 2013
Published by the Royal Society of Chemistry, www.rsc.org

fuels, typically requiring only natural gas to operate the lime kilns that are part of the recausticizing process.

In addition to dissolved lignin and degraded carbohydrates, the waste liquor from pulping contains a mixture of fatty acids and resin acids. While evaporating excess water to reach high enough organic solids concentration for the liquor to burn, this mixture of organic acids separates from the remaining liquor and is collected to provide products like the rosin used on the bows of violins or other stringed instruments, and diverse products including soap and the anticholesterol steroid sitostanol used in some fortified foods. In addition, while heating the wood chips at the beginning of the pulping process, a mixture of volatile products distil from the wood to provide turpentine.

Whereas the kraft process has benefited from this diverse range of byproducts for decades, the second most common type of pulping process has had a limited range of byproduct options. Mechanical pulping, chiefly thermomechanical pulping (TMP) and stone groundwood pulping, produces wood pulp used primarily in newsprint for newspapers and coated paper for magazines. These high yield pulps contain nearly all the lignin, hemicellulose and cellulose found in the original wood. High-yield pulp mills still burn bark and other wood wastes to provide steam for power and process heat, and newer processes recover additional waste heat from the thermomechanical pulping process and often with it, turpentine. Mechanical pulp mills are unable to recover the other navel stores chemicals and have very large electrical energy needs that are usually not practical to meet with available wood wastes. One way to improve recovery of potential byproducts at mechanical pulp mills is to carry out an early-stage "prehydrolysis" of wood in order to recover hemicellulose sugars, resin acids, and fatty acids.

The general concept of wood prehydrolysis is not new to the wood pulp industry. The prehydrolysis kraft process was developed commercially around 1950 in order to produce chemically pure cellulose for the dissolving pulp industry: rayon and acetate.[1,2] As shown in Figure 6.1, the prehydrolysis kraft pulping process could yield kraft pulp with nearly all the hemicelluloses and much of the amorphous cellulose removed. The companies that have used this process to produce dissolving pulps have all evaluated a fermentation product from the prehydrolysis solution, but ultimately decided that loss of heat in the digester and loss of fuel for the recovery boiler were too costly to be offset by the value of ethanol that could be produced. Mechanical pulping processes have evaluated various chemical pretreatment processes, and there are mills providing pulps prepared by first pretreating the wood with sodium sulfite or hydrogen peroxide. The thermomechanical pulping process includes a brief preheating stage that hydrolyzes some of the acetate esters in the hemicelluloses and generates an acidic pulping environment and dilute solution of hemicellulose fragments. This treatment must be minimized as extensive prehydrolysis produces lower brightness pulps that cannot be used effectively in most paper grades. The key discovery leading to a prehydrolysis capability for mechanical pulps was that the use of oxalic acid reduced this loss in brightness that rendered the

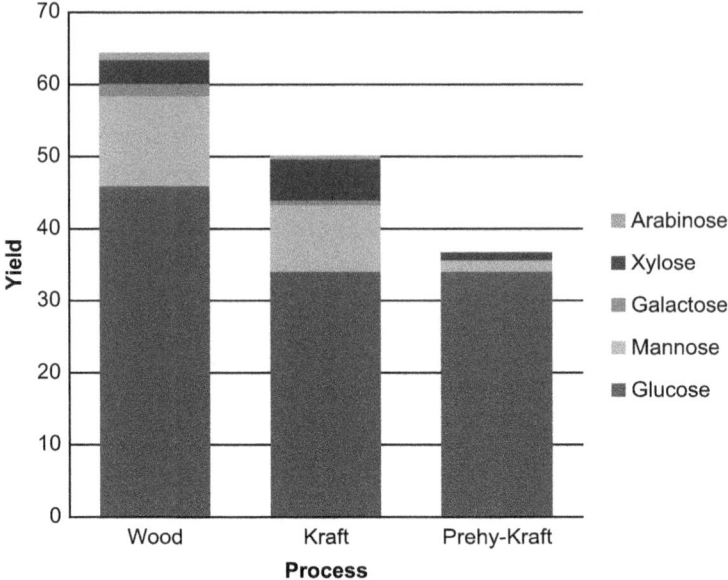

Figure 6.1 The carbohydrates in softwood and wood pulp.

pulp unsuitable for paper, and also provided a significant reduction in the amount of electrical energy needed to produce the pulp.

With pulp and paper mills struggling to remain profitable, and with incentives of potentially higher revenues from fuel and chemical byproducts along with government incentives for ethanol production and R&D grant funds for research, the American Forest and Paper Association (AF&PA), through their research committee, the Agenda 2020 Technology Alliance, initiated a research project several years ago to evaluate a range of prehydrolysis process and processing alternatives. Assembling a management and funding consortium of more than 20 corporate sponsors and with over $1 200 000 in US Department of Energy supporting grant funds, $200 000 in grant funds from the Wisconsin Department of Natural Resources, and over $400 000 in funding support from the member companies, the consortium sponsored research at five universities and two government laboratories to evaluate pretreatment conditions, impacts on wood pulp production and quality, fermentation of the prehydrolysis filtrate, and process economics. Known as the VPP (or "Value Prior to Pulping") Consortium, it conducted research to identify potentially profitable operating conditions for prehydrolysis with both hardwood and softwood kraft pulp and TMP.

VPP Consortium research evaluated alkaline and acid pretreatments and helped to explore potential business concepts for prehydrolysis and fermentation in the context of kraft pulping for both hardwood and softwood fiber, and with TMP for use in lightweight coated magazine paper (as described below in more detail). Potential business opportunities and the feasible operating space for this technology vary with each of the options. Alkaline treatments remove carbohydrates without affecting kraft pulp yield, but the resulting prehydrolysis solution has a low concentration of sugars. Moreover, much of the sugar is present as nonfermentable carbohydrate oligomers and the high ionic strength inhibits many fermentation microbes. Acid treatments for softwoods provide acceptable yields of sugars, but have to be carefully controlled since softwood fiber is the high-strength component in most paper products and the pretreatment affects both the strength of the fibers and their ability to naturally bond, as needed to produce paper strength. The prehydrolysis treatment of hardwoods has potentially more operating space since the pulp is not relied on as heavily for paper strength. Also, hardwoods contain more hemicellulose, and it is more easily removed than for softwoods. The hardwood chips are also easier to pulp following treatment than untreated wood chips.[3]

The operating space for pretreatment in association with thermomechanical pulping is very different from that for kraft pulping. Where only about half the mass removed in pretreating wood chips for kraft pulp ends up as a yield loss of pulp with the rest as a loss of waste organics as energy, in TMP nearly 100% of the mass removed in the pretreatment is a loss of product yield. However, production of TMP requires a very large input of electrical energy in refining – nominally 2000 kWh/ton of product, and the acid (or alkaline) pretreatment can reduce this energy requirement by 25% or more. The pretreatment yield and energy savings need to be carried out in a manner that does not reduce the brightness or the strength of the pulp.

Subsequent sections of this chapter describe the research on prehydrolysis with TMP and fermentation of sugars in the hydrolysate. The research on prehydrolysis-TMP focuses on the pilot-plant studies and the process and economic modeling research. The description of fermentation research focuses on the fermentation of hemicellulose sugars to ethanol as part of the VPP research program.

6.2 Prehydrolysis Thermomechanical Pulping

Prehydrolysis-TMP can be viewed as a derivative of research that started in 1980 using fungi and fungal enzymes to enhance or replace traditional mechanical pulping methods such as TMP.[4,5] It was discovered that treating wood chips with a white rot fungus and providing sufficient time and conditions for the fungus to degrade the chips significantly decreased the electrical refining energy needed to produce pulp for newsprint or other mechanical pulp paper grades.[6] In a follow-up project to learn more about the fungal mechanism, one of the compounds exuded by the fungal hyphae was found to be oxalic acid. Although the initial thought was that the oxalic acid recruited iron from the environment and initiated a Fenton radical attack on lignin moieties, it was ultimately determined that oxalic acid alone (or diethyl oxalate as an alternative chemical for introducing the acid) was sufficient to obtain the energy savings.[7,8]

McDonough at the Institute of Paper Chemistry had previously shown that mineral acids like sulfuric acid and hydrochloric acid had much the same effect, but in that case the energy savings were not of much interest to the industry because it resulted in significant darkening of the pulp that made it impossible to recover the brightness needed for paper grades (unpublished work: Master's Thesis, Cheryl Rueckert, Institute of Paper Science and Technology, 1993). The oxalic acid treatment, discovered originally by U.S. Forest Products Laboratory and BioPulping International, did not significantly reduce pulp brightness, at least at low treatment levels. The possibility of combining energy-saving oxalic acid pretreatment with a fermentation distillation plant to process the carbohydrate residuals provided the concept of a new opportunity to create added value at TMP mills.

Another study speculated that pulp brightness stabilization with oxalic acid treatment was due either to the pK_a of the acid providing a pH 2 buffer, or by oxalic acid acting as a reducing chemical and preventing some lignin condensations. Experiments showed that sulfurous acid (H_2SO_3) also preserved the brightness of wood veneer pieces and provided somewhat lower yield loss under otherwise similar conditions.[9]

The VPP Consortium then sponsored research to evaluate the effects on TMP of oxalic acid pretreatment and sulfurous acid pretreatment relative to untreated controls. The overall objective was to evaluate three wood species, spruce, the most common species used at TMP mills in the Northeast and Midwest regions; aspen, a hardwood often used because it provides high starting brightness and high opacity; and red pine, a common and readily available wood species in the Midwest that normally requires high energy and provides slightly lower brightness than spruce. Process modeling and economic evaluation focused on the case of a mill producing lightweight coated paper that consisted of about 26% bleached softwood kraft pulp (used mainly for sheet strength), 30% TMP that provides both smoothness and opacity, 30% groundwood pulp, and 15% coating material that provides a gloss surface for printing. Such paper is made to meet strength requirements for high-speed printing presses (used to publish weekly magazines).

Typically, mills with wood species that provide higher strength can increase the amount of TMP, coatings or fillers, whereas mills with lower pulp strengths will require more softwood kraft to maintain critical product qualities. The bleached softwood kraft pulp is the most expensive major component in the paper, considered as purchased market pulp for this analysis, followed by the TMP produced on-site at approximately half the cost, with the clay-coating materials and groundwood pulp as lower-cost components. If prehydrolysis causes a loss of strength in the TMP, it would have to be compensated by purchase of additional kraft pulp. Thus, the research became an effort to identify prehydrolysis conditions where there is no loss in TMP strength and brightness, but significant savings in electrical energy, while also considering the potential for fermentation of prehydrolysis sugars.

6.2.1 Experimental Prehydrolysis-TMP

Experimental work was carried out using a pilot-plant scale Sunds Defibrator CD-300 woodchip refiner operating at 1.2 kg per minute feed rate. The CD-300 consists of a chip hopper with steaming bin and metering screw followed by a plug screw feeder with a compression ratio of 4 to 1. The plug screw feeder compresses the wood chips into a plug of compressed and macerated wood chips to form a steam seal for the pressurized reactor, and also feeds the wood into the PREX chemical mixing system. The PREX is a pair of vertical feeder screws that receive and break up the wood plug from the plug screw feeder. The PREX operates with the wood plug immersed in the chemical treatment solution. As the plug breaks, the wood chips expand, filling the fiber lumens with the treatment solution and providing a very uniform distribution of the acid throughout the wood chips. Typically, one kg of wood (dry weight) absorbs about one kg of water and chemicals in the PREX immersion zone. The PREX screws lift the wood chips out of the treatment solution and drop them into the pressurized preheater, which is operated to maintain a steady chip level and residence time. Pressurized treatment was at 130 °C for 10 min. Acid strength was varied to control the yield.

The pressurized preheater discharges *via* a second metering screw at the bottom of the reactor, which feeds a plug-screw discharger, again with a 4:1 compression ratio. Chips are dewatered to 65–70% solids in the plug screw feeder before the PREX immersion zone, and again dewatered to 65–70% solids by the discharge plug screw following the preheater. Liquor take-up in the PREX is 2–3 liters per minute providing a reactor concentration of about 28% solids and a filtrate (sugar) recovery by the plug screw discharge of 75%. The plug from the discharger is broken up in a pressurized transfer auger that feeds the first-stage pulp refiner. The CD-300 refiner is a pressurized disk mill with a 12″ (300 mm) flat disc zone plus a 4″ (100 mm) conical refining zone at the periphery of the plates. The refiner plate gap was adjusted to provide a first-stage specific refining energy loading of 1.1 kWh/kg. Pulp was blown from the refiner to an atmospheric cyclone and collected in 55-gallon drums. Second-stage refining was carried out in a Sprout Waldron 12″ atmospheric disk refiner using adjustments in plate gap, and multiple passes to provide a variety of specific refiner energy loadings, and degrees of fiberization. The initial quality control test for TMP is a drainage test known as freeness with a target refiner freeness value of 150 mL. Experimental results indicated that the spruce TMP was quite responsive to the oxalic acid treatment, providing a 25% reduction in specific refiner energy (SRE) using just 0.15% oxalic acid on starting wood (dry weight basis). Red pine was much less responsive to the prehydrolysis treatment, and aspen responded poorly and suffered a noticeable brightness loss before achieving even a 15% reduction in SRE. Similar experiments carried out using sulfurous acid did not provide as much energy savings and did not protect brightness as well as anticipated.

The carbohydrate removal was largely independent of species, although aspen on average provided slightly more sugar for a given treatment severity. Two aspects of aspen contribute to this outcome. The most readily removed hemicelluloses are the arabinogalactans followed by xylan. Arabinogalactan is a minor constituent of both hardwoods and softwoods, but aspen contains about 16% xylan compared to just 7% in white spruce.[10] The pretreatment had little impact on the tear strength of the pine pulp, actually resulting in a slight increase that persisted to treatment with an oxalic acid charge of 1% on wood. Spruce also showed a minor increase in tear strength, but went through a maximum with the optimum treatment at about 0.2% oxalic acid on wood. Aspen lost 0.5 to 1 mN m^2/g tear strength under all pretreatment conditions. Hardwood TMP does not contribute much tear strength to a sheet, but this is about a 30% loss from the tear strength of untreated aspen and would have a significant effect on the paper furnish.

6.2.2   Experimental Fermentation of Hydrolysate Sugars

Fermentation research at the U.S. Forest Products Laboratory (FPL) includes biomass conversion to value-added products such as ethanol biofuel from carbohydrates, as well as the production of polyhydroxyalkanoate (PHA) bioplastics from lipids and carbohydrates by fermenting bacteria, shown in Figure 6.2, natural and synthetic fermentation enzymes, and fermentation inhibitors in extractives that could be used as antibacterial, antifungal, or pesticide agents.

Scientists at FPL have conducted pioneering research on fermentation of hemicellulose to ethanol for over 20 years, promoting lignocellulosic biomass as a way to meet increasing global energy demands. The research led initially to the adoption of a unique xylose fermenting yeast, *Pichia stipitis*, shown in Figure 6.3, commonly found in the gut of a bark beetle. The research has included the isolation of enzymes and genes responsible for fermentation of xylose by *Scheffersomyces stipitis*, the complete genome sequencing of *S. stipitis*,[11] metabolic engineering of *S. stipitis* for improved fermentation and development of patented strains of *S. stipitis* for commercial applications. The research has included evaluation of ethanol production from biomass hydrolysates.[12] An example of experimental laboratory research is described here involving the fermentation of hemicellulose to ethanol as part of the VPP

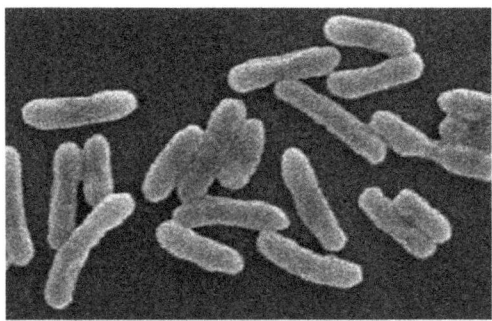

Figure 6.2   PHA producing bacteria *Pseudomonas pseudoflava*.

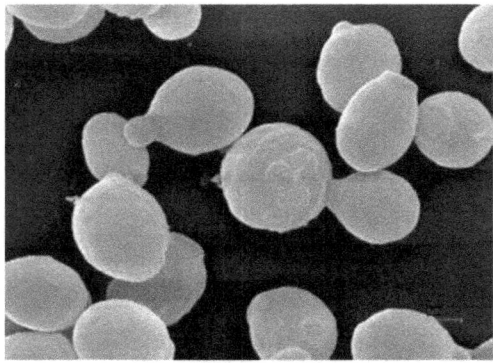

Figure 6.3 Ethanol producing yeast *Pichia stipitis*.

Table 6.1 Total sugars and pH of hydrolysate samples.

| *Hydrolysate* | *pH* | *Sugars concentration* | |
		before hydrolysis	*after hydrolysis*
Maple	∼2.14	∼80 g/L	∼132 g/L
Loblolly Pine	∼1.3	∼60 g/L	∼106 g/L
MSHW	∼4.16	∼60 g/L	∼136 g/L

research program. In this case the experimental hydrolysate was typical of that which could be obtained in prehydrolysis kraft pulping.

Hemicellulose was extracted from pulpwood of maple (a northern hardwood species), southern pine (loblolly), and mixed southern hardwood (MSHW) species by a hot-water extraction method followed by a membrane separation process to remove water and acetic acid. The resultant hemicellulose fraction was hydrolyzed with 4% H_2SO_4 and autoclaved for 15 min. Total sugar content was determined by high-pressure liquid chromatography (HPLC) techniques, and measures of the total sugar content before and after acid hydrolysis and the final pH of hydrolysates are shown in Table 6.1. Individual monosaccharide concentrations are given in Tables 6.2, 6.3 and 6.4, along with concentrations of the microbial inhibitors furfural and HMF, as determined also by HPLC.

In the fermentation experiment, *S. stipitis* fermented 100% of the xylose, glucose, galactose and mannose to ethanol in the maple, pine and MSHW hydrolysates. Complete fermentation of the xylose required about four days. *S. stipitis* successfully fermented monomeric sugars in the pine and MSHW extracts, although the hydrolysates contained such low amounts of sugar that ethanol yields were low.

In addition to the sugar monomers, some oligomers in the maple extract were hydrolyzed and the resultant sugars fermented to ethanol by *S. stipitis*, which is known to have an enzyme with the potential for hydrolysis of oligomers. Nutrient requirements were low and are expected to be inexpensive. Biotin and thiamine are the only known essential vitamins for the cultivation of these

Table 6.2   Analysis of maple hydrolysate.

Hydrolysate ID:	*Hot-water pretreated maple hydrolysate*						

Sugar Analyses, Unit: g/L

Original	*Glc*	*Xylo*	*Gal*	*Rham*	*Arab*	*Man*	*Total CHO*
HP87P HPLC	10.48	63.65	7.31	N.A.	N.A.	3.84	85.28
HP87H HPLC	1.15	71.3	ER	ER	3.3	ER	75.75
NREL HPLC	8.73	72.81	8.63	N.A.	4.47	8.63	107.52
NREL H^1-NMR	3.95	104.5	4.65	3.8	3.65	8.6	129.15
Posthydrolysis	*Glc*	*Xylo*	*Gal*	*Rham*	*Arab*	*Man*	*Total CHO*
NREL HPLC	6.00	99.65	8.60	N.A.	4.72	13.33	132.27

Inhibitor analyses, Unit: mg/L			
	HMF	*Furfural*	*Acetic acid*
Original	N.A.	1.25	1.22

Glc = glucose; Xylo = Xylose; Gal = Galactose; Rham = Rhamnose; Arab = Arabinose; Man = Mannose; Total CHO = total carbohydrate.
ER – galactose, rhamnose, mannose cannot be separated from xylose peak.

Table 6.3   Analysis of loblolly pine hydrolysate.

Hydrolysate ID:	*Loblolly Pine*						

Sugar Analyses, Unit: g/L

Original	*Glc*	*Xylo*	*Gal*	*Rham*	*Arab*	*Man*	*Total CHO*
HP87P HPLC	4.18	9.88	*low*	22.70	2.66	17.42	56.84
HP87H HPLC	3.20	35.43	ER	ER	17.77	ER	58.10
Dionex IC	3.11	15.95	9.15	0.44	17.22	9.94	55.81
4% H$_2$SO$_4$, 15 mins							
HP87P HPLC	21.7	33.14	31.1	16.0	15.62	23.24	140.8
HP87H HPLC	10.2	70.32	ER	ER	15.3	ER	95.82
Dionex IC	13.17	25.82	17.47	0.73	18.14	31.33	106.66

Cation – ICP Unit: mg/L							
	Al	*Ca*	*Cr*	*Cu*	*Fe*	*Mg*	*Ni*	*Zn*
Original	1.17	6.82	52.70	1.39	185.83	91.99	2.66	0.29

Inhibitor analyses, Unit: mg/L			
	HMF	*Furfural*	*Acetic acid*
Original	5	2.5	1.33

Glc = glucose; Xylo = Xylose; Gal = Galactose; Rham = Rhamnose; Arab = Arabinose; Man = Mannose; Total CHO = total carbohydrate.
low – below the detection limit.
ER – galactose, rhamnose, mannose cannot be separated from xylose peak.

Table 6.4 Analysis of mixed southern hardwood hydrolysate.

Hydrolysate ID:	Mixed Southern Hardwood						
Sugar Analyses, unit: g/l							
Original	*Glc*	*Xylo*	*Gal*	*Rham*	*Arab*	*Man*	*Total CHO*
HP87P HPLC	18.3	51.74	*low*	*low*	6.76	10.94	87.74
HP87H HPLC	7.8	44.2	ER	ER	4.7	ER	56.7
Dionex IC	11.16	30.81	7.15	3.09	1.31	6.62	60.14
4% H_2SO_4, 15 min							
HP87H HPLC	18.4	107.1	ER	ER	10.5	ER	136
4% H_2SO_4, 30 min							
HP87P HPLC	18.1	101.6	ER	ER	10.4	ER	130

Cation Analyses, ICP-MS unit: mg/l								
	Al	*Ca*	*Cr*	*Cu*	*Fe*	*Mg*	*Ni*	*Zn*
Original	13.1	1418.8	2.9	0.2	1061.8	807.7	3.4	11.0

Inhibitor analyses, Unit: mg/L			
	HMF	*Furfural*	*Acetic acid*
Original	*low*	*low*	*low*

Glc = glucose; Xylo = Xylose; Gal = Galactose; Rham = Rhamnose; Arab = Arabinose; Man = Mannose; Total CHO = total carbohydrate.
low – below the detection limit.
ER – galactose, rhamnose, mannose cannot be separated from xylose peak.

organisms. During the ethanol fermentation process, *S. stipitis* also produced a byproduct xylitol, a sweetener used in the food industry.

In summary, laboratory experiments showed that *S. stipitis* can successfully ferment monomeric sugars in maple, pine and MSHW hydrolysates, although in the experiments pine and MSHW hydrolysates contained such low amounts of sugar that ethanol yields were low. The xylose-metabolizing yeast *S. stipitis* strain CBS 6054 is a good candidate for commercial development of hot-water extracted maple hydrolysates, because the strain has promise for oligo-saccharide hydrolysis while fermenting subsequent monomeric sugars to ethanol. For a commercial scale-up of ethanol production, a minimum of 10–15% sugar concentration in hydrolysate would be needed for best ethanol production, at least 6% (wt) ethanol with 1 g ethanol/L per hour. A xylose fermenting microorganism like *S. stipitis* is required to reach these levels.

6.2.3 Modeling Prehydrolysis-TMP and Fermentation Process Concept

Addition of a prehydrolysis and fermentation process to an existing pulp and paper mill would impact overall mill operations, so we take the "whole-mill" approach to modeling prehydrolysis pulping and fermentation process concepts. For example, at a typical mill producing lightweight coated paper using TMP blended with purchased kraft pulp, the pretreatment of wood chips will remove hemicellulose and increase the moisture content of chips fed to the TMP refiners, while reducing specific energy required in the refiners, which in turn will allow the production of higher quality and higher volumes of TMP pulp at the same energy input, offsetting some of the more expensive purchased kraft pulp input.

To model this prehydrolysis-TMP and fermentation process concept, FPL researchers developed models of overall pulp and paper mill operations, using energy and mass balance equations. The energy and mass balance equations were designed with sufficient detail to predict operational impacts of pre-hydrolysis on overall mill-wide operations. While the energy and mass balance equations were implemented in Microsoft Excel, the results for several cases were validated by comparing with more complete WinGEMS models. Win-GEMS is the pulp and paper industry's premier process simulation tool that has been used for many years to model pulp and paper production processes.

The FPL models were constructed using hypothetical but realistic mill parameters and scales that were thought to be "typical" of commercial operations. The objective of the modeling was to determine conditions under which an investment in prehydrolysis pulping might be a good business decision. The models include capital costs that are considered to be accurate within perhaps $\pm 30\%$, and in general there remains some uncertainty about actual production performance of the process itself, so the analysis is accompanied by sensitivity analysis and stochastic risk analysis, recognizing the variability and uncertainties related to underlying parameters. The models are tools that can help to identify opportunities for commercial development, but actual investment decisions will require more detailed analysis supported by more precise engineering estimates of capital costs and operating parameters.

FPL researchers recently developed preliminary estimates of financial performance for various prehydrolysis-TMP investment scenarios with projected financial cash flow worksheets linked to process energy and mass balance equations that were developed in Microsoft Excel.[13,14] The resulting VPP-TMP process and economic model is also available for download at the VPP website maintained at SUNY-ESF (www.esf.edu/pbe/vpp/). In addition, stochastic simulation was incorporated into the VPP model as a way to show variability in the predicted financial returns in relation to probability distributions of specified engineering and financial parameters.

The VPP-TMP model computes an incremental investment analysis based on differences in projected cash flows for whole pulp and paper mill operations, between what is called the "process" case with VPP, and the "base" case without VPP (*i.e.* the existing mill without prehydrolysis and fermentation). The differences or "incremental" cash flows are used to compute measures of financial performance for the hypothetical VPP investment, including internal rate of return and discounted net present value. These measures may be used to evaluate the financial feasibility of the business concept. The Excel model computes the financial performance measures in three ways: before tax and finance, before tax, and after tax (the latter two options include the effects of capital financing or borrowing). Internal rates of return are calculated both on a "nominal" basis, which includes inflation, and on a "real" basis, which removes inflation.

For VPP/TMP the base case mill is assumed to be currently processing 300 oven dry metric tons per day (OD MT/day) of chips into thermomechanical pulp, with a specific refiner energy input of 2.1 MWh/MT. This TMP is blended with groundwood pulp and bleached kraft pulp. The blended pulp is used to make lightweight coated paper. The boundary of the VPP-TMP model starts with clean chips entering the primary TMP refiner and ends with the dried coated paper product. Full steam recovery from the refiners is implemented. The financial model suggests that if the base case TMP mill's initial capital investment was $620 million (the cost of a new mill) the after-tax internal rate of return over a 20-year lifetime would be 6.7%, but more typically an existing mill would be an older mill with smaller and depreciated capital costs.

The FPL VPP-TMP model uses engineering parameters estimated from VPP-TMP experimental results.[13] In addition, the model incorporates recent historical price data for pulp, wood, chemicals and electricity. The projected financial performance of the VPP business concept was estimated for aspen, spruce, and pine. Oxalic acid was presumed to be used as the primary pretreatment chemical. The level of oxalic acid was chosen to be the maximum level that would not significantly degrade the strength or optical properties of the wood pulp.

In the cases of aspen and pine the pretreatment was estimated to improve the strength properties of the resulting TMP pulp. The power load of the refiner was assumed to be limited at its current level, and the paper production also limited to its current level by market demand. For all the wood species analyzed the specific energy for refining would be reduced as a result of pretreatment. Thus, while maintaining the total electric power load on the refiners, the mill could increase the throughput of TMP. Since the overall paper production was also held constant it was determined that a lower amount of groundwood pulp and/ or bleached kraft pulp would be used for blending with TMP. Table 6.5 displays the major operating parameters for the base case (the existing mill without prehydrolysis and fermentation) and the three different process cases involving prehydrolysis pulping and fermentation for each of three different wood species.

In the model, the cash-flow analysis worksheet provides calculations of the after-tax net present values and internal rates of return for the three process cases. We assumed a pretax nominal discount rate of 11% with Federal and state income taxes at a combined effective rate of 41.5%. Table 6.6 shows key model results and assumptions, including projected financial performance for the three process cases (after-tax net present value, NPV, and internal rates of return, IRR, on both nominal and real dollar basis), ethanol production, and the capital costs of ethanol production and pretreatment facilities. Ethanol capital costs vary among the three cases because of varying levels of wood-sugar recovery and fermentation. Projected real IRRs of 16–27% across the three cases indicate positive expected financial performance. The main source

Table 6.5 Operating parameters for TMP cases.

Process Parameters	Base	Aspen	Spruce	Pine
Wood moisture content (% total weight)	50	50	50	50
OA loading (% on wood weight)	0.00	0.85	0.07	0.90
Sulfur Dioxide (% on wood weight)	0.0	1.0	0.0	0.0
Hydrogen Peroxide (% on pulp weight)	2.0	1.5	2.0	3.0
Sodium Bisulfite (% on pulp weight)	0.50	0.00	0.50	0.75
Hemis removed, total (% total available)	0	10	3	20
Extractives removed (% total available)	0	50	50	50
Acetic acid removed (% wood treated)	0	1.5	0.25	0.25
Primary specific energy (MWh/m.t.)	1.12	0.83	0.78	0.78
Secondary specific energy (MWh/m.t.)	0.70	0.52	0.49	0.49
Rejects specific energy (MWh/m.t.)	0.70	0.52	0.49	0.49
Kraft pulp (% fraction total pulp)	30	24	30	24
Groundwood pulp (% fraction total pulp)	35	30	21	28
LWC coating weight (% of paper weight)	15	15	15	15
Conversion monosaccharide to ethanol (%)	0	46	46	46
TMP mill process labor (workers/day)	24	24	24	24
Ethanol plant process labor (workers/day)	0	12	12	12

Table 6.6 Financial performance of incremental investment for various VPP/ TMP cases.

	Aspen	Spruce	Pine
After-tax NPV ($ million)	48.2	32.7	26
After-tax nominal IRR	30%	26%	20%
After-tax real IRR	27%	23%	16%
Ethanol production (million gal/yr)	1.01	0.59	1.92
Ethanol capital cost ($ million)	10.1	5.9	19.2
Pretreatment capital cost ($ million)	25.4	25.9	26.9

of the positive financial performance is a combination of significant specific energy savings and strength improvement of the TMP pulp, which allows for a reduction in expensive kraft pulp.

In an effort to further understand the factors that control the financial performance in the model, sensitivity analysis was performed by sequentially increasing various process parameters by 1%, one at a time. The results of sensitivity analysis varied across the different process cases, but in general the parameters to which the model was most sensitive are shown in Table 6.7, along with "reasonable" expected values of those parameters, and a range in their expected values from more "pessimistic" to more "optimistic". When the ranges of parameter values that are shown in Table 6.7 are introduced into the model it determines the sensitivity of projected financial performance (*e.g.* the sensitivity of NPV) to expected variation in process parameters. The parameters associated with the pulp blend are found to be the most important in terms of their impact on financial performance in the model. Pretreatment capital cost is next. Absent from this list is electrical power cost because in this

Table 6.7 Parameter values used for sensitivity analysis.

	Pessimistic	*Reasonable*	*Optimistic*
Per cent kraft pulp (%)	26	24	22
Per cent groundwood pulp (%)	32	30	28
Kraft pulp price ($/MT)	700	743	800
Groundwood pulp price ($/MT)	400	423	450
Pretreatment capital cost ($MM)	23	21.5	20
Primary specific energy (MW hr/MT)	0.90	0.83	0.70
Wood chips price ($/MT)	70	64	60

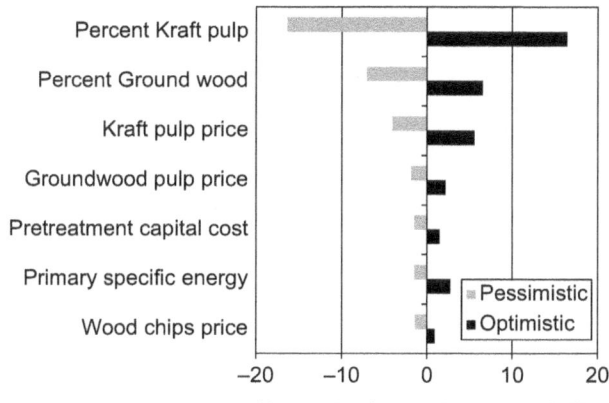

Figure 6.4 Sensitivity analyses of NPV for the aspen prehydrolysis TMP process case.

analysis the total electrical demand was held constant and thus removed from consideration in sensitivity analysis.

Figure 6.4 shows the results of the sensitivity analyses of NPV for the aspen prehydrolysis-TMP process case. The figure shows for the important process parameters how the estimated after-tax Net Present Value varies as the parameters are varied from "pessimistic" to "optimistic" values. Similar results are obtained for the other two wood species. For aspen, the after-tax incremental NPV was $48.2 million, and changing the kraft pulp blend by just 2% can change the NPV by $15 million. Changing the kraft blend has by far the greatest impact on NPV when compared to the other parameters. However, even if all the parameters are set to their "pessimistic" values, the NPV still remains positive.

A more holistic way to evaluate the financial risk is to assign probability distributions to the values of multiple engineering and financial parameters in the model (*e.g.* ethanol yield, washing efficiency, pulp prices, *etc.*) and then run model simulations by choosing values randomly from these distributions and using a large number of simulations to derive estimated probability distributions of financial performance. The results of such multivariate stochastic

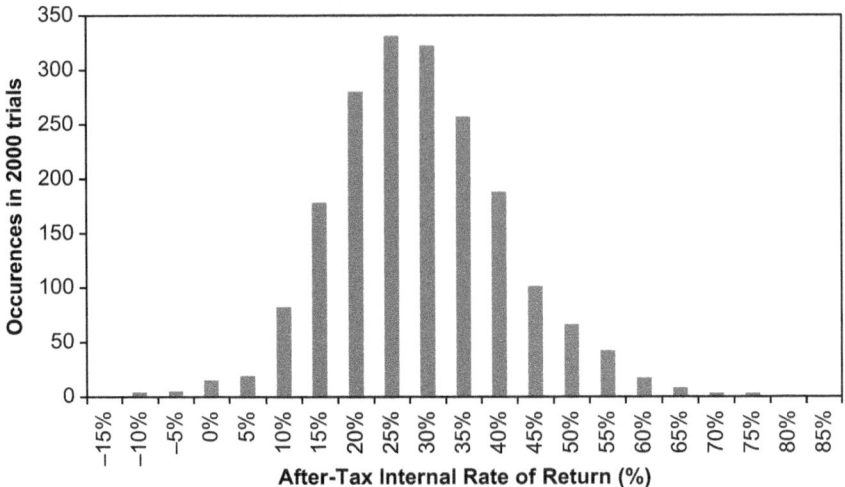

Figure 6.5   Distribution of aspen TMP IRR based on 2000 trials of a stochastic simulation.

simulations can provide the estimated distribution of after-tax real IRR for example, that naturally combine the estimated probability distributions of underlying model parameters with model sensitivities. Figure 6.5 shows the results of such a stochastic simulation of the IRR in the aspen TMP case. As expected, the distribution is centered at about 27%, as indicated in Table 6.6. Analysis of this distribution indicates that a 95% confidence interval for after-tax IRR is between 5% and 55%.

Finally, as a method for comparing distributions, the cumulative probability distributions were calculated for the three wood species, with results shown in Figure 6.6. The curves in Figure 6.6 can be interpreted as the probability of achieving an after-tax IRR value or better. For example, for the pine case the probability of achieving an IRR greater than 10% is about 70%, while the probability of achieving an IRR greater than 20% is about 30%.

In summary, all of the prehydrolysis-TMP cases were estimated to offer positive financial performance, with incremental after-tax IRRs of 16–27%. However, the stochastic risk analysis indicated that there is a relatively wide range of possible IRRs. For example, the 95% confidence interval for the aspen case includes estimated IRRs that range from 5% to 55%. It is also interesting to note that the IRR in all the cases goes up by 3% when the ethanol plant is removed and only the pretreatment is implemented, indicating that the primary economic benefit of prehydrolysis pulping in the case of TMP is the improvement in pulping and papermaking (chiefly reduced purchase of kraft market pulp as TMP output increases).

These results suggest that at current and projected ethanol prices (based on projections from the 2010 U.S. Annual Energy Outlook), the ethanol plant revenue cannot support its capital investment as well as the pretreatment

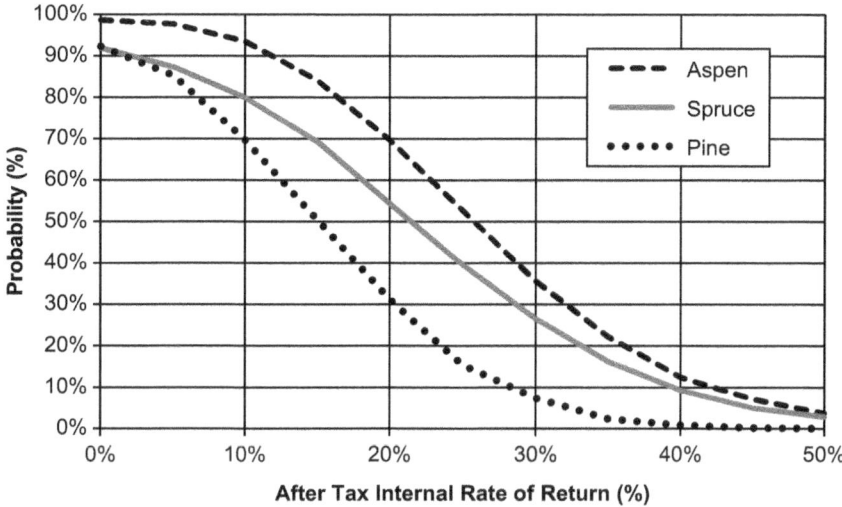

Figure 6.6 Cumulative distribution of TMP IRR based on 2000 trials of a stochastic simulation of each of the cases.

process itself. In other words, the VPP pretreatment process may be desirable by itself (without ethanol fermentation) based on the cost reductions for the pulp mill, and given the reductions in specific energy consumption, which may provide some mitigation of rising energy prices.

6.3 Summary and Path Forward

The research on prehydrolysis pulping and fermentation has identified important signposts that point the path forward in biorefining and forest products research. One important finding is that the quantities of carbohydrates that can be extracted from wood chips prior to pulping are limited, and recovery may not be justified if the carbohydrates are fermented primarily into low commodity value fuel ethanol. Thus, one signpost on the path forward is to seek higher-value carbohydrate derivatives or fermentation products to improve the commercial feasibility of prehydrolysis pulping concepts. The ability of Pichia to produce xylitol as well as ethanol is a notable example. Another important finding is that process synergies with an existing pulp and paper mill can vary substantially depending on the type of mill, and can vary even within one particular type of mill, depending on process assumptions, as shown in the VPP-TMP results. Thus, another signpost on the path forward is to seek process and mill synergies that enable biorefining to enhance overall mill performance, and to explore the role of prehydrolysis pulping in various mill contexts, such as various types of kraft pulp and paper or paperboard mills. Lastly, the finding in the case of VPP-TMP that the estimated IRR goes up by a further 3% when the ethanol plant is removed and only the pretreatment is implemented clearly suggests that it may be better in some cases to focus more attention on the improvement of existing pulping and papermaking processes than trying to repurpose existing facilities unless clearly higher financial rewards can be obtained from such repurposing. Although the pulp and paper industry may have entered an era of slower growth or declining markets, nevertheless there is an enormous global demand for paper and paperboard products, which still have high value, and thus the challenge for biorefining technology is to develop an ability to extract yet higher value from pulpwood than produced by pulp, paper, paperboard or existing biorefining concepts.

References

1. F. A. Simmonds, *et al.*, *TAPPI*, 1955, **38**, 178.
2. L. J. Bernardin, *TAPPI*, 1958, **41**, 491.
3. G. A. Richter, *TAPPI*, 1956, **39**(4), 193.
4. K.-E. Eriksson and L. Vallander, *Svensk Papperstid.*, 1982, **85**(6), R33.
5. G. C. Meyers, G. F. Leatham, T. H. Wegner and R. A. Blanchette, *TAPPI J.*, 1988, **71**(5), 105.
6. G. F. Leatham, G. C. Meyers and T. H. Wegner, *TAPPI J.*, 1990, **73**(5), 197.
7. M. Akhtar, R. E. Swaney, E. G. Horn, M. J. Lentz, G. M. Scott, C. C. Black, C. J. Houtman and T. K. Kirk, US Patent, 2002, WO 02/075043 A1.
8. (a) W. Kenealy, E. Horn and C. J. Houtman, *Holzforschung*, 2007, **61**, 223; (b) W. Kenealy, E. Horn, M. Davis, R. Swany and C. Houtman, *Holzforschung*, 2007, **61**, 230.
9. A. W. Rudie, R. Reiner, N. Ross-Sutherland, and W. Kinealy, TAPPI Engineering, Pulping and Environmental Conference Proceedings, TAPPI Press, Atlanta, October 21–23, 2007.
10. T. E. Timell, *TAPPI*, 1957, **40**, 749.
11. T. W. Jeffries and J. R. Van Vleet, *FEMS Yeast Res.*, 2009, **9**(6), 793.
12. J. Y. Jae-Won Lee, D. Zhu, Scordia and T. W. Jeffries, *Appl. Biochem. Biotechnol.*, 2010, **165**(3–4), 814.
13. C. J. Houtman and E. Horn, *TAPPI J.*, 2011, **10**(5), 21.
14. E. M. Bilek, C. J. Houtman and P. J. Ince, *TAPPI J.*, 2011, **10**(5), 31.

CHAPTER 7

Niche Position and Opportunities for Woody Biomass Conversion

THOMAS E. AMIDON,*[1,2] BILJANA BUJANOVIC,[2] SHIJIE LIU,[2] ASIF HASAN[2] AND JOEL R. HOWARD[1]

[1] Applied Biorefinery Sciences, LLC, Syracuse, NY. www.abs-llc.us;
[2] Department of Paper and Bioprocess Engineering, SUNY ESF, Syracuse, NY. www.esf.edu/pbe
*Email: teamidon@esf.edu

7.1 The "Business" of Transforming Plant Biomass for Human Use

Unlike the centuries of evolution up through which humans lived in a primarily agrarian society, today's world population is significantly larger (more than doubled in the past 50 years), and tends to dwell in urban or suburban settings where it is impractical for most people to produce their own food and fuel. As with many products, food and fuel are now provided through an economically viable, industrial manufacturing and supply chain. As part of that chain, the "biorefinery" will be a manufacturing facility that transforms raw plant biomass into products for human use.

From a manufacturing perspective, economic viability is enhanced by "efficiency"; *i.e.*, reducing production costs and also capturing all potential product value from a raw material. In this regard, similar to a petroleum refinery, a biorefinery should be designed with processes that extract as much high value product from raw material as possible. For biorefineries, woody biomass

RSC Green Chemistry No. 18
Integrated Forest Biorefineries
Edited by Lew Christopher
© The Royal Society of Chemistry 2013
Published by the Royal Society of Chemistry, www.rsc.org

presents both challenge and opportunity since it is heterogeneous with three major constituents: cellulose, hemicelluloses, and lignin, and a diverse assembly of minor components and extractives.[1] Extractives and hemicelluloses are the most amenable components to degrade either chemically or biologically, while cellulose and lignin are significantly more recalcitrant. Therefore, to optimize value from, and economic viability of, lignocellulosic biomass conversion, biorefinery processing of woody material into multiple usable products will necessarily require not only different science-based process designs or sequences to address the differing component requirements, but also to have carefully chosen process conditions to enhance the effectiveness of later operations.

7.2   The Science Behind the Technology: Woody Biomass Conversions

Autotrophic assimilation of energy, water, and carbon dioxide to create "sugar" is only the initial step in the creation of a complex suite of chemicals that plants and animals need to support life. In nature, simple sugars are quickly oxidized back to carbon dioxide and water. However, lignocellulosic plants, particularly woody plants (trees) evolved over eons to include complex assemblies of carbohydrates, proteins, and lipids that make them extraordinarily recalcitrant to decomposition back into carbon dioxide and water. This is evidenced by a tree branch laying on a forest floor that is still identifiable as a tree branch for five years or longer. To accelerate woody biomass conversion into useful products, industry has developed processes for compressing into hours the woody biomass chemical decomposition that takes nature years to accomplish. Farmers have long recognized that utilizing all possible portions of plant and animal biomass reduces time and energy investment per unit of substance derived. However, the age of increasing industrial specialization, occurred during a time when the Earth's carrying capacity was not viewed as "limited", and therefore conventional farm wisdom favoring advantages of optimizing efficient resource use became secondary to speed of production. Metrics for industrial woody biomass conversion favored time and cost minimization to achieve the singular objective of freeing (releasing) cellulose from all other chemical constituents. In the process of striving toward that objective, materials or compounds with significant potential use and value were often destroyed or discarded.

Mature chemical pulping technologies based on lignocellulosic materials are focused on the production of cellulose, while lignin, hemicelluloses, and extractives are considered either as an energy source or as waste. For example, kraft delignification results in black liquor, which is burned in a recovery boiler during the regeneration of pulping chemicals. Due to a high heating value of 23.26–25.58 kJ/g,[2] lignin represents the main source of energy in this process. Meister[3] however, describes the incineration technology as having "little thermodynamic or economic sense. With two-fifths of the plants' absorbed energy being used to make the one quarter of its dry mass that is lignin, lignin represents a large investment of biochemical effort by the plant. Reducing this macromolecule to CO_2 destroys much of that investment." Consistent with this view is a growing demand for the development of alternative technologies, which would use all lignocellulosic constituents to produce fuels, materials, and chemicals as substitutes for fossil fuels.[4–6] In accordance with the "biomass refining" philosophy, these new technologies may be envisioned to follow sequential treatments to create separate streams of relatively clean lignocellulosic components.[7,8] After purification, carbohydrate-rich streams may undergo chemical or biological conversion to produce fuels, materials, and chemicals.[1,5,9] Streams rich in lignin and extractives have a high potential for the production of high-value low-volume products to enhance the profitability of this type of biorefinery.[10] Different products including thermoplastics, carbon fibers, adhesives, and chemicals with a wide range of application (*e.g.*, oxygenated aromatic compounds) have been proposed.[11–15] A short introduction into the amazing world of lignin's nature and role in lignocellulosic biomass follows, whereas the potential use of lignin will be discussed later in this chapter.

<u>Lignin</u>: Lignin is associated with the development of vascular systems in plants. Lignin provides resistance to biodegradation and environmental stresses, such as changes in the balance of water and humidity. Derived primarily from *p*-hydroxycinnamyl alcohols (*p*-HCAs: coumaryl, coniferyl, and sinapyl alcohol; when incorporated into lignin HCAs convert into the *p*-hydroxyphenyl (H), guaiacyl (G), and syringyl (S) unit, respectively), lignin results from the radical-coupling polymerization reactions in the wall after polysaccharides have been laid down. Lignin provides a matrix in which polysaccharides are embedded and possibly crosslinked (lignin–carbohydrate complex, LCC). The result is a naturally occurring composite material, which imparts strength and rigidity to the cell wall and this strength allows organized plants to take advantage of it to grow upward.[16–18] Lignin, however, is a major barrier in chemical pulping and processing of lignocellulosics to fuels and chemicals. Presently, lignin is separated by chemical pulping processes, primarily the kraft process. The vast majority of this lignin is burned (heating value of 23.3–25.6 kJ/g for lignin *vs.* 13.6 kJ/g for hemicelluloses)[2,19] in recovery boilers during the regeneration of kraft chemicals (NaOH and Na_2S) providing relatively low-cost fuel for the pulp mill. It has been reported that less than 2% of the total available lignin is used in higher-value products in spite of its natural abundance (lignin is the most abundant natural aromatic polymer and has numerous potential specialty and commodity applications).[11,20]

Lignin polymerization proceeds with the coupling of different resonance structures of phenoxy radicals resulting from enzyme-initiated dehydrogenation of *p*-HCAs. The phenoxy oxygen and the C_β are considered the most reactive in these resonance structures, readily coupling into aryl–ether linkages. This may account for the high abundance of the β-O-4 interunit linkages in lignin, estimated to be as high as 50% in softwoods and almost 60% in hardwoods.[18] More than two-thirds of the phenylpropanoid units in lignins are linked by ether bonds and the rest by carbon–carbon bonds. Hardwood and softwood lignin structural models have been suggested and may be found elsewhere.[17] Different contributions of *p*-HCAs in the biosynthesis of hardwood (SG-lignin) and softwood (G-lignin) lignins cause significant differences in the lignin structure, including the character and content of different bonds between phenylpropanoid units and the main functional groups (for example, methoxyl (OCH_3), phenolic hydroxyl (PhOH), aliphatic hydroxyl, carbonyl, and carboxyl groups). The differences in structure cause differences in the reactivity among different lignin types with hardwood lignin being less resistant against microbiological attack, and more reactive and less prone to condense during chemical processing.

A few percent of the building blocks in lignin are not phenylpropanoid units. These units may be aromatic aldehydes (*e.g.*, vanillin and coniferaldehyde) or aromatic carboxyl acids (*e.g.*, vanillic and ferulic acid). They are proposed to be linked to lignin by ether bonds as lignin end units. Some lignins are esterified with phenolic acids; for example, aspen lignin is esterified with *p*-hydroxybenzoic acid. Also, dihydrocinnamyl alcohols, arylpropane-1,3-diol, and arylglycerol are sporadically incorporated into the lignin structure.[16,17,21] Lignin representative of protolignin is isolated by the MWL (milled-wood lignin) method that is, after more than 50 years since it was recommended, still the most broadly accepted procedure for isolation of lignin for structural studies and studies of lignin reactivity.[22,23] The yield is usually ∼25%, based on the Klason lignin content (the most commonly used method in determination of the content of lignin) using a two-step acid hydrolysis with 72% followed by 3% H_2SO_4. In spite of purification (lignin extracted in dioxane:water from ball-milled wood is purified by precipitation in water from an acetic acid solution), MWL is not carbohydrate-free. This is an important indicator of the presence of lignin–carbohydrate bonds in wood. It is notable that MWL from hardwoods contains larger amounts of carbohydrates (a few percent) than MWL from softwood (traces).[16,22,23]

Distribution of lignin: Even though the lignin concentration is high in the middle lamella (ML) and decreases in the lumen direction, at least 70% of the total lignin in wood is located in the much thicker secondary wall (SW). For example, the lignin concentration in SW, ML, and cell corners of tracheids in the earlywood of black spruce is 23%, 50%, and 80%, respectively, with more than 70% of the total lignin located in SW, which is 87% of the total wood tissue volume.[24]

There are species and morphological variations in lignin distribution. For example, ray parenchyma cells, which constitute about 5% by weight of xylem in softwoods, contain significantly higher lignin concentration than the whole wood, whereas in hardwood species, ray parenchyma cells have lower lignin concentration than fibers that are, in turn, of lower lignin content than vessels. The S/G ratio of lignin is different in different morphological regions of hardwoods. Principally, the SW of fibers contains predominantly S units, whereas SW of vessels consists mostly of G units.[16,24] Compression wood is of higher lignin content than normal softwood and contains more H- and fewer G-units.[25] In contrast to compression wood, hardwood tension wood is of lower lignin content than normal hardwood as the gelatinous layer in tension fibers is almost devoid of lignin.[26]

Lignin-carbohydrate complex, LCC: The difficulties in separation of lignin and carbohydrates during isolation of MWL and production of pulp have frequently been attributed to the presence of lignin–carbohydrate bonds in native wood.[27] The question of whether lignin is chemically bound to polysaccharides in the plant cell wall has been critical to understanding the architecture of the cell wall and the hierarchy of its biosynthesis. The results of extensive studies have shown that covalent bonds between lignin and carbohydrates (LC bonds) are present in lignified cell walls. Four types of native LC bonds have been proposed: benzylether, benzylester, phenylglycoside and acetal bonds; of these, benzylether and benzylester have been considered to be the most probable LC bonds in native wood.[28] There are different methods in the isolation of LCC from wood, and all of them aim at preparing representative LCC; that is, they are aimed at excluding both the cleavage of the native bonds and the formation of the artificial bonds.[27–31] Isolated LCC samples have been analyzed to define the type of LC bonds by a wide variety of indirect methods, including acid and alkali treatment, sodium borohydride reduction, and methylation, and lignin-specific methods such as DDQ oxidation (DDQ: 2,3-dichloro-5,6-dicyano-1,4-benzoquinone).[28] Recently, powerful 2D NMR techniques have been suggested as an efficient method of LC bond analysis since NMR analysis avoids tedious analytical steps and the possibility of chemical change to the LCC.[31]

7.3 Pretreatment Processes

Extensive research has been performed in the evaluation of pretreatments, which is the envisaged first step of sequential refining of lignocellulosics. Pretreatment is intended to remove hemicelluloses and lignin, reduce cellulose crystallinity, and increase porosity and the accessible surface area of particles to facilitate processability of the cellulose-enriched residue.[1,4,5,32–36] Typically, in the absence of pretreatment cellulose hydrolysis yield is <20% of theoretical yields, whereas after pretreatment it may exceed 90% of theoretical.[37] Different pretreatment methods have been suggested including physical, such as milling and irradiation (γ-ray, electron beam, microwave), physicochemical (steam explosion, ammonia fiber explosion, CO_2 explosion, SO_2 explosion), hydrothermal (autohydrolysis), chemical (alkali, dilute acid, gas, oxidizing agents, organic solvents, ionic liquids), biological (white-rot fungi), and electrical methods or combinations of these.[4,38–43] Pretreatment technology must be selected in accordance with the specific requirements of the lignocellulosic biomass used and, at present, there is no single feasible method optimal for pretreatment of all types of lignocellulosics. Pretreatment may result in compounds known for their inhibitory effect on the following steps of enzymatic hydrolysis and fermentation. Production of these compounds ranging from carbohydrate constituents (*e.g.*, acetic acid) and their degradation products (*e.g.*, furfural and hydroxymethyl furfural) to solubilized extractives (*e.g.*, phenolic compounds, terpenes, sterols) and lignin degradation products (phenolic compounds) must be avoided or diminished; a careful optimization of pretreatment to reduce the production of inhibitors increases efficiency and yields of hydrolysis/fermentation.[44] At present, different methods are under research and development to improve efficiency and lower the cost. Overall, a successful pretreatment should be scalable to industrial size, minimizing the use of energy, chemicals, and capital investment, minimizing the loss of sugars and the production of chemicals toxic to the enzymes or fermenting micro-organisms, and maximizing the enzymatic convertibility and the production of valuable byproducts such as lignin.[42] A few selected pretreatment processes will be described in the following sections.

7.3.1 Acid Pretreatment

The main objective of acid pretreatment is removal of hemicelluloses, which in turn makes cellulose more accessible to enzymes. Among hemicelluloses, xylans are more hydrolyzable than glucomannans, which are more acid stable. One important drawback of acid pretreatment is acid-catalyzed degradation of monosaccharides, pentoses and hexoses, resulting from acidolytic cleavage of glycosidic bonds (the generally accepted mechanism of the acidic cleavage of glycosidic bond is explained in detail elsewhere).[45] Under acid conditions, pentoses and hexoses are dehydrated to furfural and hydroxymethyl furfural (HMF), respectively. Furthermore, HMF may be degraded to levulinic acid and formic acid, whereas levulinic acid may be rearranged into α- and β-angelica lactones.[24] These compounds are well-documented inhibitors of fermentation of lignocellulosic hydrolysates.[44] Conversion of carbohydrates to polyphenolic compounds has been also observed, especially under more severe acid conditions.[46] These aromatic compounds, detected as a new Klason lignin material in acid-treated biomass, may appear in the form of granules on the surface of fibers alone or in combination with lignin, carbohydrates, and/or LCC.[47] It has been suggested that acid treatment leads to a disruption of lignin structure, which also helps to increase cellulose accessibility.[34] In acid conditions, lignin may participate in two types of competing reactions resulting in lignin cleavage (depolymerization) and condensation (repolymerization). Electrophilic furfural/HMF, containing aldehyde groups and carbon–carbon double bonds in the furan ring, can participate in lignin condensation reactions.[48] Therefore, furfural and HMF formed during acid-catalyzed degradation of monosaccharides affect the subsequent operations not only by decreasing the hydrolysis/fermentation yields but also by increasing lignin insolubility due to participation in lignin condensation reactions.

The most frequently investigated agent of acid pretreatment is sulfuric acid, but hydrochloric acid, phosphoric acid, nitric acid, and sulfur dioxide have also been tested for this purpose. The most promising seems to be dilute sulfuric acid (from very dilute 0.05% to 0.5–1.5%) which removes most of the hemicelluloses effectively at temperatures between 140 °C and 205 °C. In addition, enhanced lignin removal of nearly 35% has been reported in very dilute sulfuric acid pretreatment of yellow poplar sawdust in a continuously flowing two-stage percolation process followed by hot washing.[49] After acid pretreatment, a conditioning step is required to reduce the toxicity of the hydrolysate prior to fermentation. A pressurized, hot separation and washing of pretreated yellow poplar have been demonstrated to increase the ethanol yield by 50% compared to the control (no washing, no separation). The process includes a pH adjustment with excess lime, which is used in many detoxification schemes and is referred to as overliming. It has been suggested that overliming removes aliphatic and aromatic compounds and that a higher pH during overliming produces a less-toxic, *i.e.* more fermentable, hydrolysate.[50] In addition to inorganic acids, organic acids such as fumaric, maleic, and oxalic acids have been investigated for use in acid pretreatment. They produced less furfural compared to sulfuric acid.[33,51]

7.3.2 Alkaline Pretreatment

The effects of alkaline pretreatment on the processability of cellulose have been investigated using bases such as sodium, potassium, calcium, and ammonium hydroxide. Except for the ammonia recycling percolation treatment (ARP), which may remove 40–60% of hemicelluloses mainly in oligomeric form,[52] alkaline-based processes lead to deacetylation of polysaccharides and different degrees of lignin removal with minor concomitant solubilization of hemi-celluloses and cellulose. Lignin removal increases enzyme efficiency by increasing accessibility of polysaccharides to enzymes and improving produc-tive adsorption of enzymes, whereas deacetylation (all acetyl groups are removed) reduces the steric hindrance of hydrolytic enzymes enhancing carbohydrate digestibility.[39,52,53] In addition, alkali brings about changes in the chemical structure and quality of cellulose, including a decrease in degree of polymerization (DP) and crystallinity, which in turn improve the effects of enzymatic hydrolysis. Alkali also induces swelling, which increases internal surface area.[4] Even though there is a loss of polysaccharides due to peeling and hydrolytic reactions,[24] alkaline processes result in less sugar degradation than acid processes. An increase in the porosity of the lignocellulosic materials caused by the alkali-catalyzed cleavage of ester and ether bonds crosslinking xylan chains and lignin *via* ferulic acid contributes to an increase in the enzy-matic susceptibility of cellulose.[4,52,54,55] Ferulate esters anchoring lignin to polysaccharides are typical for herbaceous species and may be one of the important reasons that alkali pretreatment is more effective for herbaceous than for woody biomass.[38] A method for recovery of ferulic acid for diverse applications has been reported.[52] To facilitate the penetration of chemicals used in the subsequent alkali pretreatment, biomass is chipped and milled or refined under pressure.[56–59] In alkali pretreatment, the alkali is consumed by biomass through the formation of irrecoverable salts and/or salts incorporated into the biomass, and this effect must be taken into account in the process design.[32,35] Process improvements, demonstrated by the positive effects on the enzymatic hydrolysis of the remaining lignocellulosic material, have been brought about by addition of different agents including urea (NaOH/urea pretreatment of spruce) or oxidizing agents such as oxygen (oxidative lime pretreatment of poplar and corn stover) or hydrogen peroxide ($NaOH/H_2O_2$ pretreatment of wheat straw and rice hulls) to the reaction mixture.[53,56,58,60,61] Combinations with irradiation, including microwave-assisted NaOH pretreatment of switchgrass and ultrasonically assisted KOH extraction of wheat straw, showed positive effects on the enzymatic hydrolysis and led to higher yields of dissolved lignin and hemicelluloses, respectively.[57,62]

7.3.3   Steam Explosion Pretreatment

Among the physicochemical processes, steam explosion has been most commonly used for the pretreatment of lignocellulosic materials. In this process, biomass is treated with high-pressure saturated steam at temperatures of 160 °C to 260 °C and corresponding pressures between 0.69 and 4.83 MPa for a short time (several seconds to a few minutes). Afterward, biomass is explosively decompressed due to an abrupt reduction in the pressure. Biomass undergoes autohydrolysis during the steam treatment resulting in the dissolution of hemicelluloses. This is the main effect of steam explosion, which leads to a significantly improved enzymatic hydrolysis of the remaining material.[38,39] Steam explosion brings about the cleavage of LC bonds, reduction in cellulose DP, and an increased lignin solubility in alkaline and organic solvents.[63] Steam explosion has been performed with the addition of sulfuric acid (or sulfur dioxide) or carbon dioxide to decrease the time and temperature of the process with concomitant improvement of hydrolysis, *i.e.* an increase in the dissolution of hemicelluloses and a decrease in the production of inhibitory compounds. The generation of compounds that are documented inhibitors of the enzymatic hydrolysis and/or fermentation processes is one of the main drawbacks of steam explosion. In the subsequent washing stage, water-soluble hemicelluloses/monosaccharides are removed along with the inhibitors, and the overall saccharification yields are reduced. Steam explosion has shown better results for agriculture residues and hardwoods than for softwoods for which an acid catalyst is required to make the substrate accessible to enzymes.[39]

7.3.4   Ammonia Fiber Explosion Pretreatment (AFEX)

The AFEX process is very similar to steam explosion because this process is also a physicochemical process in which biomass is treated at high temperature and pressure for a short time, after which the pressure is suddenly reduced. In contrast to steam explosion, the AFEX process is performed in liquid ammonia (*e.g.*, 1–2 kg ammonia/kg of dry biomass, 90 °C, 30 min). The AFEX has been demonstrated to be a very efficient pretreatment for various herbaceous crops and grasses for which fermentation rates increase significantly after the AFEX process. This beneficial effect may be attributed as in other alkali treatments to the accumulation of alkali sensitive ferulates in these plants.[54,55] However, the process is less efficient for lignocellulosic materials of higher lignin content such as wood.[39]

7.3.5 Hydrothermal Pretreatment

Pretreatment based on hydrothermal treatment, *i.e.* use of water and heat only, has been recommended to create a cleaner, more environmentally benign process with low corrosion risk of equipment and lower quantities of hydro-lyzate neutralization residues.[64] A water prehydrolysis stage has been also studied and used in some commercial pulping operations prior to kraft pulping, with the same goal of hemicellulose removal.[65–70]

The effect of high-temperature water on lignocellulosic/wood constituents has intrigued numerous research groups since the 1930s.[1,8,48,71–81] A majority of these studies have been conducted using hardwood species because it has been demonstrated that softwoods require more severe conditions to achieve the same yields of removed hemicelluloses.[82]

In the context of sequential fractionation of lignocellulosics for the production of energy, materials, and chemicals, water-based pretreatment methods have been investigated in numerous studies and designated as aqueous pretreatment,[83,84] hydrothermal processing,[8,75] autohydrolysis,[34,77,85,86] hot-compressed water treatment,[76,87,88] liquid hot-water treatment,[32,33,35,64,89–91] mild autohydrolysis,[7] hot-water treatment,[1,36,92] and pressurized hot-water extraction.[78,93,94]

These studies have been conducted under different conditions including temperature, pressure, and time of the treatment. Ground biomass of small particle size[36,76,78,82,84,85,90,92] and chip size have been used.[1,7,77,78,91,95] It has been demonstrated that extraction of hemicelluloses from wood chips results in lower yields than from ground wood.[78] A variable water-to-solid ratio has been used ranging from 2:1 to 20:1.[8] In a few studies, to eliminate the interference of extractives on the analysis of carbohydrate and lignin in extracted wood/extracts, wood has been pre-extracted with organic solvents such as acetone and dichloromethane.[77,85] Different reactor types have been used including: percolator/flow-through reactor,[76,82,87,89] batch: Parr,[7,80,81,84] custom-built,[64,90] tube reactors,[36,91] and accelerated extractors.[78,85,95]

Even though studied under a broad range of conditions, most of these pretreatments may be categorized as subcritical water treatment ($150 < T < 370\,°C$, $0.4 < p < 22$ MPa), most frequently performed at temperatures lower than $230\,°C$ to avoid cellulose degradation.

Hydrothermal treatment under subcritical conditions and at temperatures lower than $230\,°C$ is consistent with "the biomass refining" concept because, in effect, it is a hemicellulose-dissolving/degrading step resulting in carbohydrate-rich extracts, which may be purified and used in polymer form[9] or fermented after hydrolysis.[96] An important difference between this type of hydrothermal treatment and hydrothermal treatment performed at higher temperatures is a solid stream of extracted lignocellulosic material (wood) that contains relatively undamaged cellulose if water temperature is maintained below $230\,°C$.[8] After delignification, the solid stream enriched in cellulose may be hydrolyzed to fermentable glucose or used in polymer form for the production of pulp, cellulose derivatives, or nanocellulose.[1,5,7,36,65–70,91,92,97,98]

Most of the work on hydrothermal pretreatment of lignocellulosics has been shown to follow the principles of autocatalyzed hydrolysis or auto-hydrolysis.[1,8,97] Autohydrolysis refers to the process of hydrolysis catalyzed by the hydronium ions (H^+) originating from water and acidic compounds generated from the treated lignocellulosic material. Autohydrolysis consists of a number of reactions including deacetylation of acetylated hemicelluloses and hydrolysis of hemicelluloses. Deacetylation follows the $A_{ac}2$ mechanism of (the letter A refers to the acid catalysis, ac refers to acyl–O cleavage and 2 refers to the possible reaction order of ester hydrolysis).[99,100] The H^+ ions from water ionization act as catalysts for these reactions initially. At high temperatures, the H^+ ions are generated at higher concentrations than in ambient liquid water, providing an effective medium for acid hydrolysis (the ion product, K_w for liquid water increases with increasing temperature and near the critical temperature it is about three orders of magnitude higher than that for ambient liquid water).[101] Subsequently, acetic acid ($pK_a = 4.76$) formed during deacetylation promotes further random cleavage of polymer chains, resulting in a mixture of oligosaccharides and monosaccharides. In general, lower yields (up to 90% of the hemicelluloses originally present in wood) of primarily oligomeric sugars have been produced under these conditions than in hydrothermal treatments at higher temperatures/pressures or in acidic pretreatments.[7,8,33,34] Even though xylan chains are most likely to retain 4-O-methyl glucuronic acid residues, if accessible to water and/or dissolved, the uronic acid residues would dissociate, increase the H^+ concentration, and enhance hydrolysis[8,78,102] (D-glucuronic acid, $pK_a \sim 2.8$[103]). Pectins containing unesterified and easily hydrolyzable methyl esterified galacturonic-acid units are expected to exhibit a similar effect (galacturonic acid $pK_a = 3.5$).[104,105] However, the effects of pectins on autohydrolysis are not well established.[78,82] While dissociation of uronic acids in side chains of hemicelluloses increase acidity of the medium, the glycuronide bonds are exceptionally stable toward acid and uronic acids stabilize the adjacent main-chain linkages.[24] Consequently, the overall rate of hemicellulose degradation is expected to decrease with an increase in the uronic acid-to-xylan ratio. This hypothesis has been confirmed in the study of four different hardwoods species.[106] Even though monosaccharides released during autohydrolysis may undergo degradation, degradation reactions are less frequent than in acid pretreatment because autohydrolysis is performed under mild acid conditions. The majority of the reported data indicate that the pH of hydrolyzate at the end of water treatment ranges from 3.0 to 3.8 (for hardwoods and softwoods under subcritical conditions and at temperatures $< 230\,^\circ$C).[1,65,66,70,73,76,78,82,84,85,91] The degradation of hexoses generates formic acid ($pK_a = 3.75$), which provides a further increase in the concentration of the H^+ ions. The extent of degradation of monosaccharides during autohydrolysis is an important factor that affects autohydrolysis, leads to a chain of undesirable polymerization reactions (self-polymerization of furfural/HMF and participation in lignin condensation reactions), and reduces the efficiency of the following operations on the extract and extracted material.[8,48,73] For example, it has been documented that furfural at concentrations $>2\,\text{g/L}$ is significantly toxic to yeast fermentation.[91] Significant hydrolysis of cellulose and degradation of the glucose has been observed with the pH below 3.5 and the temperature above 200 °C.[84]

Water treatment of wood at high temperatures has been investigated for the effect on lignin by studying water extracts, extracted wood, and lignin isolated from wood and extracted wood.[8,48,71–73,77,79–81,83,107–109] The amount of lignin extracted from wood during autohydrolysis at temperatures lower than 230 °C is species dependent and may be as high as 25% of the amount of lignin originally present in wood. It also depends on the temperature and time of treatment.[8,71] The reaction mechanism is expected to follow homolytic/acidolytic cleavage (depolymerization) and $C_2(C_6)$–C_α crosslinking (polymerization) illustrated in Figure 7.1.[77,79,80,110,114]

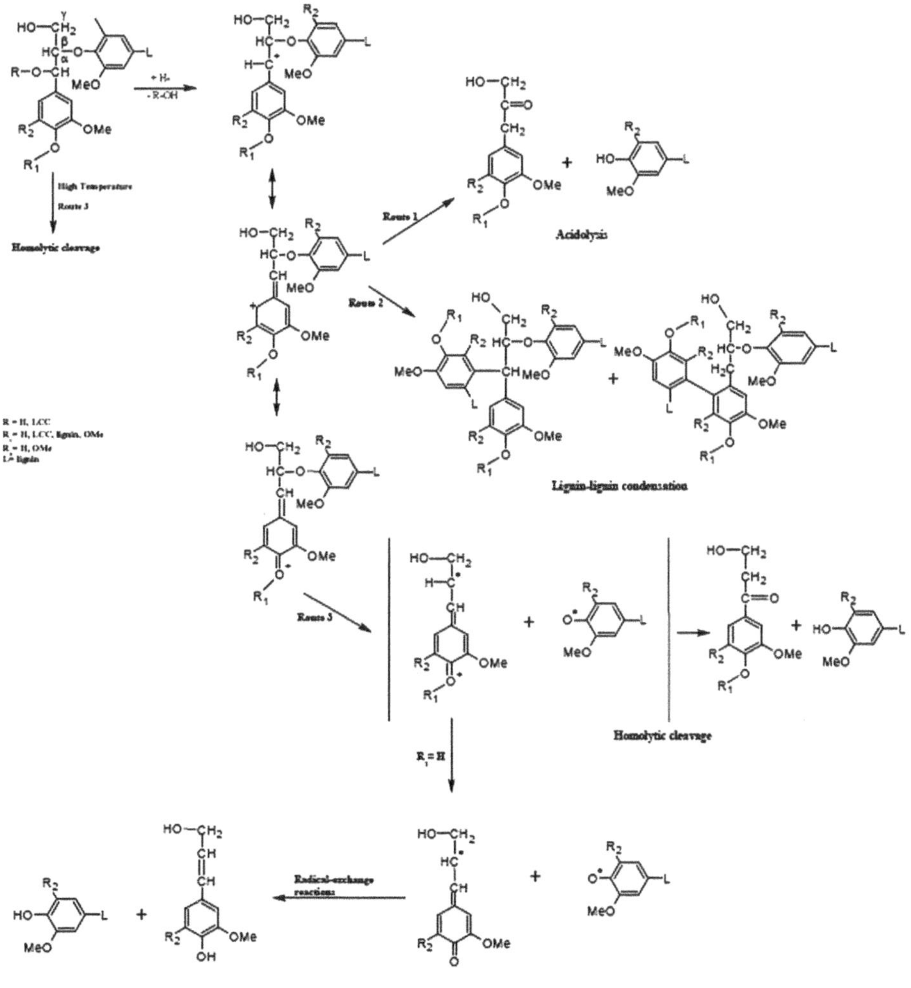

Figure 7.1 Proposed lignin reactions during autohydrolysis (adapted from Li, and Gellerstedt; and Li and Lundquist).[77,114]

The relative importance of these reaction routes depends on the conditions of autohydrolysis and the chemical properties of treated lignocellulosic biomass. MWL was subjected to conditions prevailing during autohydrolysis to evaluate modifications that lignin may endure in wood during autohydrolysis.[74] MWL containing fewer components of high molecular weight than MWL of the original *Eucalyptus globulus* wood has been isolated from the corresponding water-extracted wood (water treatment at 170 °C).[80] A decrease in molecular weight of lignin is consistent with the cleavage of the β-O-4 bonds reported in this study. With respect to the differences in reactivity of SG hardwood and G softwood lignins, Guo and Lai[111] proposed that the β-aryl ether units present in aspen lignin are more readily acid-hydrolyzable than the β-aryl ether units present in spruce lignin.

Recent studies have postulated that during hydrothermal treatment lignin melts, migrates through the cell wall, coalesces to form spherical droplets, and deposits on the cells upon cooling.[36,112] The lignin droplets formed on this way have been considered as an important obstacle of the following enzymatic saccharification of the biomass.[91]

The stability of LC bonds in acid conditions is of interest while studying the effect of water treatment of wood performed at elevated temperature (up to 230 °C) because cleavage of these bonds would bring about more reactive lignocellululosic material in the following phase of processing. Model experiments have suggested that even though LC bonds are chemically labile in acid,[29,113] the degradation rate varies depending on the type of structure.[29,113] For example, while studying the stability of benzylether LC bonds, Košikova *et al.*[29] concluded that etherified benzyl models were more stable than nonetherified counterparts.

Similar to the H-factor, which has been developed by Vroom[115] for kraft pulping, factors aimed at encompassing the effects of temperature and time of hydrothermal treatment in a single number have been introduced. Based on the same principles, the use of a P-factor to express time and temperature of water treatment has been suggested.[116] The P-factor may be calculated by taking into consideration the activation energy typical for the cleavage of glycosidic bonds in hardwood xylan (*Eucalyptus saligna*, $E_a = 125.6 \, \text{kJ/mol}$); P-factor $= \int_{t_0}^{t} \exp(40.48 - 15106/T) dt$, T (K) and t (h). It has been shown that an increase in the P-factor correlates well with the amount of residual xylan in wood at various temperatures.[116] The P-factor should also include the effects of heating and cooling periods. However, if these periods were short, their contribution to the P-factor is minor and may be ignored.

7.4   Bringing the Science to Commerce: *ABS Process*™ Biorefinery Technology

ABS Process™ Biorefinery Technology (BT) is designed to be a commercially deployable, economically viable, overarching biorefinery sequence that encompasses steps from pretreatment of lignocellulosic raw material through to purification and preparation of multiple materials for commercial sale. As can be seen in Figure 7.2, *ABS Process*™ BT main subsystems include screening and cleaning of incoming raw plant biomass, followed by *Hot Water Extraction*™ cooking, followed by numerous steps to separate, recover, and concentrate output products for sale and transport. The economic viability of the process stems, in part from multiple output products, and in part from the efficiency of the process design. *ABS Process*™ BT "cooks" (*Hot Water Extraction*™, or "*HWE*™") lignocellulosic biomass, in water only, to extract and purify multiple naturally occurring chemicals that can serve as renewable alternatives to fuels and industrial chemicals currently derived from petroleum, coal, and natural gas. However, in addition to producing multiple commercial chemicals, *HWE*™ cooking leaves the structural portion of the extracted biomass still intact, but improved, through creation of several key properties which add significant

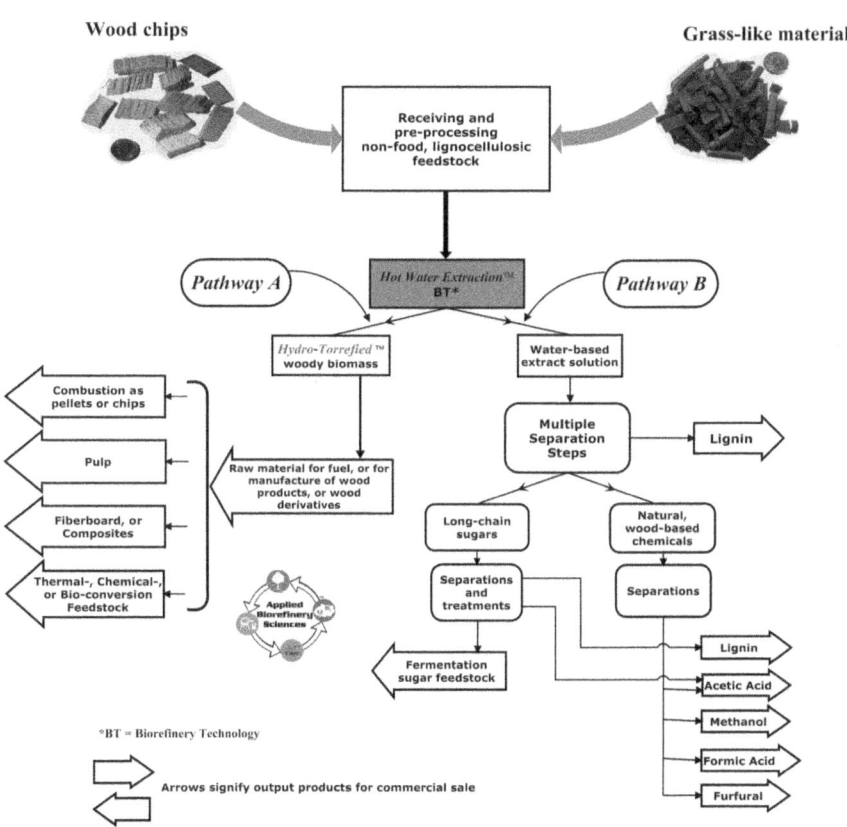

Figure 7.2 ABS Process: generalized process flow.

value. For example, wood chips subjected to *HWE*™ cooking are more energy dense (Btu/ton), lighter (per unit of equal volume), lower in ash content, and markedly less hydrophilic. These improvements have significant implications for *HWE*™ application in the manufacture of wood-derived products such as paper, fiberboard, chipboard, and fuel pellets. Similarly, agricultural residues become more chemically reactive following *HWE*™ cooking, and may be more readily transformed into nanocellulose or sugars available for nutritional or chemical conversion. By virtue of these multiple, commercial-product pathways, *HWE*™-based biorefinery technology can serve at least three global industrial market opportunities: wood products, sugar-based products, and chemicals.

Drawing from Section 7.3 Pretreatment Processes, *HWE*™ cooking, as a hydrothermal pretreatment performed in the absence of acids and bases, is preferred because of its ability to generate multiple high-value output products without chemical input, recovery, or disposal costs. Instead of added chemicals in the cooking phase, *ABS Process*™ BT relies upon an autocatalytically driven reaction initiated by deacetylation of hemicelluloses and release of acetic acid into the extract solution (explained in detail in Section 7.3.5). The resulting acidic conditions (final pH ∼ 3.5) permit further solubilization and diffusion of biomass constituents and derived compounds, including hemicelluloses of different DP, oligomers, monosaccharides, phenolic compounds and other trace compounds into the extract solution (Figure 7.2, Pathway B). Then, membrane fractionation and other chemical separation techniques are applied to the extract solution to separate hemicelluloses of different DP and carbohydrate oligomers and monomers from acetic acid, furfural, formic acid, methanol, and lignin/lignin-related compounds for sale commercially and to purify the mixture into fermentable sugars. In Figure 7.2 Pathway A, extracted lignocellulosic biomass, with reduced hemicellulose content, carries the advantage of expanded product opportunities, as well as enhanced product properties at reduced energy and chemical manufacturing costs.

As stated above, the commercial chemical and material outputs will serve at least three global, industrial market opportunities: wood products, sugar-based products, and chemicals. Following are specific examples of potential products in each of these pathways.

7.5   Output Products from the *ABS Process*™

7.5.1   Products from Extracted Wood and Nonfood Agricultural Materials

Many conventional wood-derived products can be improved when produced from woody biomass subjected to *HWE*™ cooking. Following hot-water extraction, the residual woody material contains predominantly cellulose and lignin. Cellulose is an excellent biopolymer of wide applications. Lignin is an aromatic polymer with a wide range of valuable commercial applications (even though in mature technologies it is used as a fuel due to its relatively high-energy content). Following are examples emphasizing the benefits of *HWE*™ cooking.

Pulp and paper: Following the removal of most hemicellulose related compounds through HWE^{TM} cooking, residual sugar maple chips are reduced in mass typically by 23%. These Hydro-Torrefied™ chips, with reduced hemicellulose content, higher porosity (lower density), and with lignin of higher reactivity and weaker LCC, can improve pulping and bleaching, improve net energy efficiencies, and can offer lower chemical use, increased pulp viscosity, as well as higher bulk, absorbance, and stiffness of pulp.[1] A lower tensile strength inherent to the paper made from this type of pulp can be increased using a conventional dry strength agent, starch, or recently evaluated polylactic acid as a potential product of a biorefinery based on lignocellulosics.[117]

Fuel Pellets: ABS research on fuel pellets produced from Hydro-Torrefied™ woody materials showed that these pellets have: 1) reduced ash content; 2) increased Btu/lb content; 3) increased hydrostability; and 4) increased physical durability. HWE-induced ash reduction (property #1) is well documented, and properties 2, 3, and 4 are consistent with higher relative lignin and lower hemicellulose contents. All these improved properties combine to create superior fuel pellets.[118]

Fiberboard: Decreased woodchip bulk density (after hot-water extraction) leads to lighter products, and reduced hemicellulose content leads to decreased hydrophilicity, increased weather resistance, increased dimensional stability, and an expected increase in mold resistance. This could be a significantly improved material for reconstituted wood products. A project is underway in collaboration between SUNY ESF and Washington State University to quantify the gains.[118]

Carboxymethylated Cellulose: Hot-water pre-extraction of wood significantly enhanced the reactivity of the cellulose component in the residual woody material and could facilitate the preparation of the carboxymethylcellulose (CMC) as the bulk of hemicelluloses are removed.[119]

Nanocellulose: Nanocellulose is a high-value product that can be obtained by acid hydrolysis of cellulose. Removal of hemicellulose material through HWE^{TM} cooking makes the residual lignocellulosic material more chemically reactive and amenable to delignification.[70,117,120] In addition, the increased relative cellulose content along with more accessible regions of amorphous cellulose, in combination, are expected to improve the recovery rate of nanocellulose.

Sugar from Cellulose – Glucose: As mentioned above, lignocellulosic materials are more chemically reactive following HWE^{TM} cooking, and they are therefore more amenable to hydrolysis by biochemical or thermochemical pathways. Research is underway to test the economic viability of depolymerizing residual amorphous cellulose from hot-water-extracted cellulose material into fermentable glucose as a coproduct with nanocellulose production.

7.5.2 Products from Extracted Sugars

Up until the Energy Independence and Security Act (EISA) in 2007, processes increasing utility of nonfood plant biomass generally, as a group, targeted depolymerization of cellulose into monomer and dimer sugars for fermentation to ethanol. Per stipulations in EISA 2007, the United States Department of Energy (DOE) began to include other sugar-based alcohol and biodiesel fuels as acceptable end products for biomass based processes, but fuels derived from sugar fermentation still dominate the R&D objectives of government grants, and therefore both public and private sector agendas.

Aside from alcohol fuels, other fermentation products have been under extensive study including butanol, propane-diol, hydrogen, higher alkanes, polyhydroxyalkanoates (PHA), lactic acid, *etc.*, as well as Xylitol fermented as a sweetener specifically from 5-carbon (xylose) extracted sugars.

7.5.3 Chemicals and Materials

As shown in Figure 7.2, ABS Process™ BT extracts, separates, and recovers naturally occurring, water-soluble hemicellulose related chemicals, such as sugar, acetic acid, formic acid, furfural, methanol, and lignin. All of the chemicals in this group are saleable as commercial chemicals. Research underway at SUNY-ESF reveals that multiple varieties of sulfur-free ("clean") lignin may be recoverable from both Pathway A and Pathway B of ABS Process™ BT. This could make lignin the single most valuable product of the biorefinery. In addition to the chemicals noted heretofore, wood extractives may also possess value worthy of commercial consideration.

<u>Lignin</u>: Lignin in *HWE*™ (160 °C, 2 h, P-factor = 540) results in solubilization of ∼15% of the total amount of lignin in sugar maple.[120] Dissolved lignin appears in the hot-water extract in two distinct fractions of lignin/lignin-like material: an insoluble fraction, Insol fraction, which readily precipitates from the hot-water extract and a soluble fraction, Sol fraction, which may be extracted from the hot-water extract with organic solvents, such as chloroform or ether.[107,108,121] To separate the precipitate, an ultrafiltration membrane (1000 Da nominal molecular-weight cut-off; N_2) has been used in recent experiments. The amount of the Insol fraction was ∼1.8% of the total amount of wood treated in *HWE*™ cooking. The lignin content of the Insol fraction was 87% (Klason + acid soluble lignin[23]). An increase in the lignin content to ∼93% was observed after mild acid hydrolysis (pH2, 2 h, 60 °C, N_2). The acid-labile LC bonds, such as benzyl–ether bonds (C_α–C_6) might have been broken during this treatment. Carbohydrate analysis demonstrated that xylose was the predominant sugar in the purified Insol fraction (more than 75% of the remaining carbohydrates were xylose based). The properties of the purified Insol fraction are presented in Table 7.1 along with the properties of hardwood kraft lignin and the Alcell lignin (hardwood organosolv lignin).[11,107–109]

The S/G ratio of ∼2.6 of extracted lignin is more than twice as high as the S/G ratio characteristic for lignin in sugar maple.[122] This result is consistent with an earlier suggested high solubility of S-units during hydrothermal treatment.[8] The high content of S-units in lignin dissolved during *HWE*™ cooking of hardwoods may also indicate that lignin in fibers is readily available because S-lignin is mostly present in fibers.[24] The relatively high content of the PhOH groups compared to MWL is lower than that registered for kraft and

Alcell lignin. This lignin is of higher polydispersity than kraft and Alcell lignin. During HWE^{TM} cooking, the PhOH content of lignin in sugar maple increased from 0.35 mmol/g lignin to 1.83 mmol/g lignin.[107,120] This considerable increase in the content of PhOH groups of lignin remaining in extracted wood is a strong indication of the cleavage of β-O-4 bonds during HWE^{TM}/autohydrolysis which is in accordance with the results of previous studies.[77,79–81]

Organic extracts (extraction with chloroform and ether) of permeates obtained during laboratory ultrafiltration of hot-water extracts of sugar maple (solvent-soluble fraction of hot-water extract, Sol fraction) were analyzed by GC-MS.[107,121] The total amount of extracted material was ~1.5% of OD wood treated in HWE^{TM} cooking, or about 6.5% of the total material removed. Typical phenolic wood extractives vanillin, guaiacol, and syringol were identified. A number of organic low-molecular weight (LMW) compounds were present only in the solvent-soluble fraction of the hot-water extracts but not in the extracts produced from extraction of sugar maple with organic solvents. Specifically, p-hydroxy benzaldehyde, syringaldehyde, p-hydroxybenzoic acid, vanillic acid, syringic acid, coniferaldehyde, sinapaldehyde, dihydroconiferyl alcohol, dihydroferulic acid, medioresinol, and syringaresinol were detected only in the solvent-soluble fraction of the hot-water extracts. These phenolic compounds may be phenolic extractives in which case their presence in hot-water extract actually corresponds to the environmental leaching of phenols from leaves or bark.[123] Identified dissolved aromatic acids contribute to autohydrolysis processes during HWE^{TM} cooking by providing additional sources of the H^+ ions; p-hydroxybenzoic acid, $pK_a = 4.3$; ferulic acid, $pK_a = 4.27$; vanillic acid, $pK_a = 4.08$; syringic acid, $pK_a = 3.86$.[124] In accordance with earlier results on aqueous hydrolysis of lignin performed on hemlock wood and Brauns' lignin,[72] these compounds may also be attributed to lignin cleavage reactions taking place during HWE^{TM} cooking. Moreover, evidence for the occurrence of homolytic cleavage of the β-aryl ether bonds was found based on the presence of syringaldehyde, sinapaldehyde, syringic acid, and syringaresinol. These compounds were proposed to be formed *via* a homolytic cleavage of an initially formed quinone methide (Figure 7.1 route 3) or result from oxidation reactions (syringic acid).[77,114] Coniferyl alcohol, which may also be expected as a result of homolytic cleavage of the phenolic β-aryl ether structure, was not identified in the Sol fraction. The absence of coniferyl alcohol may be accounted for by an inherent low stability of this compound at high temperature and in acid conditions.[125]

Table 7.1 Properties of hardwood kraft lignin, Alcell lignin, and purified Insol fraction of sugar maple.

Lignin	PhOH mmol/g	S/G	MW_w	MW_n	Polydispersity	Ref.
HKL	4.3	1.2	2400	–	1.8	11
Alcell	4.3	2.3	2200	–	1.7	11
Purified Insol	3.0	2.6	1600	5200	3.1	107–109

Depending on the conditions of HWE^{TM} cooking, the total amount of Insol and Sol fraction of hot-water extract may amount to more than 3.5% of OD wood used in HWE^{TM} or 15% of the total material dissolved during HWE^{TM} (160 °C, 2 h; P = 540). At the proposed biorefinery capacity of 1000 tons of wood per day,[1] the annual production of these two fractions may equal 10 000 tons. This is a significant amount of valuable material with tremendous potential to increase economic returns from HWE^{TM} cooking of lignocellulosic hardwoods and enhance the desirability of hardwood biorefineries.

In contrast to kraft lignin (which is the most readily available source of lignin today) the lignin released during HWE^{TM} cooking has the significant advantage of being almost entirely sulfur-free. Due to its considerable sulfur content, the utilization of kraft lignin is limited despite an annual worldwide production of $\sim 100 \times 10^6$ tons.[11] Even though as a fuel source lignin is not very energy dense, producing less than 1/4 as much energy as middle distillate fuels, most kraft lignin is used for energy production and is burned in the kraft recovery process. Less than 2% of lignin is used for other purposes.[11,126]

7.5.4   Insol Fraction

Adhesives: The Insol fraction containing lignin can be evaluated for the production of adhesives.[3,13,20] An extensive condensation of lignin and furfural/HMF, has been observed in earlier studies.[48,127] Accordingly, the Insol fraction containing both lignin and sugars appears to be an excellent candidate to produce resin under acid conditions. This type of lignin-based thermosetting resin would be a green alternative for environmentally unfavorable phenol-formaldehyde resins.

Lignin-based polymer blends: Even though lignins of different origin vary in chemical configuration and physicochemical properties, which may cause pronounced differences in mechanical behavior and thermal mobility, one important area of study of lignin application has been the production of thermoplastics.[11] To improve the processability and mechanical properties of lignin fibers, lignin may be blended with various synthetic or biorenewable polymers, which decrease the glass-transition temperature and increase the plastic response to mechanical deformation. Thermoplastics have been made by the blending of different technical lignins, such as kraft and Alcell lignin with different synthetic polymers, among which polyethylene oxide (PEO) showed the most promising results.[11] Lignin recovered after HWE^{TM} cooking is expected to have similar thermal properties and processability as Alcell lignin. Being of similar S/G ratio, PhOH content, and MW/polydispersity, these lignins are probably of similar condensation level, which mainly governs lignin thermal mobility.[11,107–109]

Lignin-based carbon fibers: From the 1950s onward, the demand for carbon fibers as a reinforcing material in composite products has continued to grow. Today carbon fiber composite products are routinely used in sports equipment, marine products, construction, aircraft and the automotive industry. Carbon fibers are produced in a two-step thermal process; the first step is a stabilization stage in which the precursor fibers are thermostabilized at 200–300 °C under tension; the second stage is thermal treatment of fibers at 1000–2000 °C in an inert atmosphere, conditions that maintain the fibrous structure. Even though expensive, polyacrylonitrile (PAN) is the most important precursor material for the production of carbon fibers and almost 80% of commercially available carbon fibers are derived from PAN. It has been estimated that the cost of carbon fibers can be significantly reduced if lignin and lignin and/or lignin synthetic polymer blends are used for carbon-fiber production instead PAN. Carbon fibers of satisfactory mechanical properties have been successively made from blends composed primarily of kraft lignin and Alcell lignin with overall yields of up to 45%. Since mechanical properties of carbon fibers are highly size dependent and increase with decreasing size diameter, the production of lignin-based carbon fibers of lower diameter (<10 μm) than achieved at present is an important goal.[126,128,129] Lignin as a biorefinery product is expected to be of similar processibility and final carbon fiber properties to Alcell lignin and more appropriate for this use than kraft lignin due to the inherent chemical structure being more similar to native lignin with tunable properties adaptable for many uses, which provide for better processibility, including lower polydispersity, less contamination, and lower condensation level.

Production of oxygenated aromatic compounds: Oxidation in alkaline medium is an established process for the depolymerization of lignins to high-value chemicals, among which vanillin and syringaldehyde are of prime interest. Vanillin with an approximate selling price of $5–9/pound (vs. $1200–4000/kg when extracted from *Vanilla* pods) finds application in different industries including chemical and pharmaceutical for the production of herbicides, anti-foaming agents and drugs such as papaverine, L-dopa, and antimicrobial agent, trimethoprim.[130] Syringaldehyde, with a selling price of tens of dollars per pound, is used in similar applications as vanillin due to the similar chemical structure and properties. Its prospective use includes the production of hair and fiber dye and drugs for obesity and breast cancer.[131] The extent of lignin oxidation may be improved by the introduction of catalysts. Thus, lignin dissolved during HWE^{TM} cooking can be examined for oxidation in alkaline media in the presence of Fe^{3+} and Cu^{2+} catalysts in the form of CuO, Fe_2O_3, and mixtures of CuO and Fe_2O_3 or metal organic frameworks.[132] Lignin can also be oxidatively depolymerized with polyoxometalates.[133]

7.5.5 Sol Fraction

The permeate obtained through ultrafiltration of sugar maple hot-water extract was treated (at bench scale) with chloroform and the Sol fraction was recovered after chloroform evaporation. This method of laboratory isolation of the Sol fraction is unacceptable in industrial conditions (chlorinated organic solvent) and a feasible way of separation in industrial conditions must be developed. Different approaches could be considered, including the use of hydrophobic membranes and/or a combination of extraction with an appropriate nonpolar solvent (hexane) and adsorption onto a neutral acrylic ester sorbent.[14,15,134] Recently, a new category of "switchable solvents" has been introduced. These solvents change from being hydrophobic, with very low miscibility with water when in air, to being hydrophilic and completely miscible in water under an atmosphere of CO_2.[135] Alternatively, the Sol fraction from *HWE*™ cooking may be recovered in a two-step process consisting of polymerization in the presence of phenoloxidases (laccase, lignin peroxidase)[136,137] followed by membrane separation. Treatment of willow hydrolysate with phenoloxidases led to the removal of monoaromatic phenolic compunds, which increased fermentability of the hydrolysate.[44] This detoxification effect of enzymes corroborated a mechanism of oxidative polymerization.

In the development of new biorefinery technologies, including those based on hot-water extraction, lignin should be carefully considered as a significant source of valuable chemicals and materials, which may be a determining factor in successful biorefinery operations. Its use for the production of chemicals and materials is expected to increase in the future because the flexible and mild conditions applied in the biorefinery operations enable the production of lignin of tunable properties close in structure and properties to the native lignin.

7.6 Summary

It is virtually axiomatic that increased value can be generated by isolating relatively pure substances from heterogeneous raw materials. Woody biomass has been widely used throughout the world to supply resources to fulfill human needs, however, the whole raw material has not necessarily been utilized efficiently. Traditional wood fractioning techniques targeted cellulose as the end product, while discarding or destroying much of the other potential value. The world's population has more than doubled to over 7.25 billion over the past 50 years. If we wish to achieve long-term sustainability for humans on Earth, such wasteful practices need to be replaced or modified to more efficiently utilize available resources. One means for contributing to that goal is to develop processes that recover a larger portion of the raw material in a broader portfolio of products. Applied Biorefinery Sciences, LLC is commercializing a value-optimization pathway (*ABS Process*™ BT) for generating a multiproduct portfolio by isolating and recovering higher total yield and value from woody biomass. By more efficiently fractionating woody biomass into usable products, *ABS Process*™ BT creates a previously unavailable, economically viable niche position for capturing sustainable, renewable value.

References

1. T. E. Amidon, C. D. Wood, A. M. Shupe, Y. Wang, M. Graves and S. Liu, Biorefinery, *Conversion of woody biomass to chemicals, energy and materials. J Biobased Mater Bio*, 2008, **2**, 100–120.
2. A. Demirbaş, Relationship between lignin contents and heating values of biomass, *Energy Conv. Manag.*, 2001, 183–188.

3. J. J. Meister, Modification of Lignin, *J Macromol Sci – Pol R*, 2002, **C42**, 235–289.
4. Y. Sun and J. Cheng, Hydrolysis of lignocellulosic materials for ethanol production: a review, *Bioresource Technol.*, 2002, **83**, 1–11.
5. S. Liu, T. E. Amidon, R. C. Francis, B. V. Ramarao, Y.-. Lai and G. M. Scott, From forest biomass to chemicals and energy, *Ind. Biotechnol.*, 2006, **2**, 113–120.
6. A. J. Ragauskas, C. K. Williams, B. H. Davison, G. Britovsek, J. Cairney, C. A. Eckert, W. J. Frederick, J. P. Hallett, D. J. Leak, C. L. Liotta, J. R. Mielenz, R. Murphy, R. Templer and T. Tschaplinski, The path forward for biofuels and biomaterials, *Science.*, 2006, **311**, 484–489.
7. G. D. Garrote, Mild autohydrolysis: an environmentally friendly technology for xylooligosaccharide production from wood, *J Chem Technol Biotechnol*, 1999, **74**, 1101–1109.
8. G. Garrote, H. Dominguez and J. C. Parajó, Hydrothermal processing of lignocellulosic materials, *Holz Roh Werkst*, 1999, **57**, 191–202.
9. A. J. Stipanovic, J. S. Haghpanah, T. E. Amidon, G. M. Scott, V. Barber and K. Mishra, Opportunities for hardwood hemicelluloses in biodegradable polymer blends, *Materials, Chemicals, and Energy from Forest Biomass,* Argyopoulos, D.S., ed., ACS Sym Ser 954, 2007, pp. 107–120.
10. S. Fernando, S. Adhikari, C. Chandrapal and N. Murali, Biorefineries: Current status, challenges, and future direction, *Energ Fuel*, 2006, **20**, 1727–1737.
11. S. Kubo, R. D. Gilber and J. F. Kadla, Lignin-based polymer blends and biocomposite materials, In *Natural Fibers, Biopolymers, and Biocomposites.*, Mohanty, A.K., Misra, M. and Drzal, L.T., ed.; CRC Taylor & Francis Group: Boca Raton, 2005, pp. 671–697.
12. T. Kaneko, T. H. Thi, D. J. Shi and M. Akashi, Environmentally degradable, high-performance thermoplastics from phenolic phytomonomers, *Nat Mater*, 2006, **5**, 966–970.
13. D. Stewart, Lignin as a base material for materials applications: Chemistry, application, and economics, *Ind Crop Prod*, 2008, **27**, 202–207.
14. H. D. Embree, T. Chen and G. F. Payene, Oxygenated aromatic compounds from renewable resources: motivation, opportunities, and adsorptive separations, *Chem Eng J*, 2001, **84**, 133–147.
15. J. A. Koehler, B. J. Brune, T. Chen, A. J. Glemza, P. Vishwanath, P. J. Smith and G. F. Payne, Potential approach for fractionating oxygenated aromatic compounds from renewable resources, *Ind Eng Chem Res*, 2000, **39**, 3347–3355.
16. E. Sjöström, R. Alén, Analytical Methods in Wood Chemistry, Pulping, and Papermaking, Springer Series in Wood Science, Springer-Verlag: Berlin Heidelberg, 1999.
17. J. Ralph, G. Brunow and W. Boerjan Lignins, In *Encyclopedia of Life Sciences.*, Osborne, K., ed.; John Wiley & Sons, Ltd: 2007; pp. 1–10.
18. D. S. Argyropoulos, S. B. Menachem Lignin, In *Advances in Biochemical Engineering/Biotechnology.*, Scheper, T., ed.; Springer-Verlag: Berlin Heidelberg, 1997; Vol. 57, pp. 128–158.

19. S. Walton, A. van Heiningen and P. van Walsum, Inhibition effects on fermentation of hardwood extracted hemicelluloses by acetic acid and sodium, *Bioresource Technol*, 2010, **101**, 1935–1940.
20. E. K. Pye, Industrial lignin production and applications, In *Biorefineries-Industrial Processes and Products*, Kamm, M., ed.; Wiley-VCH Verlag GmbH&Co. KGaA: Weinheim, 2006; Vol.2, pp. 165–200.
21. H. Kim, J. Ralph, N. Yahiaoui, M. Pean and A.-M. Boudet, Cross-coupling of hydroxycinnamyl aldehydes into lignins, *Org Lett*, 2000, **2**, 2197–2200.
22. A. Björkman, Studies on finely divided wood. Part 1. Extraction of lignin with neutral solvents, *Sven Papperstidn*, 1956, **59**, 477–485.
23. S. Y. Lin, C. W. Dence, *Methods in Lignin Chemistry*. Springer: Berlin Heidelberg, 1992.
24. E. Sjöström, Wood Chemistry. *Fundamentals and Applications. 2nd edn.* Academic Press: San Diego, 1993.
25. T. E. Timell, Recent progress in the chemistry and topochemistry of compression wood, *Wood Sci Technol*, 1982, **16**, 83–122.
26. T. E. Timell, The chemical composition of tension wood, *Sven Papperstidn*, 1969, **72**, 173–181.
27. A. Björkman, Studies on finely divided wood. II Extraction of lignin-carbohydrate complexes with neutral solvents, *Sven Papperstidn*, 1957, **60**, 243–251.
28. T. Koshijima, T. Watanabe, *Association between Lignin and Carbohydrates in Wood and other Plant Tissues*. Springer Berlin: Heidelberg, 2003.
29. B. Košikova, D. Joniak and L. Kosáková, On the properties of benzyl ether bonds in the lignin saccharide complex isolated from spruce, *Holzforschung*, 1979, **33**, 11–14.
30. Y.-. Kim, K. Iiyama, A. Kurahashi and G. Meshitsuka, Structural feature of lignin in cell walls of normal and fast-growing poplar (*Populus maximowiczii* Henry), *Mokuzai Gakkaishi*, 1995, **4**, 837–843.
31. M. Y. Balakshin, E. A. Capanema and H.-M. Chang, MWL fraction with high concentration of lignin carbohydrate linkages: Isolation and 2D NMR spectroscopic analysis, *Holzforschung*, 2007, **61**, 1–7.
32. A. T. W. M. Hendriks and G. Zeeman, Pretreatments to enhance the digestibility of lignocellulosic biomass, *Bioresource Technol*, 2009, **100**, 10–18.
33. P. Alvira, E. Tomás-Pejó, M. Ballesteros and M. J. Negro, Pretreatment technologies for an efficient bioethanol production process based in enzymatic hydrolysis: A review, *Bioresource Technol*, 2010, **101**, 4851–4861.
34. C. E. Wyman, B. E. Dale, R. T. Elander, M. Holtzapple, M. R. Ladisch and Y. Y. Lee, Coordinated development of leading biomass pretreatment technologies, *Bioresource Technol*, 2005, **96**, 1959–1966.
35. N. Mosier, C. Wyman, B. Dale, R. Elander, Y. Y. Lee, M. Holtzapple and M. Ladisch, Features of promising technologies for pretreatment of lignocellulosic biomass, *Bioresource Technol*, 2005, **96**, 673–686.
36. M. Zeng, N. S. Mosier, C. P. Huang, D. M. Sherman and M. R. Ladisch, Microscopic examination of changes of plant cell structure in corn stover

due to hot water pretreatment and enzymatic hydrolysis, *Biotechnol Bioeng*, 2006, **97**, 265–278.

37. L. R. Lynd, Overview and evaluation of fuel ethanol from cellulosic biomass: Technology, economics, the environment, and policy, *Annu Rev Energ Env*, 1996, **21**, 403–465.

38. M. J. Taherzadeh and K. Karimi, Pretreatment of lignocellulosic wastes to improve ethanol and biogas production: A review, *Int J Mol Sci*, 2008, **9**, 1621–1651.

39. P. Kumar, D. M. Barrett, M. J. Delwiche and P. Stroeve, Methods for pretreatment of lignocellulosic biomass for efficient hydrolysis and biofuel production, *Ind Eng Chem Res*, 2009, **48**, 3713–3729.

40. S. Zhu, Perspective use of ionic liquids for the efficient utilization of lignocellulosic materials, *J Chem Technol Biot*, 2008, **83**, 777–779.

41. S. H. Lee, T. V. Doherty, R. J. Linhardt and J. S. Dordick, Ionic liquid-mediated selective extraction of lignin from wood leading to enhanced enzymatic cellulose hydrolysis, *Biotechnol Bioeng*, 2009, **102**, 1368–1376.

42. H. Jørgensen, J. B. Kristensen and C. Felby, Enzymatic conversion of lignocellulose into fermentable sugars: challenges and opportunities. *Biofuel Bioprod Bio*, 2007, **1**, 119–134.

43. H. L. Chum, S. K. Black, D. K. Johnson, K. V. Sarkanen and D. Robert, Organosolv pretreatment for enzymatic hydrolysis of poplars: isolation and quantitative structural studies of lignins, *Clean Technol Envir*, 1999, **1**, 187–198.

44. L. J. Jönsson, E. Palmqvist, N.-O. Nilverbrant and B. Hahn-Hägerdal, Detoxification of wood hydrolysates with laccase and peroxidase from the white-rot fungus *Trametes verisocolor*, *Appl Microbiol Biot*, 1998, **49**, 691–697.

45. D. Fengel, G. Wegener Wood: Chemistry, *Ultrastructure, Reactions.* Walter de Gruyter: Berlin, 1989.

46. P. Sannigrahi, D. H. Kim, S. Jung and A. Ragauskas, Pseudo-lignin and pretreatment chemistry, *Energ Environ Sci*, 2011, **4**, 1306–1310.

47. B. S. Donohoe, S. R. Decker, M. P. Tucker, M. E. Himmel and T. B. Vinzant, Visualizing lignin coalescence and migration through maize cell walls following thermochemical pretreatment, *Biotechnol Bioeng*, 2008, **101**, 913–925.

48. M. G. S. Chua and M. Wayman, Characterization of autohydrolysis aspen (*P. tremuloides*) lignins. Part 3. Infrared and ultraviolet studies of extracted autohydrolysis lignin, *Can J Chem*, 1979, **57**, 2603–2611.

49. N. J. Nagle, R. T. Elander, M. M. Newman, B. T. Rohrback, R. O. Ruiz and R. W. Torget, Efficacy of a hot washing process for pretreated yellow poplar to enhance bioethanol production, *Biotechnol Progr*, 2002, **18**, 734–738.

50. A. Mohagheghi, M. Ruth and D. J. Schell, Conditioning hemicellulose hydrolysates for fermentation: Effects of overliming pH on sugar and ethanol yields, *Process Biochem*, 2006, **41**, 1806–1811.

51. J.-. Lee, R. C. L. B. Rodrigues, H. J. Kim, I.-. Choi and T. W. Jeffries, The roles of xylan and lignin in oxalic acid pretreated corncob during separate enzymatic hydrolysis and ethanol fermentation, *Bioresource Technol*, 2010, **101**, 4379–4385.
52. F. Carvalheiro, L. C. Duarte and F. M. Gírio, Hemicellulose bio-refineries: a review on biomass pretreatments, *J Sci Ind Res India*, 2008, **67**, 849–864.
53. S. Kim and M. T. Holtzapple, Effect of structural features on enzyme digestibility of corn stover, *Bioresource Technol*, 2006, **97**, 583–591.
54. M.M.d.O. Buanafina, Feruloylation in grasses: Current and future perspectives, *Mol Plant*, 2009, **2**, 861–872.
55. J. Ralph, R. D. Hatfield, J. H. Grabber, H.-J. G. Jung, S. Quideau and R. F. Helm, Cell wall cross-linking in grasses by ferulates and diferulates, In *Lignin and Lignan Biosynthesis*, Lewis, N. G. and Sarkanen, S. ed., ACS Symposium Series, Vol. 697, Washington DC, 1998, pp. 209–236.
56. V. S. Chang, M. Nagwani, C. H. Kim and M. T. Holtzapple, Oxidative lime pretreatment of high-lignin biomass poplar wood and newspaper, *Appl. Biochem. Biotechnol.*, 2001, **94**, 1–28.
57. Z. Hu and Z. Wen, Enhancing enzymatic digestability of switchgrass by microwave-assisted alkali pretreatment, *Biochem Eng J*, 2008, **38**, 369–378.
58. Y. Zhao, Y. Wang, J. Zhu, A. Ragauskas and Y. Deng, Enhanced enzymatic hydrolysis of spruce alkaline pretreatment at low temperatures, *Biotechnol. Bioeng.*, 2008, **99**, 1320–1328.
59. K. Mirahmadi, M. M. Kabir, A. Jeihanipour, K. Karimi and M. J. Taherzadeh, Alkaline pretreatment of spruce and birch to improve bioethanol and biogas production, *BioResources*, 2010, **5**, 928–938.
60. B. C. Saha and M. A. Cotta, Enzymatic saccharification and fermentation of alkaline peroxide pretreated rice hulls to ethanol, *Enzyme Microb. Tech.*, 2007, **41**, 528–532.
61. B. C. Saha and M. A. Cotta, Ethanol production from alkaline peroxide pretreated enzymatically saccharified wheat straw, *Biotechnol. Prog*, 2006, **22**, 449–453.
62. R. C. Sun and J. Tomkinson, Characterization of hemicelluloses obtained by classical and ultrasonically assisted extractions from wheat straw, *Carbohydr. Polym.*, 2002, **50**, 263–271.
63. J. Li, G. Henriksson and G. Gellerstedt, Lignin depolymerization/repolymerization and its critical role for delignification of aspen wood by steam explosion, *Bioresource Technol*, 2007, **98**, 3061–3068.
64. M. Laser, D. Schulman, S. G. Allen, J. Lichwa, M. J. J. Antal and L. R. Lynd, A comparison of liquid hot water and steam pretreatments of sugar cane bagasse for bioconversion to ethanol, *Bioresource Technol*, 2002, **81**, 33–44.
65. D. J. F. Brasch, Prehydrolysis-kraft pulping of *Pinus radiata* grown in New Zealand, *TAPPI*, 1965, **48**, 245–248.

66. R. L. Casebier and J. K. Hamilton, Chemistry and mechanism of water prehydrolysis on southern pine wood, *TAPPI*, 1969, **52**, 2369–2377.
67. J. H. Lora and M. Wayman, Delignification of hardwoods by auto-hydrolysis and extraction, *TAPPI*, 1978, **61**, 47–50.
68. S. H. Yoon, K. Macewan and A. R. P. van Heiningen, Hot-water pre-extraction from loblolly pine (*Pinus taeda*) in an integrated forest product biorefinery, *TAPPI*, 2008, **7**, 27–31.
69. S. H. Yoon and A. R. P. van Heiningen, Kraft pulping and papermaking properties of hot-water pre-extracted loblolly pine in an integrated forest product biorefinery, *TAPPI*, 2008, **7**, 22–27.
70. W. W. Al-Dajani, U. Tschirner and T. Jensen, Pre-extraction of hemi-celluloses and subsequent kraft pulping Part II: Acid and autohydrolysis, *TAPPI*, 2009, **8**, 30–37.
71. S. I. Aronovsky and R. A. Gortner, The cooking process I-Role of water in the cooking of wood, *Ind & Eng Chem*, 1930, **22**, 264–274.
72. O. Goldschmid, Aqueous hydrolysis of lignin, *TAPPI*, 1955, **38**, 728–732.
73. M. G. S. Chua and M. Wayman, Characterization of autohydrolysis aspen (*P. tremuloides*) lignins. Part 1. Composition and molecular weight distribution of extracted autohydrolysis lignin., *Can J Chem*, 1979, **57**, 1141–1149.
74. J. H. Lora and M. Wayman, Autohydrolysis of aspen milled wood lignin, *Can J Chem*, 1980, **58**, 668–676.
75. G. Garrote, H. Domínguez and J. C. Parajó, Study on the deacetylation of hemicelluloses during the hydrothermal processing of Eucalyptus wood, *Holz Roh Werkst*, 2001, **57**, 191–202.
76. H. Ando, T. Sakaki, T. Kokusho, M. Shibata, Y. Uemura and Y. Hatate, Decomposition behavior of plant biomass in hot-compressed water, *Ind Eng Chem Res*, 2000, **39**, 3688–3693.
77. J. Li and G. Gellerstedt, Improved lignin properties and reactivity by modifications in the autohydrolysis process of aspen wood, *Ind Crop Prod*, 2008, **27**, 175–181.
78. T. Song, A. Pranovich, I. Sumersky and B. Holmbom, Extraction of galactoglucomannan from spruce with pressurized hot water, *Holz-forschung*, 2008, **62**, 659–666.
79. M. Leschinsky, H. K. Weber, R. Patt and H. Sixta, Formation of insoluble components during autohydrolysis of *Eucalyptus globulus*, *Lenzinger Berichte*, 2009, **87**, 16–25.
80. M. Leschinsky, G. Zuckerstätter, H. K. Weber, R. Patt and H. Sixta, Effect of autohydrolysis of *Eucalyptus globulus* wood on lignin structure. Part 1: Comparison of different lignin fractions formed during water prehydrolysis., *Holzforschung*, 2008, **62**, 645–652.
81. M. Leschinsky, G. Zuckerstätter, H. K. Weber, R. Patt and H. Sixta, Effect of autohydrolysis of *Eucalyptus globulus* wood on lignin structure. Part 2: Influence of autohydrolysis intensity, *Holzforschung*, 2008, **62**, 653–658.

82. K. Leppännen, P. Spetz, A. Pranovich, K. Hartonen, V. Kitunen and H. Ilvesniemi, Pressurized hot water extraction of Norway spruce hemicelluloses using a flow-through system, *Wood Sci Technol*, 2011, **45**, 223–236.

83. R. P. Overend, E. Chornet and J. A. Gascoigne, Fractionation of lignocellulosics by steam-aqueous pretreatments, *Philos T Roy Soc A*, 1987, **321**, 523–536.

84. J. Weil, M. Brewer, R. Hendrickson, A. Sarikaya and M. R. Ladisch, Continuous pH monitoring during pretreatment of yellow poplar wood sawdust by pressure cooking in water, *Appl Biochem Biotech*, 1998, **70–72**, 99–111.

85. M. S. Tunc and A. R. P. van Heiningen, Autohydrolysis of mixed southern hardwoods: Effect of P-factor, *Nord Pulp Pap Res J*, 2009, **24**, 46–51.

86. A. Mittal, S. G. Chatterjee, G. M. Scott and T. E. Amidon, Modeling xylan solubilization during autohydrolysis of sugar maple wood meal: Reaction kinetics, *Holzforschung*, 2009, **63**, 307–314.

87. S. G. Allen, L. C. Kam, A. J. Zemann and M. J. J. Antal, Fractionation of sugar cane with hot, compressed liquid water, *Ind Eng Chem Res*, 1996, **35**, 2709–2715.

88. Y. Yu, X. Lou and H. Wu, Some recent advances in hydrolysis of biomass in hot-compressed water and its comparison with other hydrolysis methods, *Energ Fuel*, 2008, **22**, 46–60.

89. G. P. van Walsum, S. G. Allen, M. J. Spences, M. S. Laser, M. J. Antal and L. R. Lynd, Conversion of lignocellulosics pretreated with liquid hot water to ethanol, *Appl Biochem Biotech*, 1996, **57/58**, 157–170.

90. S. G. Allen, D. Schulman, J. Lichwa and M. J. J. Antal, A Comparison between hot liquid water and steam fractionation of corn fiber, *Ind Eng Chem Res*, 2001, **40**, 2934–2941.

91. Y. Kim, N. Mosier and M. R. Ladisch, Enzymatic digestion of liquid hot water pretreated hybrid poplar, *Biotechnol Progr*, 2009, **25**, 340–348.

92. I. Hasegawa, K. Tabata, O. Okuma and K. Mae, New pretreatment methods combining a hot water treatment and water/acetone extraction for thermo-chemical conversion of biomass, *Energ Fuel*, 2004, **18**, 755–760.

93. K. Hartonen, J. Parshintsev, K. Sandberg, E. Bergelin, L. Nisula and M.-L. Riekkola, Isolation of flavonoids from aspen knotwood by pressurized hot water extraction and comparison with other extraction techniques, *Talanta*, 2007, **74**, 32–38.

94. C. C. Teo, S. N. Tan, J. W. H. Yong, C. S. Hew and E. S. Ong, Pressurized hot water extraction (PHWE), *J Chromatogr A*, 2010, **1217**, 2484–2494.

95. M. S. Tunc and A. R. P. van Heiningen, Characterization and molecular weight distribution of carbohydrates isolated from autohydrolysis extract of mixed southern hardwoods, *Carbohyd Polym*, 2010, **83**, 8–13.

96. J. L. Xu, S. Liu, Optimization of ethanol production from hot-water extracts of sugar maple chips, *Renew Energ*, 2009, **34**, 2353–2356.

97. S. Liu, Woody biomass: Niche position as a source of sustainable renewable chemicals and energy and kinetics of hot-water extraction/ hydrolysis, *Biotechnol Adv*, 2010, **28**, 563–582.

98. A. Gandini, Polymers from renewable resources: A Challenge for the future of macromolecular materials, *Macromolecules*, 2008, **41**, 9491–9504.

99. K. Yates, Mechanisms of ester hydrolysis in aqueous sulfuric acids, *J Am Chem Soc*, 1967, **89**, 2686–2692.

100. P. V. Krammer and H. Vogel, Hydrolysis of esters in subcritical and supercritical water, *J Supercrit Fluid*, 2000, **16**, 189–206.

101. N. Akiya and P. E. Savage, Roles of water for chemical reactions in high-temperature water, *Chem Rev*, 2002, **102**, 2725–2750.

102. S. Caparrós, Garrote, G.; Ariza, J.; López, F. Autohydrolysis of Arundo donax L., a kinetic assessment, *Ind Eng Chem Res*, 2006, **45**, 8909–8920.

103. H. M. Wang, *Determination of the pKa of glucuronic acid and the carboxyl groups of heparin by 13C-nuclear-magnetic-resonance spectroscopy, Biochem J*, 1991, **278**, 689–695.

104. I. Fraeye, T. Duvetter, I. Verlent, D. N. Sila, M. Hendrickx and A. V. Loey, Comparison of enzymatic de-esterification of strawberry and apple pectin at elevated pressure by fungal pectinmethylesterase, *Innov Food Sci Emerg*, 2007, **8**, 93–101.

105. J. Hafrén and U. Westermark, Distribution of acidic and esterified polygalacturonons in sapwood of spruce, birch and aspen, *Nord Pulp Pap Res J*, 2001, **16**, 284–289.

106. E. L. Springer and L. L. Zoch, Hydrolysis of xylan in different species of hardwoods, *TAPPI*, 1968, **51**, 214–218.

107. B. Bujanovic, M. Goundalkar, T. Amidon, Non-carbohydrate-based products extracted during hot-water extraction of sugar maple, 30th SETAC North America Annual Meeting: Green Biorefinery, November 19–23, 2009, *New Orleans, LA*.

108. M. J. Goundalkar, B. Bujanovic and T. Amidon, Lignin in the hot-water extract of sugar maple-isolation, characterization and potential use, International Biorefinery Conference, IBC 09, October 6–7, 2009, Syracuse, NY, IBC Book of Abstracts: 3A, oral presentation.

109. M. J. Goundalkar, B. Bujanovic, C. Gong and T. Amidon, Characterization of Organic Precipitate from Hot-Water Extraction of Hardwoods, The 37th Northeast Regional Meeting of the ACS, NERM2010, Poster #39, June 2–5, 2010, SUNY Potsdam, NY, 2010.

110. K. Lundquist and R. Lundgren, Acid degradation of lignin, *Acta Chem Scand*, 1972, **26**, 2005–2023.

111. X. P. Guo and Y. Z. Lai, Variation of softwood and hardwood lignin in the reactivity of aryl ether units under acidic conditions, 14th International Symposium on Wood Fibre Pulping Chemistry, June 25–28, 2007, Durban, South Africa, Proceedings: oral presentation #141.

112. M. J. Selig, S. Viamajala, S. R. Decker, M. P. Tucker, M. E. Himmel and T. B. Vinzant, Deposition of lignin droplets produced during dilute acid

pretreatment of maize stems retardsenzymatic hydrolysis of cellulose, *Biotechnol Progr*, 2007, **23**, 1333–1339.

113. M. Poláková, D. Joniak and M. Ďuriš, Preparation of some benzyl D-glucoronates from 4-methoxybenzylidene derivatives of D-glucuronic acid, *Monatschefte für Chemie*, 2000, **131**, 1197–1205.
114. S. Li and K. Lundquist, Cleavage of arylglycerol b-aryl ethers under neutral and acid conditions, *Nord Pulp Pap Res J*, 2000, **15**, 292–299.
115. K. E. Vroom, The H factor: A means of expressing cooking times and temperatures as a single variable. *Pulp Paper Mag Can*, 1957, **58**, 228–231.
116. H. Sixta *Handbook of Pulp*. Wiley-VCH: Weinheim, Germany, 2006.
117. A. Hasan, B. Bujanovic and T. Amidon, Strength properties of kraft pulp produced from hot-water extracted wood chips within the biorefinery, *J Biobased Mater Bio*, 2010, **4**, 46–52.
118. T. E. Amidon, S. Liu, J. Howard and V. Yadama, Biofuels from woody biomass and potential integration into existing wood fuel, wood products, and pulp and paper products, Energy, Utility & Environment Conference (EUEC), February 1–3, 2010, Phoenix, AZ.
119. S. N. Chien, H. Ren, M. Aoyagi, T. E. Amidon and Y. Z. Lai, Fractionation of wood polymers by carboxymethylation: Influence of reaction conditions, *J Biobased Mater Bio*, 2010, **4**, 40–45.
120. C. Gong, M. J. Goundalkar, B. Bujanovic and T. Amidon, Evaluation of different sulfur-free delignification methods for hot-water extracted hardwood, *J Wood Chem Technol*, 2012, **32**, 93–104.
121. M. J. Goundalkar, B. Bujanovic and T. Amidon, Analysis of Non-Carbohydrate Based Low-Molecular Weight Organic Compounds Dissolved During Hot-Water Extraction of Sugar Maple, *Cell Chem Technol*, 2010, **44**, 27–33.
122. S. K. Bose, R. C. Francis, M. Govender, T. Bush and A. Spark, Lignin content versus syringyl to guaiacyl ratio amongst poplars, *Bioresource Technol*, 2009, **100**, 1628–1633.
123. D. R. Bedgood, A. G. Bishop, P. D. Prenzler and K. Robards, Analytical approaches to the determination of simple biphenols in forest trees such as *Acer* (maple), *Betula* (birch), *Coniferus*, *Eucalyptus*, *Juniperus* (cedar), *Picea* (spruce) and *Quercus* (oak), *Analyst*, 2005, **130**, 809–823.
124. F. Z. Erdemgil, S. Şanli, N. Şanli, G. Özkan, J. Barbosa, J. Guiteras and J. L. Beltrán, Determination of pKa values of some hydroxylated benzoic acids in methanol-water binary mixtures by LC methodology and potentiometry, *Talanta*, 2007, **72**, 496.
125. U. Westermark, B. Samuelsson and K. Lundquist, Homolytic cleavage of the b-ether bond in phenolic b-O-4 structures in wood lignin and in guaiacylglycerol-b-guaiacyl ether, *Res Chem Intermediat*, 1995, **21**, 343–352.
126. J. F. Kadla, S. Kubo, R. D. Gilberrt and R. Venditti Lignin-based carbon fibers, In *Chemical modification, properties, and usage of lignin.*, 1st edn; Hu, T.Q., ed.; Kluwer Academic/Plenum Publishers: New York, 2002; pp. 121–137.

127. G. H. van der Klashorst, Modification of lignin at the 2- and 6-positions of the phenylpropanoid nuclei, In *Lignin Properties and Materials*, Glasser, W.G. and Sarkanen, S., ed.; ACS Symposium Series 397: Washington DC, 1989; pp. 346–360.
128. J. F. Kadla, S. Kubo, R. A. Venditti, R. D. Gilbert, A. L. Compere and W. Griffith, Lignin-based carbon fibers for composite fiber applications, *Carbon*, 2002, **40**, 2913–2920.
129. A. L. Compere, W. L. Griffith, C. F. Leitten and S. Petrovan, Improving the fundamental properties of lignin-based carbon fiber for transportation applications, *Int Sampe Tech Conf*, 2004.
130. R. L. Howard, E. Abotsi, E. L. Jansen van Rensburg and S. Howard, Lignocellulose biotechnology: issues of bioconversion and enzyme production, *Afr J Biotechnol*, 2003, **2**, 602–619.
131. C. Eckert, C. Liotta, A. Ragauskas, J. Hallett, C. Kitchens, E. Hill and L. Draucker, Tunable solvents for fine chemicals from the biorefinery, *Green Chem*, 2007, **9**, 545–548.
132. Q. Xiang, Production of oxychemicals from precipitated hardwood lignin, *Appl Biochem Biotech*, 2001, **71**, 91–93.
133. B. Bujanovic, S. A. Ralph, R. S. Reiner, K. Hirth and R. H. Atalla, Polyoxometalates in oxidative delignification of chemical pulps: Effect on lignin, *Materials*, 2010, **3**, 1888–1903.
134. N. J. Walton, A. Narbad, C. B. Faulds and G. Williamson, Novel approaches to the biosynthesis of vanillin, *Curr Opin Biotech*, 2000, **11**, 490–496.
135. P. G. Jessop, L. Phan, A. Carrier, S. Robinson, C. J. Dürr and J. R. Harjani, A solvent having switchable hydrophilicity, *Green Chem*, 2010, **12**, 809–814.
136. A. Marjasvaara, M. Torvinen, H. Kinnunen and P. Vainiotalo, Laccase-catalyzed polymerization of two phenolic compounds studied by MALDI-TOF and ESI-FTIC, *MS with CID experiments, Biomacromolecules*, 2006, **7**, 1604–1609.
137. C. Regalado, B. E. Garcia-Almendarez and M. A. Duarte-Vazquez, Biotechnological applications of peroxidases, *Phytochem Rev*, 2004, **3**, 256.

CHAPTER 8

Lignin Recovery and Lignin-Based Products

GÖRAN GELLERSTEDT, PER TOMANI,
PETER AXEGÅRD AND BIRGIT BACKLUND*

Innventia AB, Box 5604, SE-114 86 Stockholm, Sweden
*Email: birgit.backlund@innventia.com; Tel: +46 8 676 7226;
Fax: +46 8 411 5518

8.1 Lignin Sources

8.1.1 Sources in Nature

Lignin is present in the tissue of vascular plants where it adds to the mechanical strength and structural integrity of the cell wall. Furthermore, its structure, being much more hydrophobic than that of polysaccharides, provides the necessary balance in water transport and swelling ability of the cell. In wood species, only small differences in the lignin content between stem, roots, branches, and tops can be observed. Softwoods such as spruce, pine or hemlock have a lignin content of the order of 25–30% by weight whereas hardwoods usually have lower amounts; 20–25%.[1] In Table 8.1, the chemical composition of some wood species is given.

Various annual plants have even lower amounts and *e.g.* flax only contains around 3% of lignin as an average amount albeit with large differences between different parts of the plant.[2] Although annual plants can contain higher amounts of lignin (up to ca 15%), the normally encountered large differences in

RSC Green Chemistry No. 18
Integrated Forest Biorefineries
Edited by Lew Christopher

Table 8.1 Chemical composition of wood species commonly employed for the production of bleached paper pulps.

Wood component	Content, % on wood				
	Pine	Spruce	Birch	Eucalyptus	Eucalyptus
	P. sylvestris	P. abies	B. pendula	E. globulus	E. urograndis
Klason lignin	28.2	27.2	19.1	17.9	26.8
Acid soluble part of lignin	0.3	0.2	4.0	6.0	3.6
Xylan	7.8	8.7	25.6	20.8	15.6
Glucomannan	18.2	19.2	3.7	4.2	3.0
Cellulose	44.2	44.4	44.8	51.0	50.6
Extractives	1.3	0.2	2.8	0.2	0.4

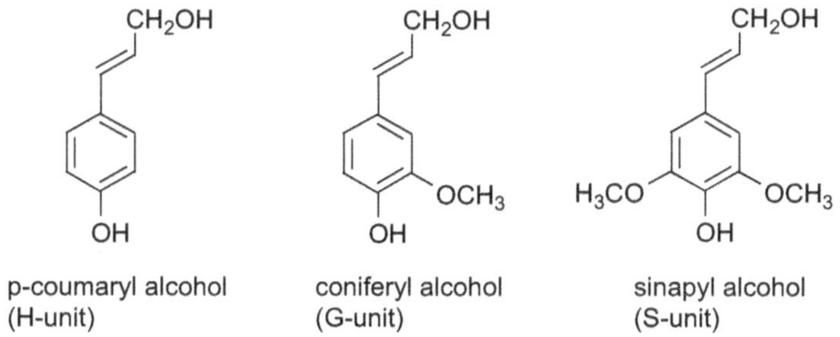

p-coumaryl alcohol
(H-unit)

coniferyl alcohol
(G-unit)

sinapyl alcohol
(S-unit)

Figure 8.1 Common precursors of lignin and their denotations.

amount and chemical structure in different parts of the plant will affect all types of lignin-isolation processes in a negative way.

From a structural point of view, differences between different types of biomass material exist. Softwood lignin is built up by oxidative coupling and chemically controlled polymerization of coniferyl alcohol with minor amounts of p-coumaryl alcohol being incorporated in compression wood lignin. Hardwood lignin, on the other hand, has both coniferyl alcohol and sinapyl alcohol as building units although the proportions may vary in different plants. In annual plants, all three alcohols serve as precursors of the lignin, Figure 8.1. A comprehensive review on lignin formation has recently been published.[3]

In the plant cell wall, the lignin is chemically linked to both cellulose and hemicelluloses (glucomannan, xylan) but with a large predominance of the latter. As a consequence, no simple extraction process for pure lignin from plant material exists and any separation process must involve chemical degradation/modification of the structure. The principle of lignin–carbohydrate network structures building up the fiber cell wall is, however, not random but arranged such that structurally different lignins are linked to individual polysaccharides.[4,5] The various networks, in turn, are arranged in repeating layers of cellulose, glucomannan (when present), and xylan.[6] For representative softwood (spruce) and hardwood (birch, eucalypt) species, it has been found that xylan preferentially is linked to a rather linear type of lignin polymer with a predominance of alkyl-aryl ether structures, whereas the lignin linked to cellulose and glucomannan is more branched and/or crosslinked. As a consequence, the xylan-bound lignin is much easier to degrade and dissolve in lignin-extraction processes (pulping), irrespective of type.

8.1.2 Industrial Sources

The forestry industry produces huge quantities of lignin in its pulping operations. Here, the main objective is to liberate the cellulosic fibers present in wood or (in some countries) in annual plants. Two major technologies, developed more than a century ago, are used, namely alkaline pulping utilizing either the kraft (aqueous sodium hydroxide and sodium sulfide) or the soda process (aqueous sodium hydroxide, sometimes together with anthraquinone) or acidic pulping with excess aqueous bisulfite together with either sodium, magnesium, calcium, or ammonium hydroxide. Kraft is the dominating process representing over 95% of all chemical pulp produced. In all cases, the used pulping liquor (black liquor) will contain the liberated lignin either as lignin phenolate (alkaline pulping) or as lignosulfonate (acidic pulping) together with degraded carbohydrates and some extractives. In most pulp mills, the black liquor is subsequently evaporated to high dryness before being burnt to generate the necessary process energy. At the same time, the inorganic chemicals are recovered and returned to the process.

Various alternative processes for the separation of cellulosic fibers from lignin have been suggested with the objective of either producing paper-making fibers or pure sugar moieties by hydrolysis of polysaccharides for further fermentation to bioethanol. Thus, the steam explosion process was developed in the 1970s as a more energy-efficient process for the production of mechanical pulp.[7–9] The conditions used in the process; treatment of the biomass material with steam at about 200 °C or higher for a short period of time followed by rapid decompression, were, however, found to result in poor pulp quality. In later development of the process, biomass disintegration in combination with polysaccharide hydrolysis and sugar fermentation has, therefore, been in focus.[10–12] The lignin can be obtained as a byproduct by extraction of the exploded fibrous material with either aqueous alkali or an organic solvent.[10,13,14]

Pulping of wood with aqueous ethanol (organosolv pulping) was first suggested in 1978 and an industrial pilot-plant unit was constructed at Newcastle, N.B., Canada in 1989.[15,16] The low flexibility, with the process only being able to utilize certain hardwood species without the addition of a catalyst (sulfuric acid), was, however, a serious drawback and the unit was shut down after a few years of operation. Recently, the technology has been revisited with the driving force being the possibility of making bioethanol from hydrolyzed wood polysaccharides.[12] At the same time, a facile isolation of the lignin is feasible.[16] A pilot-plant unit is currently being constructed in Grand Junction, CO, USA.

A direct method for the separation of carbohydrates and lignin in wood (biomass) is by hydrolysis using strong acid like sulfuric acid or hydrochloric acid.[17] Thereby, the polysaccharides are hydrolyzed to monomeric sugars (and byproducts) that can be further fermented to bioethanol. In order to obtain an optimal yield of sugars, the process is usually done in two stages with the first mild stage taking care of hemicelluloses and amorphous cellulose, while the second is done to degrade the crystalline portion of cellulose. In the process, the lignin portion is condensed into a highly insoluble polymeric material. The wood hydrolysis process is currently developed in pilot-plant scale by Sekab in Domsjö, Sweden.

8.2   Lignin Production and Process Integration

8.2.1   Lignins from Alkaline Pulping

Kraft Pulping.   The potentially predominant source of lignin is that available in black liquors from kraft pulping of wood. Approximately 40–50% of the organic material present in such liquors is lignin, with the remainder being various degradation products from polysaccharides and minor amounts of extractives (Table 8.2).

Since the annual production of kraft pulp (softwood and hardwood) in the world is of the order of 130 million tonnes, this corresponds to a total release of approximately 55 million tonnes of kraft lignin.

The lignin in black liquor is chemically modified in comparison to the original wood lignin, with a high concentration of phenolic hydroxyl groups being a characteristic feature. These are ionized due to the high alkalinity, thus providing water solubility to the lignin. Consequently, by lowering the pH in the liquor, lignin can be precipitated.

Soda Pulping.   In the production of paper pulp from annual plants such as rice and wheat straw, and flax and bagasse, soda pulping is usually employed. There are also a few examples of soda pulping of hardwoods. Frequently, anthraquinone (AQ) is also added as a delignification booster. Similarly to kraft black liquors, lignin can be isolated from the soda black liquor

Table 8.2   Main components in black liquor from softwood (pine) and hardwood (birch).[29]

Component (kg/ton of pulp)	Pine	Birch
Lignin	490	330
Carbohydrate derived		
– Hydroxy acids	320	230
– Acetic acid	50	120
– Formic acid	80	50
Turpentine	10	not present
Resin and/or fatty acids	50	40
Misc. products	60	80

by addition of acid, filtration, and washing. Structurally, soda lignin should be very similar to kraft lignin, but without any sulfur. Some 6000 tonnes of soda lignins from annual plants are currently being produced annually and commercialized through GreenValue SA in Switzerland, which is one of the largest producers of these qualities.

8.2.1.1   *Lignin Removal from Pulping Liquors*

In kraft or soda pulping, approximately 90–95% of the original lignin in the raw material fed to the process is degraded and solubilized in the black liquor. The predominant reaction during pulping, a cleavage of alkyl–aryl ether linkages with the concomitant formation of new phenolic hydroxyl groups, renders the lignin fragments soluble in alkali through neutralization of such groups. In addition, some condensation of lignin fragments and the release of malodorous sulfur-containing low molecular mass products (kraft pulping) take place. Altogether, the dissolved lignin has a low average degree of polymerization, although the spectrum of products will cover a broad molecular mass range. As a result, kraft and soda lignins have rather high degrees of polydispersity with the presence of low molecular mass phenols as well as of high molecular mass lignin attached to carbohydrate residues.[18–20]

In the LignoBoost process,[21–23] recently developed in cooperation between Innventia and Chalmers, partially evaporated kraft black liquor (30–50% DS) is acidified to pH-values about 10 with carbon dioxide (150–250 kg/tonne lignin) followed by ageing of the precipitated lignin phase, separation by filtration and dewatering of the precipitated lignin, resuspension of the lignin filter cake in dilute aqueous sulfuric acid at pH ~ 2.5 (150–250 kg H_2SO_4/tonne lignin; spent acid from chlorine dioxide generators is acceptable), separation by filtration, and, finally, washing and dewatering including air blowing of the filter cake, Figure 8.2.[24]

It has been found that the addition of carbon dioxide must be done under well-controlled conditions, especially with respect to the temperature (different for different wood species), in order to get good lignin filtration properties. If these criteria are not met, the precipitated lignin forms particles close to the colloidal state and subsequent dewatering will become exceedingly difficult.[24,25] In the acidic step (Filtration 2 in Figure 8.2), all acidic groups (phenols and carboxyl acids) present in the lignin become protonated and, after filtration, washing and dewatering, an almost ash-free lignin can be collected (0.02–1% ash; 65–70% dryness) and, if needed, dried. The lignin has been shown to be of high purity with only minor contamination of carbohydrates and ash. In addition, some 2–3% of sulfur is present with about half being chemically linked to the lignin. Currently, softwood as well as hardwood kraft lignin, produced according to the LignoBoost principle, is available industrially in pilot-plant scale at Bäckhammars Bruk in Sweden.

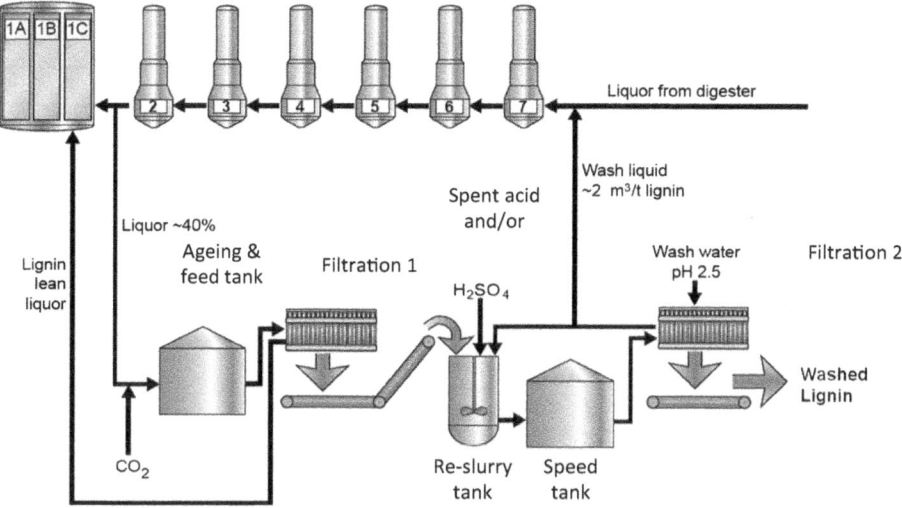

Figure 8.2 Schematic view of the LignoBoost process.

The major advantages compared to previous lignin removal technology are that:

- The filter area can be kept smaller resulting in lower investment cost.
- The sulfuric acid addition and acidic wash volume water can be kept lower resulting in lower operating costs.
- The yield of lignin is higher.
- The lignin has a lower content of ash and organic impurities.

In a similar type of process, using earlier lignin removal technology, kraft lignin from softwood and hardwood is commercially produced in the US (Mead-Westvaco, Charleston, S.C.) with an annual production estimated to about 30 000 ton. The process configuration is, however, not public in this case. Most of the product is subsequently sulfonated and used as dispersant while the parent unsulfonated lignin has uses such as matrix for controlled release of agrochemicals.

8.2.1.2 Analytical Data on Kraft and Soda Lignins

In alkaline pulping, hydrolysis of alkyl-aryl and dialkyl ether linkages in lignin takes place resulting in a depolymerization reaction and a concomitant formation of new phenolic and aliphatic hydroxyl groups together with some thiol groups (kraft pulping). The latter types of groups are to a large extent eliminated together with the terminal side-chain carbon (Cγ) during the cook, whereas the phenolic groups contribute to successively increased lignin solubility in the alkaline pulping liquor. With prolonged pulping time,

Table 8.3   Analytical data for kraft lignins from softwood (spruce) and hardwood (birch, *Eucalyptus globulus*).[20]

Functionality	Spruce	Birch	E. globulus
Carbohydrates, %	1.5	4.1	1.7
Inorganics (ash), %	1	0.5	0.8
Elemental analysis, %			
– carbon	64–65	62–63	61
– hydrogen	5.8–5.9	5.8–5.9	5.8–5.9
– oxygen	29–27	29–30	30–31
– sulfur	1.4–1.6	2.2–2.4	1.9–2.0
Aromatic OH, mmol/g	4.1[a]	4.3	3.3
Aliphatic OH (total), mmol/g	3.1[a]	1.7	1.5
Sec. (benzylic) OH, mmol/g	1.0[a]	n.a.	n.a.
Carboxyl groups, mmol/g	0.5	0.5	0.2
Molecular mass, M_w	4500	1600	2300
Polydispersity, M_w/M_n	4.5	3.6	4.3
Glass transition temp. T_g, °C	148	119	133
Free aromatic C5, mmol/g	3.1[b]	1.2[b]	n.a.

[a]Data based on quantitative ^{13}C NMR degradation.[26]
[b]Calculated from data based on analysis of kraft lignins by oxidative.[27,28]

radical-induced recondensation reactions between lignin fragments of the guaiacyl (G) type become abundant. The resulting lignin present in the black liquor thus contains a large variety of phenolic fragments ranging from monomeric simple phenols to complex irregular polymeric structures with high molecular mass. Minor or small amounts of sugar residues are usually also present and chemically attached to the lignin. By lowering the pH in the pulping liquor to values around or below the pK_a-value for phenols, the lignin precipitates out as a dark-colored solid powder. Some analytical data for softwood and hardwood kraft lignin have been collected in Table 8.3.[26–28]

8.2.1.3   *Process Integration and System Aspects*

A major advantage of lignin extraction is that a pulp mill with a capacity limitation in the recovery boiler can increase its pulp capacity, Figure 8.3. Most recovery boilers are capacity limited on the flue-gas side. The logical design basis is then to keep the flue-gas volume constant while taking out lignin. This means that the input of black liquor is increased so that, when lignin is taken out, the flue gas volume will remain unchanged.

   Lignin is also a new byproduct with the potential to give the pulp mill additional revenues from the wood. The produced lignin can be used in the mill's lime kiln to replace fossil fuels, with the potential for the mill to become completely fossil-free in its operations. Another opportunity is to sell lignin to external customers and by that also discharge sulfur (2–3% in kraft lignin) and minor amounts of sodium (adjustable by washing). Although some lignin has to be used for the internal energy production in the mill, the successively higher

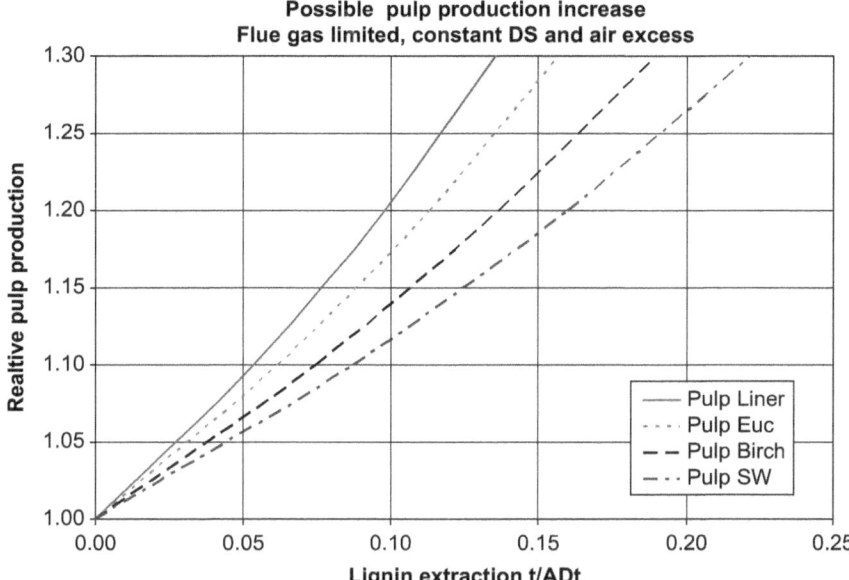

Figure 8.3 Possible pulp production increase *vs.* lignin extraction for liner and bleached pulp grades. The calculation is made for constant liquor concentration and air surplus.

energy efficiency in the complete pulping operation will allow for a large partial out-take of lignin in many modern mills. Pulp mills with extra biomass boiler capacity can of course introduce lignin removal and compensate the reduced steam production in their recovery boiler with increased steam production in their biomass boiler.

When lignin is precipitated from the black liquor, the most obvious, direct, consequences for a kraft pulp mill are:

- changes in the mill steam balance;
- changes in the Na/S balance;
- changes in black liquor properties and heating value.

Steam balance and effects on the recovery boiler. The steam balance differs between pulp mills, especially between old and new mills. Modern pulp mills offer opportunities to utilize large amounts of excess energy (Table 8.4) for power production to the grid or for lignin removal. Older pulp mills may need internal energy-efficiency measures in the pulp mill or more steam production in the bark boiler to create opportunities for lignin removal. Soft-wood-based kraft pulp mills usually have a larger amount of lignin available for extraction compared to hardwood-based mills, since softwood contains more lignin. However, there are Eucalyptus clones with as high lignin content as in softwood, Table 8.1.

Table 8.4 Example of the steam and power balances for state-of-the-art kraft market pulp mills 2010.

	Eucalyptus urograndis	Eucalyptus grandis	Loblolly pine
STEAM BALANCE (GJ/ADt)			
Total steam generation*	17.4	16.5	20.7
Total mill consumption	10.1	10	11.8
Excess energy	7.3	6.5	8.9

*Recovery boiler, power boiler and secondary heat.

Table 8.5 Influence of lignin extraction on the heat of combustion of softwood and hardwood black liquors. In the "lignin-extracted black liquor" case, the extracted lignin corresponds to 0.14 t DS/ADt.

Heat of Combustion		Black liquor	Lignin	Lignin-extracted black liquor
Dry solids gross	MJ/kg DS	14.5	27.4	13.3
Dry solids net	MJ/kg DS	12.4	26.1	11.1
Wet fuel (@ 80% DS) net	MJ/kg DS	11.8	25.5	10.5
Wet fuel (@ 72% DS) net	MJ/kg DS	11.4	25.2	10.2

Lignin removal can also be used to increase the power production per produced ADt pulp. The normal situation is that the excess steam is used in a condensing turbine to produce power. However, this type of power production has relatively low efficiency (about 25%) compared to high-efficiency power boilers (about 45%). The total power production from a kraft mill per ADt pulp can be increased considerably if lignin is removed and used in this type of high efficiency power boiler instead of power production at the mill site. This can be of special interest for pulp mills located in remote areas where the grid is not available for power delivery.

Lignin is an energy-rich component in black liquor. Extracting lignin from the black liquor therefore decreases the heat value of the black liquor (Table 8.5), and the evaporation and recovery boiler operation will consequently be influenced.

It is difficult to predict exactly how much lignin can be extracted without adversely affecting the combustion in the recovery boiler. According to earlier experiments, the swelling behavior of the black liquor does not decrease markedly until almost all the lignin is removed. We therefore believe that the adiabatic combustion temperature is a good measure of the combustion performance. Normal kraft black liquor can be safely burned without support fuel down to about 60% DS. This corresponds to an adiabatic combustion temperature of about 1430 °C (Figure 8.4).

In a kraft recovery boiler, the SO_2 emission increases drastically when the combustion temperature drops to a low level. Special considerations are

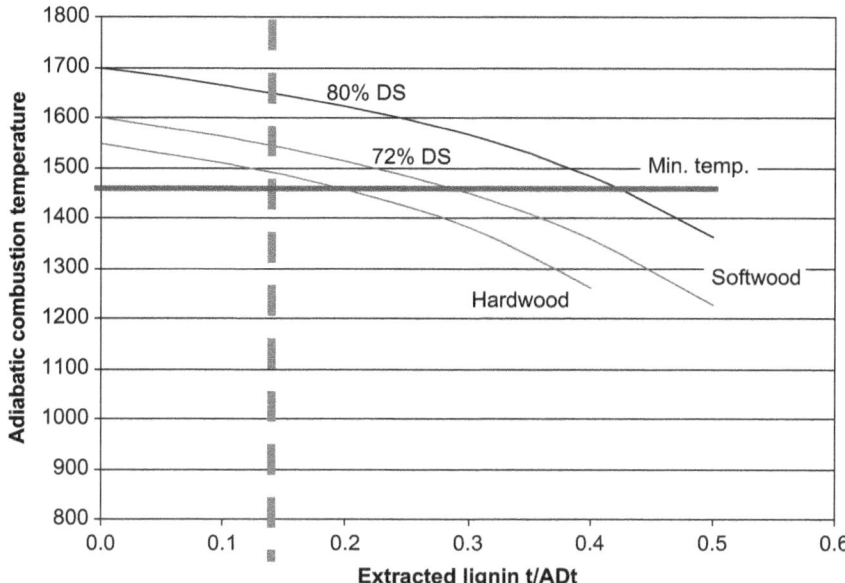

Figure 8.4 Adiabatic combustion temperature for lignin-extracted black liquor and the minimum temperature needed for reliable combustion in the recovery boiler. Normal kraft black liquor can be safely burned without support fuel down to about 60% DS.

therefore necessary at high levels of lignin extraction. At the moderate level of extraction no extra measures should, however, be necessary.

Boiling point elevation and black liquor viscosity. The boiling point elevation (BPE) seems to be relatively unaffected by the lignin extraction at concentrations below 50% DS. At higher concentrations there is a decrease and at 70% DS it seems to be a couple of degrees lower for a 50/50% mix of black liquor and lignin-lean liquor.

Figure 8.5 shows that the temperature influence on the viscosity is the same as for normal black liquor. The viscosity reduction is, however, rather different for different liquors.

The reduced viscosity and BPE increase the heat transfer markedly. If it is assumed that the viscosity is reduced by 50% in the lignin-lean liquor, then the viscosity in the mixture will probably be reduced by about 25% compared to that of normal black liquor in effect 1. This would result in an increased evaporation capacity of about 1–2% in the evaporation plant.

If it is assumed that the boiling point elevation is reduced by 3 °C at 75%, then the mixture will have the BPE reduced by about 1.5 °C at 75%. This will increase the capacity of the evaporation plant by about 3%.

In total, the decreased BPE and viscosity could thus yield a capacity increase of up to 5%. This would almost completely compensate for the increased

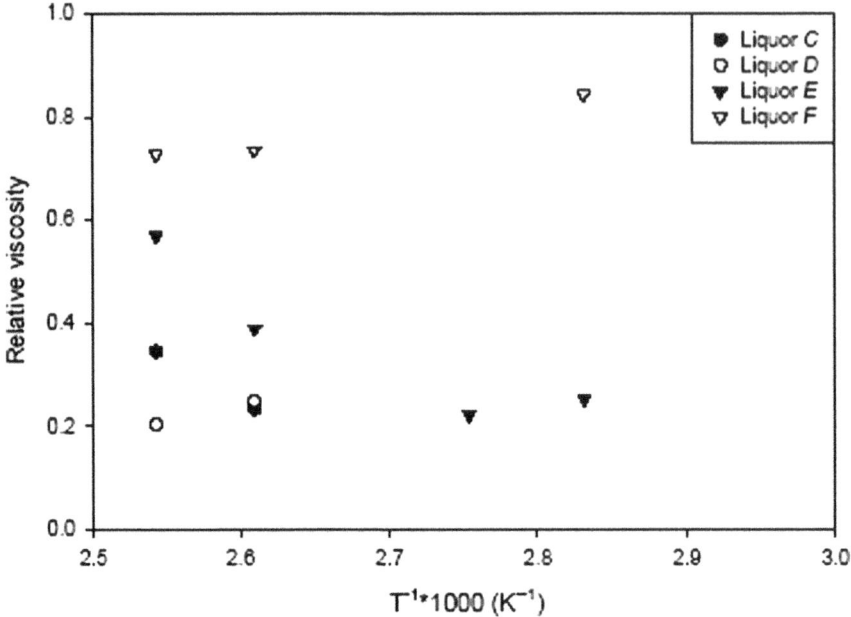

Figure 8.5 Relative viscosity (viscosity in relation to normal black liquor) as a
function of temperature at 70 w/w% dry content for different lignin-lean
liquors. C, D, E = softwood. F = hardwood.

evaporation demand due to the water added in the LignoBoost process. The
lignin extraction as such might therefore not require any increased heating
surfaces in the evaporation plant.

Existing evaporation plants have, however, often been expanded and modified over the years and the effect of lignin extraction on the evaporation plant
must therefore be studied for each specific evaporation plant and liquor.

Na/S balance. Lignin precipitation affects the Na/S balance in the mill in
various ways depending on the charge of fresh sulfuric acid and on whether
the lignin is reused into the pulp mill as a lime-kiln fuel or not. Recovery
boiler ash (ESP dust, mainly Na_2SO_4 (s) and Na_2CO_3 (s)) is purged to
remove sulfur, but at the same time valuable sodium is lost, which means
that make-up NaOH or $NaCO_3$ needs to be added to control the Na/S
balance.

With a normal sulfur level chemically bonded to the extracted lignin (20–30
kg S/tonne lignin) and a normal need for acid (150–200 kg H_2SO_4/tonne lignin)
reuse of lignin as biofuel in the lime kiln will result in a net input of sulfur of
0–45 kg/tonne lignin to the pulp mill. The range depends on:

• the amount of spent acid available and already included into the Na/S-
balance as alternative to fresh sulfuric acid;

- the sulfur intake with the lime kiln fuel before replaced with lignin;
- modifications in the LignoBoost operation that lower the demand of sulfuric acid.

Recycled inorganics. Another aspect to consider regarding the return of lignin-lean liquor is the risk of precipitation of burkeite [$Na_2(CO_3) \cdot 2Na_2(SO_4)$] and/or dicarbonate [$2Na_2(CO_3) \cdot Na_2(SO_4)$] in the evaporator train.

8.2.2 Lignin from Sulfite Pulping

The production of chemical pulp through pulping with aqueous bisulfite is successively declining due to the higher versatility of the kraft process. The more complex chemical recovery system and lower pulp strength of sulfite processes are other important factors. Today, less than 5% of the total chemical pulp production is based on the sulfite process. The dissolution of lignin from the cellulosic fibers takes place through a combination of sulfonation of lignin and acid-catalyzed hydrolysis of lignin–lignin and lignin–carbohydrate linkages. The resulting lignosulfonate is completely water soluble at all pH-values and constitutes approximately 50% (for softwood) of the total organic material in the spent pulping liquor; the remainder being various monosugars together with some acids and extractives (Table 8.6).

For a long time, many sulfite mills using calcium hydroxide as cooking base have been forced to remove their spent pulping liquor from the system for use as a lignosulfonate source, because of the lack of a chemical recovery system. Furthermore, the sugars present in the liquor can be fermented into bioethanol. Thereby, a rather complete biorefinery system is made possible with specialty pulp, lignosulfonate and ethanol as major products.

The major reactions encountered in sulfite pulping; sulfonation of lignin and acid-induced hydrolysis of lignin and carbohydrate ether linkages, and to a smaller extent lignin condensation; result in a highly polydisperse lignosulfonate, usually with an average molecular mass similar to that of kraft lignin, but with a

Table 8.6 Main components in spent sulfite liquor from softwood (spruce) and hardwood (birch).[29]

Component (kg/ton of pulp)	Spruce	Birch
Lignosulfonate	480	370
Carbohydrates	280	375
– Arabinose	10	10
– Xylose	60	340
– Mannose	120	10
– Galactose	50	10
– Glucose	40	5
Aldonic acids	50	95
Acetic acid	40	100
Extractives	40	40
Misc. compounds	40	60

broader distribution.[30] Lignosulfonates cannot be precipitated through change in pH of the spent liquor and are instead isolated through complete evaporation of the water. Thereby, a crude lignosulfonate, containing substantial amounts of carbohydrate-derived products, can be isolated in powder form as a sodium, calcium, magnesium, or ammonium salt depending on process. Currently, about 1 000 000 ton of lignosulfonate is produced annually in the world (with Borregaard-Lignotech as the dominant producer) with major uses as dispersant and binders in a variety of applications.

8.2.2.1 *Analytical Data on Sulfite Lignins*

In acid sulfite pulping, three different types of reaction are contributing to the overall behavior of lignin. Sulfonation of lignin starts early in the cook with the predominant reaction taking place at the benzylic side-chain carbon (C_α). A successive acid-induced hydrolysis of alkyl-aryl and dialkyl ether linkages results in the dissolution of highly water-soluble lignosulfonate ions. Lignosulfonate fragments with a broad range of molecular weights are formed, having an average degree of sulfonation of about one sulfonate group per two phenylpropane units. By an improper choice of pulping conditions, secondary recondensation reactions between lignin fragments can become abundant and even result in incomplete delignification. Although lignosulfonates can be precipitated from water by the addition of certain amine salts, the normal way of isolation is by spray drying.

8.2.3 Lignin from Other Liquors

Steam Explosion Lignin.   Steam explosion of wood or other biomass involves treatment of the material with water at high temperature. In the process, acetic acid is released from the biomass and most of the hemicelluloses are degraded by acid hydrolysis of glucosidic linkages, giving rise to the formation of monomeric and oligomeric sugars together with conversion products such as furfural and hydroxymethyl furfural (HMF).[31] These products can be removed by water washing of the lignocellulosic residue. The further separation of lignin from cellulose must be done by extraction of the lignin, *e.g.* with aqueous alkali followed by neutralization and filtration. The process has been found to be very sensitive to the type of biomass used and, in practice, only hardwood and annual plant material can be successfully processed unless additional chemicals are added to the wood.[14]

Solvent Pulping Lignin.   In solvent pulping with aqueous ethanol at high temperature (190–200 °C), wood acids like acetic acid are released, thus rendering the pulping liquor slightly acidic. Due to the presence of ethanol, lignin is directly solubilized through acid hydrolysis of lignin–lignin and lignin–carbohydrate linkages and partial evaporation of the spent liquor is sufficient for lignin precipitation. This process is sensitive to the type of biomass used and softwood can only be processed if a strong acid like sulfuric acid is added.[16] As an alternative to ethanol, solvent pulping in acetic acid or formic acid has been suggested.[32,33] In either of these acids, lignin is directly dissolved, but due to the low acidity, an additional stronger acid must be used for softwoods in order to achieve a substantial delignification.

Hydrolysis Process Lignin. Wood saccharification was used for several decades in the former Soviet Union for production of glucose directly from wood.[17] By treatment of wood with strong acid, all carbohydrates can be solubilized through acid hydrolysis of glucosidic linkages, leaving lignin as a solid residue. In more advanced forms, the process has been further developed for the production of bioethanol (the CASH process) through a two-step hydrolysis utilizing aqueous sulfur dioxide at high temperature (160–170 °C) for 10–15 min to hydrolyze the hemicelluloses in the first step. Addition of sulfuric acid and raising the temperature further results in hydrolysis of the crystalline cellulose. The lignin that is released in the process as a solid residue is heavily condensed and still contains substantial amounts of unprocessed material and products of carbohydrate origin.

Structural Changes. In steam explosion as well as in organosolv pulping, the process conditions are weakly acidic resulting in acid hydrolysis of alkyl-aryl ether linkages as the major lignin degradation reaction. In the absence of a high concentration of an external nucleophile such as ethanol or acetic acid, condensation reactions may also play a dominant role.[13] Consequently, the latter reaction takes place in steam explosion pulping and, in particular, in direct wood hydrolysis. Organosolv processes, on the other hand, result in lignins having a lower degree of polydispersity and such lignins will also have a low average molecular mass.[34]

8.3 Lignin Upgrading and Products

8.3.1 Situation Today

The dispersing ability of sulfonated lignin (lignosulfonates), obtained from the spent liquor after sulfite pulping, has been known for a long time. Originally, the lack of a recovery system for inorganic chemicals in calcium-based sulfite mills served as a driving force towards isolation of lignosulfonates, but the rising demand of dispersants, *e.g.* in the construction industry, has paved the way for production of lignosulfonates also from mills based on sodium or ammonium bisulfite. Despite the successively diminishing production of sulfite pulp in the world, the consumption of lignosulfonates has remained relatively stable with Borregaard-Lignotech being the predominant producer. In recent years, several new production facilities have been installed and the rapid economic growth in countries like China and India has resulted in expanding new markets.

In many applications, the crude evaporated and dried spent sulfite liquor, still containing monosugars and other carbohydrate-derived degradation products, cannot be used as such for quality reasons. Thus, upgrading through fermentation of sugars to ethanol and/or ultrafiltration to separate low molecular mass material, sometimes in combination with cation exchange are commonly used for production of more well-defined lignosulfonates. These find uses as dispersant in a variety of applications, with addition to concrete currently being the largest. Thereby, a reduction in water content can be achieved without adversely affecting the mixing and transport of the concrete and the final product becomes stronger. Similar effects can also be obtained in the production of gypsum wallboard, thereby reducing the energy used in drying. On oxygen oxidation of (softwood) lignosulfonate in alkali, part of the material is converted into vanillin, while the remainder can be obtained as a partially oxidized lignosulfonate that finds use as a dye dispersant.

The annual production of kraft and soda lignin is much smaller than that of lignosulfonates; about 30 000 tons of kraft lignin (Mead-Westvaco) and 6000 tons of soda lignin (GreenValue SA). A major portion of the kraft lignin is sulfonated with careful control of molecular mass, degree of sulfonation, and number of phenolic hydroxyl groups. Such lignins are predominantly used in the dyestuff industry as dispersants, with minor amounts being consumed in the agrochemical industry. Unmodified kraft lignin, on the other hand, shows good emulsifying properties and is used to prepare temperature-stable asphalt emulsions.[35]

In recent years, soda lignins from annual plants have found increasing markets. The lack of sulfur in the product results, together with a higher concentration of carboxyl groups (about twice as high as kraft lignin), in a dispersing ability that can be further increased by sulfonation. Other uses include soda lignin as partial replacement for phenol in phenol–formaldehyde resins.

8.3.2   Applications for Polymeric Lignin

The development of an efficient method for the large-scale industrial production of kraft lignin (see Section 8.2.1.1) as well as the foreseen increase in availability of new types of lignin, such as organosolv-based, will facilitate the development of new technical uses besides energy production. During a long period of time, research efforts have been directed to the development of various applications for lignin, *e.g.* as coreactant in phenol–formaldehyde resins and in materials such as polyurethanes, polyesters and epoxy resins. This work was recently summarized by Gandini and Belgacem.[36]

8.3.2.1   Carbon Fibers

Carbon fiber (CF) is a low-weight and high-strength material that is used as reinforcing fiber in composites where strength, stiffness, and low weight are important parameters. The aircraft industry, engineering industries, and sporting goods producers are major consumers of CF, which, currently, is produced from polyacrylonitrile (PAN) and from petroleum- or coal-based pitch. The high cost of CF production is a major drawback for an increased use in large-volume applications such as replacement of steel by CF-reinforced composites in the automotive industry. At present, only the very high cost segment of vehicles takes advantage of CF-composites, but if a major cost reduction could be achieved, a substantial portion of the steel ($> 60\%$) in the average vehicle could be replaced with a concomitant large reduction in fuel consumption and in greenhouse-gas emissions.[37]

The possibility of using lignin as a renewable precursor for CF has been explored starting in the 1960s with the use of lignosulfonate together with polyvinyl alcohol as plasticizer.[38] In later attempts, steam explosion lignin, organosolv lignins, as well as kraft lignin have been used. Provided that the lignin has been purified and has suitable molecular characteristics such as low mean molecular mass, low polydispersity, and thermoplasticity, all types of lignin can be used for making CF by melt spinning, albeit in most cases with a plasticizer like PEO (polyethylene oxide) or PET (polyethylene terephtalate) added.[39–42] So far, the strength properties of lignin-based CF are, however, in all cases inferior to those of commercial CF, at best around 50% of those obtained from pitch-based CF.

The production steps in the manufacturing of CF involve fiber extrusion, thermostabilization, carbonization, graphitization, and surface treatment. In addition, the precursor, whether lignin or other, must have a high purity. For lignin, this can be achieved by solvent fractionation,[43] extensive acid washing[41] or by the use of membrane filtration prior to precipitation.[20] Once extruded, the fiber must be thermally stabilized to avoid any self-fusion on further heating. Normally, this is done by treatment with air (oxygen) at temperatures around or above 200 °C, whereby the thermoplastic fiber is converted into a thermoset. It has been demonstrated that this part of the CF process is a major bottleneck since the stabilization reaction requires very long times (several hours) to proceed to completion. On further heating in an inert atmosphere, the stabilized lignin fiber is carbonized through the loss of water and other volatile fragmentation products. At a temperature of 1000 °C, about 50% of the original lignin can be recovered as CF corresponding to a yield of carbon of about 80%. At this temperature, the carbon atoms are still somewhat unordered but a true graphite structure can be reached through further heating to temperatures in the region of 2000 °C or higher. The production steps in the manufacturing of CF are outlined in Figure 8.6.

8.3.2.2 Activated Carbon

The use of activated carbon as an adsorbent has been known for a long time and both technology and a market exist for activated carbon products today. The presence of a very porous structure with large internal surface area (500–2000 m^2/g) results in excellent adsorption characteristics towards a variety of substances. Major applications include the removal of organic and inorganic

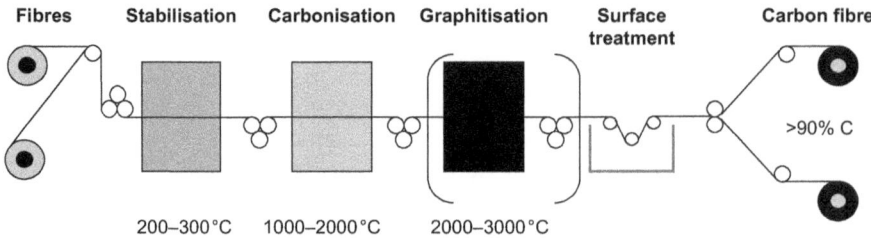

Figure 8.6   Production steps in carbon-fiber manufacturing.

pollutants from liquid or gaseous phases such as flue gases or municipal waste. Many materials can be used for the production of activated carbon with common precursors being coal, coconut shells, and wood. Consequently, isolated lignin such as kraft lignin, with a carbon content of 60–65% could constitute an interesting alternative.[44] The advantage of using lignin is not obvious, however, although development of new application areas and/or niche markets can be envisioned where specific surface properties of activated carbon from lignin can be tailor-made, *e.g.* by extrusion of lignin fibers followed by carbonization and activation at high temperature.

The preparation of activated carbon (from *e.g.* lignin) involves two processes; carbonization of the starting material into a char and subsequent activation to form a micropore structure with high surface area. These processes can be achieved using either physical or chemical activation. In the former case, the material is first pyrolyzed in inert atmosphere at temperatures in the range 600–900 °C, followed by activation with either carbon dioxide or steam at 600–1200 °C.[45–48] In chemical activation, the starting material is premixed with chemicals such as phosphoric acid, potassium hydroxide or sodium hydroxide, and heated at temperatures of 450–900 °C. In this case, carbonization and activation proceeds simultaneously.[49–52] Obviously, both activation methods are energy consuming but the latter less so. The cost of added chemicals will instead be higher. Irrespective of the preparation method (physical or chemical) highly active carbon with surface areas and pore volumes approaching 2000 m^2/g and 1 cm^3/g, respectively, can be obtained from lignin.

8.3.2.3   *Polyurethanes*

The possibility of utilizing lignin as a polyol component in polyurethanes was first recognized some 50 years ago.[53] The reactivity of the hydrolysis lignin used was, however, very poor and it was concluded that no product of technical value could be assumed. Other early attempts to utilize lignin in polymer systems have also been made with little success and it was concluded that a major limitation concerns its polydisperse and multifunctional characteristics.[54] In a comprehensive work on lignin in polyurethanes and other polymer systems, it was recognized that a more uniform reactivity of the lignin should be advantageous. Since virtually all lignins contain both aliphatic and aromatic hydroxyl

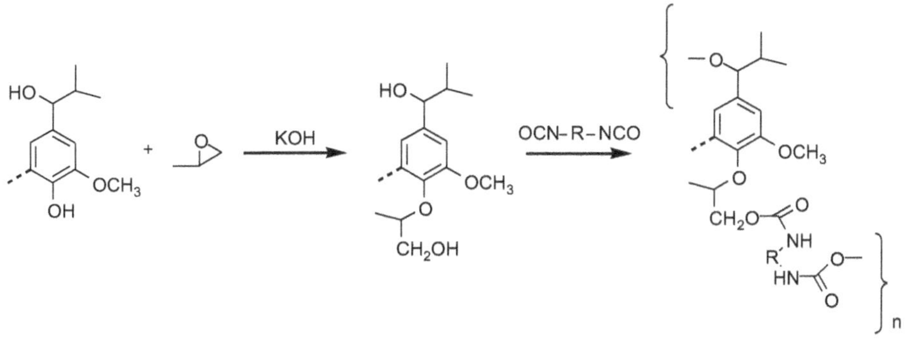

Figure 8.7 Formation of polyurethane from lignin *via* hydroxypropylation of phenolic units. (The reaction between lignin and propylene oxide can be vigorous unless the experimental conditions are followed rigorously)

groups with a large variation in reactivity, a structural modification leading to only one type of functionality should be introduced. Etherification of phenolic hydroxyl groups by reaction of the lignin with propyleneoxide, employing potassium hydroxide as catalyst, was therefore introduced and used in subsequent work, Figure 8.7.[55,56]

The high polydispersity and comparatively low reactivity encountered in many technical lignins was, however, found to affect the degree of substitution in lignin and the concomitant formation of homopolymers from propylene oxide. Consequently, lignins having low polydispersity and/or low molecular weight have been found to result in more homogeneous polyurethanes.[57]

The influence of lignin structure on the hydroxypropylation reaction as well as on the ability to produce polyurethanes with appropriate properties were later studied.[36] Both soda and organosolv lignins were found to behave more uniformly and with higher reactivity than either kraft or oxidized organosolv lignin, presumably due to a lower molecular weight. In all cases, however, substantial amounts of homopolymerization took place in addition to the substitution reaction but, nevertheless, well-performing polyurethane foams were prepared with methylenediphenyl diisocyanate (MDI) together with either oxypropylated organosolv or soda lignin.

In an alternative approach, softwood kraft lignin, fractionated to eliminate both low and high molecular weight material, was used directly together with a polyether triol and MDI for preparation of polyurethanes.[58,59] Improved mechanical properties were obtained when up to about 20% lignin was used, whereas at lignin contents higher than 30%, only hard and brittle polyurethanes formed. Similar results have also been obtained with hardwood organosolv (Alcell[R]) lignin.[60] A pronounced influence of the molecular weight of the lignin has been noted and in a direct comparison, low molecular weight kraft lignin was superior to either medium or high molecular weight fractions.[59] For Alcell lignin, on the other hand, the medium and high molecular weight fractions were superior.[61] These results clearly indicate that in order to obtain lignin-containing polyurethanes with good mechanical properties, a pre-fractionation of the original lignin, in order to reach a suitable molecular weight range, is necessary. In work with softwood kraft lignin, using hexamethylene diisocyanate (HDI) as a condensation partner, it was also concluded that the limited lignin addition achieved was due to the heterogeneous structure of the lignin. Consequently, the purification (fractionation) should be improved.[62]

8.3.2.4 Adhesives

In resins of the phenol–formaldehyde type (PF-resins), as well as in other phenol-based resins used in the manufacturing of particleboard, plywood, and fiberboard, lignin could partially replace the phenol, see Figure 8.8.[36]

The potential advantage of lignin in this type of application has prompted several investigations using a variety of technical lignins such as kraft lignin, steam explosion lignin, hydrolysis lignin, and organosolv lignin.[63–65] The very low reactivity of lignins, and the low level of reactive sites in comparison to phenol or natural polyphenols such as condensed tannins, result, however, in long reaction times and may also give inferior quality of the final product, see Figure 8.9.[66]

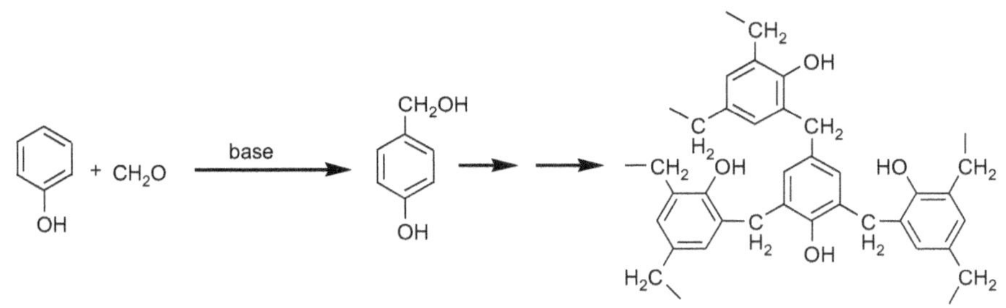

Figure 8.8   Classic phenol–formaldehyde condensation.

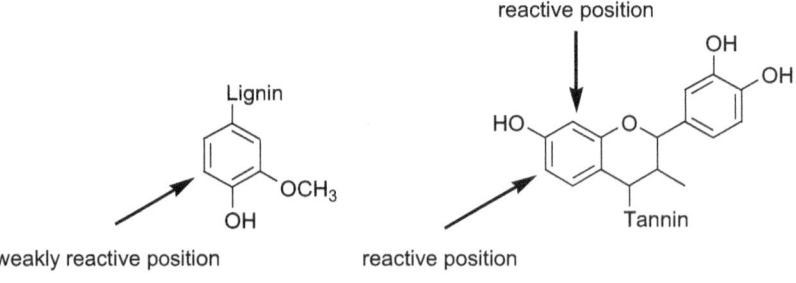

Figure 8.9   Difference in reactivity between lignin and tannin towards formaldehyde.

Various attempts to overcome these difficulties have been made, such as fractionation of the lignin,[65] prereaction with formaldehyde to methylolated lignin followed by reaction with phenol-formaldehyde,[65,67] demethylation,[68] or addition of small amounts of resorcinol.[69] Other modifications involve the addition of diisocyanate (MDI) together with the PF-resol and methylolated lignin, whereby the reaction time can be considerably reduced and good quality boards are produced.[70,71] In studies on novolac resins, based on PF resin with partial substitution of the phenol with lignin, it could also be shown that the addition of MDI as curing agent was superior to hexamethylenetetramine (HMTA) since the curing time could be reduced.[72] Although there are no true breakthroughs in the use of lignin in PF-based resins, it has been shown that the environmentally problematic formaldehyde can be replaced by glyoxal, while still maintaining the reactivity of the system.[73] The superior reactivity of phenol itself or of phenols such as resorcinol with meta-substitution of the hydroxyl groups is, however, a drawback when using lignin and new innovative approaches are needed to improve its reactivity.

8.3.2.5 Complexing Agents

The possibility of using kraft lignin as well as other types of technical lignin for sorption of (heavy) metal ions in contaminated water streams has been studied for several metals in model systems.[74] Two different mechanisms have been suggested, namely ion exchange in which phenolic hydroxyl groups and/or phenolates are the active sites,[75–77] and adsorption through particle or film diffusion.[78] Both mechanisms have been supported by experimental data and it seems reasonable to assume that, depending on the experimental conditions chosen, each of these may contribute. By converting kraft lignin into its calcium salt, and using this product as an ion-exchange resin, a stoichiometric amount of Pb(II) could be released.[77] Using purified eucalypt kraft lignin at slightly acidic conditions, on the other hand, gave data for Cu(II) and Cd(II) consistent with an adsorption mechanism. In most studies to date, Cu(II) has been the target metal ion and it could be shown that purified eucalypt kraft lignin was able to adsorb 1.37 mmole of Cu(II)/gram at room temperature. For Cd(II), a corresponding value of 1.22 mmole/g was found.[78] The rather high solubility of kraft lignin in water that increases with pH is, however, detrimental when used for water-purification purposes. Therefore, attempts have been made to modify the lignin by reaction with synthetic resin in order to obtain a product insoluble in water. For this purpose, poly(methyl methacrylate) as well as polysulfone resins have been tested.[76,77]

Only little information is available concerning the sorption of organic materials by lignin. It has, however, been shown that kraft lignin, modified through treatment with aqueous sulfuric acid, can be used to bind N-nitrosamines and bile acids.[79] The same type of modified kraft lignin was later also shown to reduce the genotoxic effects of the carcinogen N-methyl-N'-nitro-N-nitrosoguanidine as well as other nitroso compounds.[80,81]

8.3.3 Applications for Monomeric Lignin

An alternative approach to the use of lignin is degradation at high temperature with formation of low molecular mass products. Such chemistry has been known for a long time and can be used to produce phenols,[82] aromatic hydrocarbons for use in diesel blends[83] or bio-oil.[84] Recently, a more direct conversion of lignin to hydrocarbons has been performed by high-temperature reduction with formic acid using ethanol (or other alcohols) as solvent. Thereby, a complete degradation was achieved with most of the oxygen being removed as water, with simultaneous formation of low molecular mass aliphatic and aromatic hydrocarbons and phenols.[85,86]

8.3.3.1 *Phenols*

By a proper choice of catalyst and reaction conditions, lignin can be converted into a mixture of phenols in good yield. A continuous reactor system has been described, in which solid particles of lignosulfonate[87] or kraft lignin[88] are mixed with catalyst and hydrogen and reacted at about 340–450 °C to monophenols and other reaction products, as summarized in Table 8.7. Since the product mixture is complex, Figure 8.10, a second distillation step would, however, be necessary to get pure compounds and, so far, this drawback has been prohibitive for any further development and commercialization.

The conversion of lignin to phenolic products can also be done by high-temperature treatment using formic acid as the hydrogen donor. Lignosulfonate as well as steam explosion lignin and hydrolysis lignin all give similar results with complex mixtures of hydrocarbons and phenols as final

Table 8.7   Mass balance with types of products formed and consumption of hydrogen on catalytic hydrogenation of kraft lignin.

Product	*Volatiles* $\rightarrow C_5H_{12}$	*Water*	*Neutrals* $<240\ ^\circ C$	*Monophenols*	*High M_w* $>240\ ^\circ C$	*Hydrogen*
% by weight	25.2	17.9	14.0	37.5	11.1	5.7

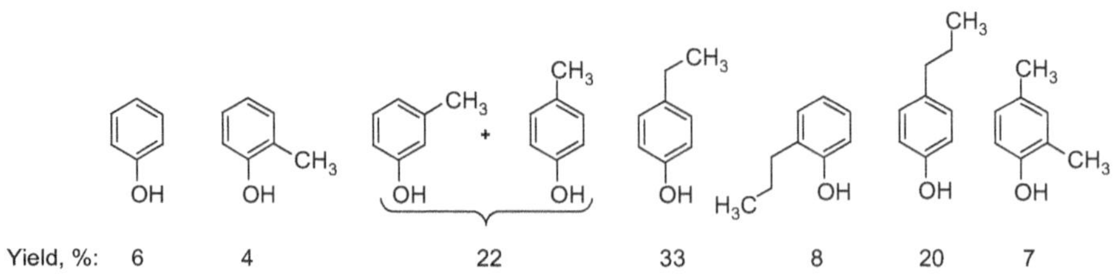

Yield, %:   6   4   22   33   8   20   7

Figure 8.10   Relative yield of individual mono-phenols after catalytic hydrogenation of kraft lignin.[57]

products.[86] Although the formation of char is minimal in this type of process, the product complexity indicates that further optimization of reaction conditions might result in a more uniform product mixture with either only hydrocarbons (biodiesel) or phenols being formed.

8.3.3.2 Other Aromatics

The presence of aromatic rings in lignin should favor a selective degradation of all side chains leaving only the aromatic skeleton as the most stable part of the molecule. Attempts to produce aromatic hydrocarbons have, however, only resulted in complex mixtures more suitable for use as biodiesel.[83,85] Therefore, the future production of large-volume products such as benzene, toluene, and xylene (BTX) should preferentially be done by further development of catalyst systems able to convert syngas into specific aromatic hydrocarbons.[89,90] Here, black liquor gasification as well as gasification using green biomass are renewable alternatives to the use of coal.

8.3.4 Fuel Applications

The moisture content of kraft lignin from the LignoBoost process is normally 65–70% DS. There is in some fuel applications a need for further drying to below 10% moisture. The energy value for kraft lignin, the higher heating value, is relatively high, 26–27.5 MJ/kg DS, compared to other biofuels (Table 8.8). Wood and bark has a higher heating value in the range of 20–22 MJ/kg DS.

Kraft lignin can be used as a biofuel or as additive in fuels. Solid lignin fuels are fired as lignin lumps, powder or pellets. The amount of sulfur in kraft lignin (2–3%) needs some consideration when lignin is fired. It can in some cases be used as an advantage. If the sulfur content cannot be utilized as an advantage, it has to be handled by, for example, addition of calcium to form $CaSO_4$, alternatively collect SO_2 in a flue gas scrubber. The advantage with a sulfur-containing biofuel is that potential corrosion problems from chlorides combined with potassium can be handled. This is the same type of effect that is marketed by Vattenfall where a sulfur-containing salt is added to the combustion to prevent corrosion in heat and power plants.

An elementary analysis of kraft lignin is shown in Table 8.9. The ash content originates mainly from sodium, which means that the ash melting point will be

Table 8.8 Example of physical properties of softwood lignins.

Property	Unit	Lignin filter cake	Lignin powder
Higher heating value	MJ/kg DS	26–28	26–28
Lower heating value	MJ/kg fuel	14–17	24–25
Ash content for fuel quality	% of DS	0.3–1.2	0.3–1.2
Moisture content	% of fuel	30–40	3–6
Bulk density	kg/m^3	*ca.* 500	*ca.* 700

Table 8.9   Example of elementary analysis and ash analysis for some softwood lignins.

Element	% dry ash-free
C (coal)	63–66
H (hydrogen)	5.7–6.2
O (oxygen)	25.9–27.5
S (sulfur)	1.8–3.2
N (nitrogen)	0.1–0.2
Cl (chlorine)	0.01

Ash component	g/kg ash
Na (sodium)	120–230
K (potassium)	25–75
Ca (calcium)	2–80
Mg (magnesium)	1–10

decreased when lignin is burnt in a boiler. The ash content is possible to control in the LignoBoost process; typical ash contents are in the range of 0.5 to 1%, but significantly lower ash contents can be obtained by additional washing.

8.3.4.1   Kraft Lignin Pellets and Kraft Lignin as Additive in Biofuel Pellets

Densification of biomass by means of pelletizing has been recognized as a well-proven technology for improving several of the properties of biomass important for various industrial applications, *e.g.* energy conversion or different material processing applications. The main drivers for pelletizing are improved transportation, storage and handling properties. Depending on the type of end user, different quality parameters are important, *e.g.* the amount of fines and crumbled pellets, ash content, heat value, *etc.*[91] From a pellet producer's point of view, it is not only the quality of the product that is important, it is also important to reduce the energy and power requirements for the pellet production process. A lot of work has been done on pelletizing of kraft lignin, and the main findings from these studies are summarized below.

Pelletizing of Kraft Lignin.   Kraft lignin with two different moisture contents, 20 and 33 wt% has been pelletized on an industrial scale. Compared to wood pellets, the energy consumption was decreased by 11% in both cases and, in addition, for the higher moisture content the production rate increased by 50%. The bulk density of both pellet qualities was about the same, around 700 kg/m^3, but due to the difference in moisture the energy density differed. Typically, the moisture content of fresh kraft lignin (raw filter cake) is 30–40 wt-% with a corresponding range of bulk and energy density of 500–600 kg/m^3 and 2.4–2.8 MWh/m^3, respectively. The kraft lignin produced in the industrial-scale trials had bulk densities close to 700 kg/m^3 and energy densities of 3.3–4.3 MWh/m^3, see Table 8.10.

Table 8.10 Typical bulk and energy density of kraft lignin.

Lignin type	Moisture wt%	Bulk density kg/m³	Energy* density MWh/m³
Raw filter cake	30–40	500–600	2.4–2.8
Dried powder	5	700	4.9
Pellets	33	700	3.3
Pellets	20	700	4.3

*Based on the lower (net) calorific value (NCV/LHV)

Kraft Lignin as Binder in Pelletization of Straw and Woody Biomass. Several studies on copelletizing of kraft lignin and various types of agricultural and woody biomass qualities have been made in which the binding properties of kraft lignin has been utilized.[92] The results show generally that the pellet strength, mechanical durability and bulk density is improved when 1–3% of kraft lignin is added. As kraft lignin increases the friction in the pellet mill operation, a suitable die (press channel length) should be chosen and the dosage system for the kraft lignin may need modification.

8.3.4.2 Kraft Lignin Fuel Slurry

Heavy fuel oils are extensively used as the main fuel in material processing operations in various industries, *e.g.* pulp and paper, cement, mining, metal. In many applications, especially in the pulp and paper industry, heavy fuel oil has been substituted with tall oil or tall oil pitch, which is a byproduct from the pulping process. Several studies have shown that fuel slurries can be made by mixing kraft lignin with heavy fuel oil or tall oil (pitch) using conventional material processing equipment. The energy share of kraft lignin in these slurries can be up to 50%, depending on the type of oil mix that is used. The resulting fuel slurry has excellent (long-term) storage properties with respect to sedimentation. The viscosity of the lignin fuel slurries can be modified by the use of additives and practical experiences show that slurries with a kraft lignin energy share of 30%, with a lower heating value of 30 MJ/kg, can be used in typical (lime) kiln oil burners with none-to-minor modifications of the firing system.

8.3.4.3 Experiences from Large-Scale Boiler Trials

LignoBoost kraft lignin has been tested extensively as a biofuel in both small-scale and large-scale commercial boilers, including a lime kiln.

Cocombustion of kraft lignin and bark. Trials with cofiring kraft lignin and bark in a 12-MWth fluidized-bed boiler[93] showed that kraft lignin can be handled easily and be cofired together with bark in a fluidized-bed boiler, Figure 8.11.

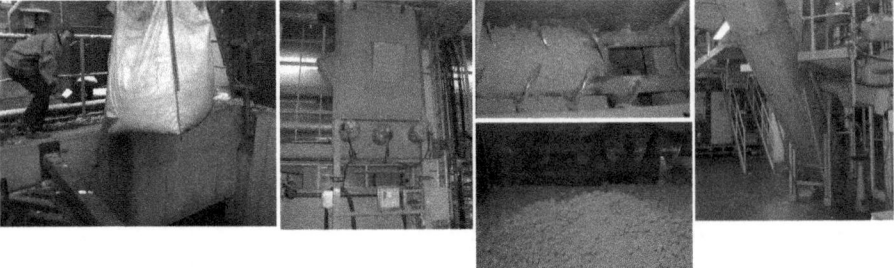

- The lignin was easy to feed into the boiler

- The lignin filter cake broke down to smaller pieces but no increased CO

- No vaulting in the fuel bin

- No clogging problems in the fuel chute

Figure 8.11   Cocombustion trial in a 12-MW research CFB boiler at Chalmers University of Technology.

The combustion performance was normal and not influenced by the lignin. The sulfur content of the kraft lignin had a significantly positive effect in reducing the alkali chloride content in the deposits, thus reducing the risk of sticky deposits and high-temperature corrosion (Figure 8.12). When kraft lignin was cofired with bark, the sulfur emission increased, compared to when bark is incinerated by itself. However, in this case, most of the sulfur was captured by calcium in the bark ash. The conventional capture of sulfur was also demonstrated with the addition of limestone to the bed. The addition of kraft lignin had no measurable effect on the sintering properties of the bed material. However, the sintering temperature of the cyclone bed material was decreased when limestone was added,

Full-scale substitution of coal in a pressurized fluidized-bed combustion (PFBC) boiler.   LignoBoost kraft lignin was mixed with coal and fired in full-scale trials in the PFBC boiler plant at Värtaverket in Stockholm,[94] with each of the two modules having a thermal capacity of 210 MW. The extensive long-term trials involving the cofiring of kraft lignin and coal showed that continuous cofiring had no effect on the important combustion parameters. A substitution rate of up to 15%, on an energy basis, was demonstrated and a substitution rate of 20–30% seems to be possible. In total, more than 1500 tonnes of kraft lignin were used in 2008. Fuel handling and preparation proved to be the most difficult tasks.

Trials in a full-scale lime kiln.   One appealing application is to replace fossil fuels fired in the lime kiln with dried kraft lignin powder, which has a higher

SO₂, NO (ppm)   NaC l+KCl (ppm)

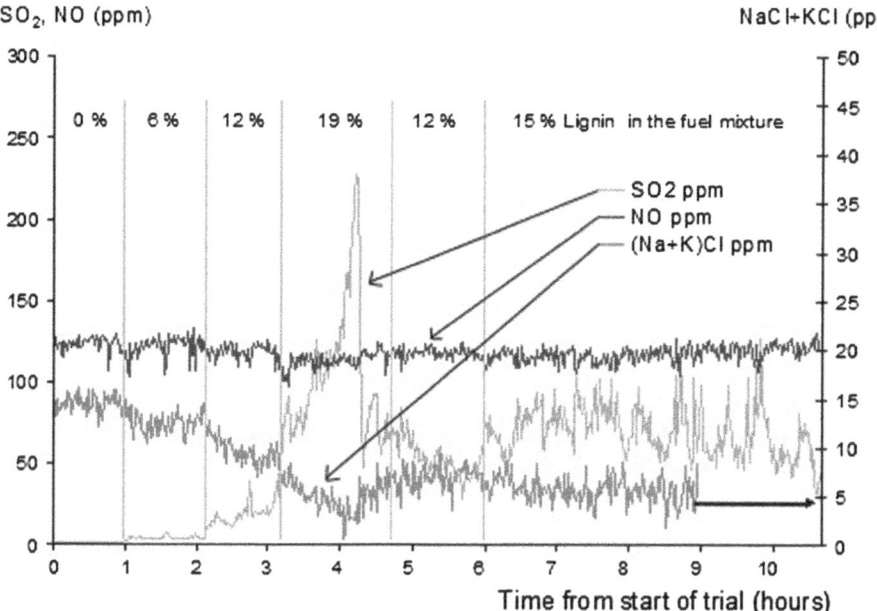

Figure 8.12 Results from cocombustion of kraft lignin and bark. The sulfur content in kraft lignin had a significantly positive effect in reducing the alkali chloride content in the deposits, thus reducing the risk of sticky deposits and high-temperature corrosion.

energy value than many other biofuels. Full-scale tests of firing a lime kiln with kraft lignin were carried out at a kraft mill in Sweden.[95] This lime kiln has a capacity of 275 tonnes of burned lime per day and the kiln is equipped with a lime mud dryer and a lime product cooler. The burner is designed for the simultaneous cofiring of oil, biomass powder, gas (NCG) and methanol. The existing powder fuel feeding system at the mill is designed to supply the two lime kilns with dried and pulverized bark or wood.

Kraft lignin powder/oil mixtures were evaluated up to a level of 100% kraft lignin. During a total of 32 h, 37 tonnes of kraft lignin were cofired with fuel oil. For approximately 15 h, the kraft lignin heat input was above 50% of the total heat input. At the end of the trial, the fossil-fuel oil was completely shut off and the kiln was operated on 100% kraft lignin for approximately 2 h. Experience gained from the full-scale combustion trial and the shorter lime kiln feeding trial shows that it is possible to achieve a stable and continuous operation of a lime kiln with kraft lignin as fuel. It was possible to use the standard powder burner and feeding equipment when firing dried and sieved kraft lignin. The temperature levels in the kiln were approximately the same as with fossil-fuel oil or wood powder. It was possible to produce lime with a consistent quality when firing with kraft lignin. The temperature reached in the burner zone was sufficient for the proper sintering of lime nodules. White liquor could be produced from the lime with the same causticizing efficiency and at the same rate as during normal operation. We saw no significant influence on the emissions of CO, H_2S, NO_x and SO_2 and, according to the plant operator, the kiln could be controlled easily when firing kraft lignin.

References

1. G. Gellerstedt and G. Henriksson in *Monomers, Polymers and Composites from Renewable Resources* (ed. M. N. Belgacem, A Gandini), Elsevier, Oxford, UK, 2008, pp. 201–224.
2. A. Day, K. Ruel, G. Neutelings, D. Cronier, H. David, S. Hawkins and B. Chabbert, *Planta*, 2008, **222**, 234–245.
3. J. Ralph, K. Lundquist, G. Brunow, F. Lu, H. Kim, P. F. Schatz, J. M. Marita, R. D. Hatfield, S. A Ralph, J. H. Christensen and W. Boerjan, *Phytochem. Rev*, 2004, **3**, 29–60.
4. M. Lawoko, G. Henriksson and G. Gellerstedt, *Biomacromolecules*, 2005, **6**, 3467–3473.
5. G. Henriksson, M. Lawoko, M. E. E. Martin and G. Gellerstedt, *Holzforschung*, 2007, **61**, 668–674.
6. L. Salmén and A.-M. Olsson, *J. Pulp Pap. Sci.*, 1998, **24**, 99–103.
7. B. V. Kokta, Process for preparing pulp for paper, Can Pat.1,230,208 (1978).
8. E. A. Delong, Method of rendering lignin separable from cellulose and hemicellulose in lignocellulosic material and the product so produced, Can. Pat. 1,096,374 (1981).
9. B. V. Kokta and A. Ahmed in *Environmentally Friendly Technologies for the Pulp and Paper Industry* (ed.Young, M. Akhtar). John Wiley & Sons Inc., New York, U.S., 1998, pp. 191–214.
10. K. Shimizu, K. Sudo, H. Ono, M. Ishihara, T. Fujii and S. Hishiyama, *Biomass Bioenergy*, 1998, **14**, 195–203.
11. A. Wingren, M. Galbe and G. Zacchi, *Biotechnol. Prog.*, 2003, **19**, 1109–1117.
12. A. Kurabi, A. Berlin, N. Gilkes, D. Kilburn, R. Bura, J. Robinson, A. Markov, A. Skomarovsky, A Gusakov, O. Okunev, A. Sinitsyn, D. Gregg, D. Xie and J. Saddler, *Appl. Biochem. Biotechnol.*, 2005, **121–124**, 219–230.
13. J. Li, G. Henriksson and G. Gellerstedt, *Biores. Technol*, 2007, **98**, 3061–3068.
14. J. Li, G. Gellerstedt and K. Toven, *Biores. Technol*, 2009, **100**, 2556–2561.
15. H. Lora and S. Aziz, *Tappi J.*, 1985, **68**, 94–97.
16. E. K. Pye and J. H. Lora, *Tappi J.*, 1991, **74**, 113–118.
17. M. L. Rabinovich, *2nd Nordic Wood Biorefinery Conference, Helsinki, Finland*. Proceedings, pp. 111–120 (2009).
18. R. Mörck, H Yoshida, K. P. Kringstad and H. Hatakeyama, *Holzforschung*, 1986, **40**, 51–60.
19. W. G. Glasser, V. Davé and C. E. Frazier, *J. Wood Chem. Technol.*, 1993, **13**, 545–559.

20. I. Brodin, E. Sjöholm and G Gellerstedt, *Holzforschung*, 2009, **63**, 290–297.
21. F. Öhman, H. Theliander, P. Tomani and P. Axegård, A method for separating lignin from black liquor, Europ. patent application EP 1794363, 2006.
22. F. Öhman, H. Theliander and M. Norgren, Method for separating lignin from a lignin containing liquid/slurry, Europ. patent application EP 1797236, 2011.
23. F. Öhman, H. Theliander, P. Tomani and P. Axegård, A method for separating lignin from black liquor, a lignin product, and use of lignin product for the production of fuels or materials, Int. patent application PCT/SE 2008/000142.
24. F. Öhman, H. Theliander, P. Tomani and P. Axegård, Method for separating lignin from black liquor, Int. patent application PCT/SE2005/001301.
25. F. J. Ball and W. G. Vardell, Continuous acidulation and coagulation of lignin in black liquor, Can. Patent 66,818, 1963.
26. D. R. Robert, M. Bardet, G. Gellerstedt and E.-L. Lindfors, *J. Wood Chem. Technol.*, 1984, **4**, 239–263.
27. G. Gellerstedt and K. Gustafsson, *J. Wood Chem. Technol.*, 1987, **7**, 65–80.
28. G. Gellerstedt, K. Gustafsson and R. A. Northey, *Nordic Pulp Pap. Res. J*, 1988, **3**, 87–94.
29. E. Sjöström in *Wood Chemistry. Fundamentals and Applications.* Academic Press, San Diego, CA, USA, 1993, pp. 236–248.
30. S. Y. Lin and W. J. Detroit in *1ˢᵗ International Symposium on Wood and Pulping Chemistry*, Stockholm, Sweden. Proceedings, Vol 4, 1991, pp. 44–52.
31. J. Li, G. Henriksson and G. Gellerstedt, *Appl. Biochem. Biotechnol.*, 2005, **125**, 175–188.
32. H. H. Nimz and R. Casten, *Holz Roh- Werkst*, 1986, **44**, 207–212.
33. M. Delmas and B. Benjelloun-Mlayah in *Appita Annual Conference and Exhibition. Proceedings*, 2008, pp. 41–45.
34. W. G. Glasser, V. Davé and C. E. Frazier, *J. Wood Chem. Technol.*, 1993, **13**, 545–559.
35. B.-A. Thorstensson, Proceedings *2008 Nordic Wood Biorefinery Conference*, Stockholm, Sweden, 11–14 March, 2008.
36. A. Gandini and M. N. Belgacem in *Lignin as Components of Macromolecular Materials: Partial or Total Oxypropylation of Natural Polymers and the use of Ensuing Materials as Composites from Renewable Resources*, Elsevier, Oxford, UK, 2008, pp. 243–272; 273–288.
37. F. S. Baker, N. C. Gallego and D. A. Baker, Extended Abstract #421 (CD), *Carbon 2010-International Carbon Conference*, Clemson, USA, 11–16 July, 2010.
38. Y. Fukuoka, *Jpn. Chem. Quarterly*, 1969, **5**(3), 63–66.
39. K. Sudo and K. Shimizu, *J. Appl. Polym. Sci.*, 1992, **44**, 127–134.
40. Y. Uraki, S. Kubo, N. Nigo, Y. Sano and T. Sasaya, *Holzforschung*, 1995, **49**, 343–350.

41. J. F. Kadla, S. Kubo, R. A. Venditti, R. D. Gilbert, A. L Compere and W. Griffith, *Carbon*, 2002, **40**, 2913–2920.
42. S. Kubo and J. F. Kadla, *J. Polym. Environ.*, 2005, **13**, 97–105.
43. S. Kubo, Y. Uraki and Y. Sano, *Carbon*, 1998, **36**, 1119–1124.
44. P. J. M. Suhas Carrott and M. M. L. Ribeiro Carrott, *Biores. Technol*, 2007, **98**, 2301–2312.
45. L. M. Cotoruelo, M. D. Marques, J. Rodriguez-Mirasol, T. Cordero and J. J. Rodriguez, *Ind. Eng. Chem. Res.*, 2007, **46**, 4982–4990.
46. P. J. M. Carrott, M. M. L. Suhas Ribeiro Carrott, C. I. Guerrero and L. A Delgado, *J. Anal. Appl. Pyrol.*, 2008, **82**, 264–271.
47. L. M. Cotoruelo, M. D. Marques, J. Rodriguez-Mirasol, J. J. Rodriguez and T. Cordero, *J. Colloid Interface Sci.*, 2009, **332**, 39–45.
48. P. J. M. Suhas Carrott and M. M. L. Ribeiro Carrott, *Carbon*, 2009, **47**, 1012–1017.
49. V. Fierro, V. Torné-Fernandez and A. Celzard, *Micropor. Mesopor. Mater*, 2006, **92**, 243–250.
50. V. Fierro, V. Torné-Fernandez, A. Celzard and D. Montané, *J. Haz. Mater*, 2007, **149**, 126–133.
51. K. Babel and K. Jurewicz, *Carbon*, 2008, **46**, 1948–1956.
52. V. Torné-Fernandez, J. M. Mateo-Sanz, D. Montané and V. Fierro, *J. Chem. Eng. Data*, 2009, **54**, 2216–2221.
53. K. Kratzl, K. Buchtela, J. Gratzl, J. Zauner and O. Ettingshausen, *Tappi*, 1962, **45**, 113–119.
54. V. P. Saraf and W. G. Glasser, *J. Appl. Polym. Sci.*, 1984, **29**, 1831–1841.
55. L. C.-F. Wu and W. G. Glasser, *J. Appl. Polym. Sci.*, 1984, **29**, 1111–1123.
56. W. G. Glasser, W. de Oliveira, S. S. Kelley and L. S. Nieh, Method of producing prepolymers from hydroxyalkyl lignin derivatives, U.S. Patent 5,102,992 (1992).
57. J. H. Lora and W. G. Glasser, *J. Polym. Environ.*, 2002, **10**, 39–48.
58. H. Yoshida, R. Mörck, K. P. Kringstad and H. Hatakayama, *J. Appl. Polym. Sci.*, 1990, **40**, 1819–1832.
59. H. Yoshida, R. Mörck, K. P. Kringstad and H. Hatakayama, *J. Appl. Polym. Sci.*, 1987, **34**, 1187–1198.
60. R. W. Thring, M. N. Vanderlaan and S. L Griffin, *Biomass Bioenergy*, 1997, **13**, 125–132.
61. M. N. Vanderlaan and R. W. Thring, *Biomass Bioenergy*, 1998, **14**, 525–531.
62. H. Cheradame, M. Detoisien, A. Gandini and F. Pla, *British Polym. J*, 1989, **21**, 269–275.
63. W. H. Newman and W. G. Glasser, *Holzforschung*, 1985, **39**, 345–353.
64. W. G. Glasser and R. C. Strickland, *Biomass*, 1987, **13**, 235–254.
65. M. Olivares, J. A. Guzman, A. Natho and A. Saavedra, *Wood Sci. Technol.*, 1988, **22**, 157–165.
66. A. Pizzi, *J. Adhesion Sci. Technol*, 2006, **20**, 829–846.
67. L.-W. Zhao, B. F. Griggs, C.-L. Chen and J. S. Gratzl, *J. Wood Chem. Technol.*, 1994, **14**, 127–145.
68. Y. Liu and K. Li, *J. Adhesion*, 2006, **82**, 593–605.

69. K. Shimatani, Y. Sano and T. Sasaya, *Holzforschung*, 1994, **48**, 337–342.
70. A. Stephanou and A. Pizzi, *Holzforschung*, 1993, **47**, 439–445.
71. A. Stephanou and A. Pizzi, *Holzforschung*, 1993, **47**, 501–506.
72. A. Tejado, G. Kortaberria, C. Pena, J. Labidi, J. M. Echeverria and I. Mondragon, *Ind. Crops Prod*, 2008, **27**, 208–213.
73. N-E. el Mansouri, A. Pizzi and J. Salvado, *J. Appl. Polym. Sci.*, 2007, **103**, 1690–1699.
74. P. J. M. Suhas Carrott and M. M. L. Ribeiro Carrott, *Biores. Technol*, 2007, **98**, 2301–2312.
75. S. B. Lalvani, T. S. Wiltowski, D. Murphy and L. S. Lalvani, *Environ. Technol.*, 1997, **18**, 1163–1168.
76. D. R. Crist, R. H. Crist and J. R. Martin, *J. Chem. Technol. Biotechnol.*, 2003, **78**, 199–202.
77. R. H. Crist, J. R. Martin and D. R. Crist, *Sep. Sci. Technol*, 2004, **39**, 1535–1545.
78. D. Mohan, C. U. Pittman Jr and P. H. Steele, *J. Colloid Interface Sci.*, 2006, **297**, 489–504.
79. B. Kosikova, D. Slamenova, M. Mikulasova, E. Horvathova and J. Labaj, *Biomass Bioenergy*, 2002, **23**, 153–159.
80. J. Labaj, D. Slamenova and B. Kosikova, *Nutr. Cancer*, 2003, **47**, 95–103.
81. B. Kosikova, E. Slavikova and J. Labaj, *Bioresources*, 2009, **4**, 72–79.
82. D. W. Goheen in *Lignin Structure and Reactions*, Adv. Chem. Ser. 59 American Chemical Society, Washington D.C, 1996, pp. 205–225.
83. J. S. Shabtai, W. W. Zmierczak, E. Chornet and D. Johnson, Process for converting lignins into a high octane blending component. U.S. Pat. Appl. 2003/0115792 A1 (2003).
84. S. Czernik and A.V. Bridgwater, *Energy Fuels*, 2004, **18**, 590–598.
85. M. Kleinert and T. Barth, *Energy Fuels*, 2008, **22**, 1371–1379.
86. G. Gellerstedt, J. Li, I. Eide, M. Kleinert and T. Barth, *Energy Fuels*, 2008, **22**, 4240–4244.
87. S. B. Alpert and S. C. Schuman, Production of chemicals from lignin, Can. Pat. CA 851708, 2007.
88. D. T. A. Huibers and H. J. Parkhurst Jr, Lignin hydrocracking process to produce phenol and benzene. US Patent 4,420,644 (1983).
89. R. L. Varma, K. Jothimurugesan, N. N. Bakhshi, J. F. Mathews and S. H. Ng, *Can. J. Chem. Eng*, 1986, **64**, 141–148.
90. R. L. Varma, N. N. Kakhshi, J. F. Mathews and S. H. Ng, *Ind. Eng. Chem. Res.*, 1987, **26**, 183–188.
91. Pellet standard EN 14961:1:2010, Solid biofuels – Fuel Specification and classes, Part 1 – General requirements, CEN (European Committee for Standardization), January 2010.
92. N. P. K. Nielsen, *Importance of raw material properties in wood pellet production*, (2009) Industrial PhD thesis, University of Copenhagen, Faculty of Life Sciences, Copenhagen, Denmark.

93. N. Berlin, P. Tomani, H. Salman, S.Herstad Swärd and L.-E. Åmand, *Tappi Engineering, Pulping & Environmental Conf.*, Portland, Oregon, Aug 25–27, 2008.
94. E.-K. Lindman, *Bioenergy 2009*, Jyväskylä, Finland, 31 Aug – 4 Sept 2009.
95. N. Berglin, P. Tomani, P. Olowson, T. Hultberg, S. Persson, *Tappi International Chemical Recovery Conference*, Williamsburg, USA, March, 2010.

CHAPTER 9

Integrated Forest Biorefineries: Gasification and Pyrolysis for Fuel and Power Production

SUSHIL ADHIKARI,*[a] SUCHITHRA THANGALAZHY-GOPAKUMAR[a,b] AND STEVEN TAYLOR[a,c]

[a] Department of Biosystems Engineering, Auburn University, Auburn, AL 36849, USA; [b] Department of Chemical Engineering, Auburn University, Auburn, AL 36849, USA; [c] Center for Bioenergy and Bioproducts, Auburn University, Auburn, AL 36849, USA
*Email: sushil.adhikari@auburn.edu; Tel.: + 1 334 844 3543; Fax: + 1 334 844 3530

9.1 Biomass Gasification

Biomass has evolved as one of the most promising sources of fuel for the future. This has spurred the growth of research and development efforts in the bioenergy field in both federal and private sectors.[1] This impetus is influenced by different factors such as dwindling fossil fuels, an increased need of energy security, environmental concerns, and the promotion of socioeconomic benefits to rural areas. Another important fact is that biomass is somewhat uniformly distributed throughout the world.[2] Previous studies[3,4] have shown the worldwide recoverable biomass residues is about 31 exajoules per year, which is almost equivalent to 10% of the commercial energy use now. The United States, for example, has about 1.3 billion dry tons of biomass available annually.[2,5] The end-use products from

RSC Green Chemistry No. 18
Integrated Forest Biorefineries
Edited by Lew Christopher
© The Royal Society of Chemistry 2013
Published by the Royal Society of Chemistry, www.rsc.org

biomass conversion can be utilized in heat and power applications, transportation fuels (biodiesel, bioethanol), or chemicals for subsequent processing.[6] To date, the only long-term solution for carbon-based fuels and chemicals is biomass, and the biomass has been successfully converted into solid, liquid, and gas fuels.[7,8] Huber *et al.*[9] opine that among all the renewable energy sources, biomass is the most optimal long-term fuel for transportation. Biomass can be converted into biofuels using either thermochemical or biochemical processes. Among thermal conversion processes, gasification has received significant interest. Gasification is the conversion of biomass into a mixture of combustible and noncombustible gases (referred to as producer gas or syngas hereafter) by partial oxidation at a high temperature around 800–900 °C in the presence of a gasifying medium, such as air, oxygen, or steam. Producer gas from biomass is a mixture of carbon monoxide (CO), carbon dioxide (CO_2), hydrogen (H_2), water (H_2O), and a small amount of methane (CH_4) along with nitrogen (N_2) if air is used as an oxidizing medium. This chemical transformation can take place in different gasifiers that will be discussed in a later section. Gasification is an attractive alternative due to its higher efficiency as compared to other processes such as direct combustion, pyrolysis, and liquefaction.[10–12]

Thermal processes for biomass to biofuels and bioenergy conversion are distinguished from each other based on the amount of air supplied, residence time, temperature, and consequently the heat-transfer rate in the process. For example, supplying excess air results in combustion, while treatment without air or oxygen results in pyrolysis products.[13] Both the gasification and pyrolysis processes will be discussed in this chapter. The use of producer gas for power generation[14] is widely accepted and considered a mature technology[14] although biomass gasification has its own set of challenges, which will be discussed later in this chapter. In this gasification section, biomass characterization, different types of gasifiers along with their advantages and disadvantages, chemical reactions involved during the gasification, and process parameters that influence the quality of producer gas will be discussed. In addition, there will be a discussion involving the applications of producer gas to power and fuels production.

9.1.1   Biomass Characterization

Biomass contains a large number of organic compounds and a small amount of inorganic impurities, also known as ash. Organic compounds are essentially made of four major elements: carbon (C), hydrogen (H), oxygen (O) and nitrogen (N). Most of the biomass feedstocks also consist of small amounts of chlorine and sulfur. Three major properties that are used to characterize biomass for gasification are: (i) proximate analysis, (ii) ultimate analysis, and (iii) energy content. Proximate analysis provides information regarding moisture content, volatile matter, ash, and fixed carbon. Moisture (M) in biomass can be classified as "free" moisture and "bound" moisture. As the name suggests, free moisture resides outside the walls, whereas bound moisture is absorbed within cell walls. High moisture content is a major characteristic of biomass, and it could be as high as 95% in certain biomass types (*e.g.*, water hyacinth and algae). Biomass that will be used for gasification is dried to a certain moisture content, typically within 10–20 wt.%. Every kilogram of moisture requires a minimum of 2260 kJ energy to vaporize water, and gasification of high moisture content biomass can be an energy-intensive process. Volatile matter (VM) is defined as condensable and noncondensable vapors released when the biomass is heated at a certain temperature. Ash is the inorganic solid residues left once the fuel is completely burned, and most biomass contains alkali and alkaline earth metals that pose some challenges during gasification. ASTM methods E871, E872, and E1755 are used to determine moisture content, volatile matter, and ash in the biomass, respectively. Fixed carbon (FC) in biomass is determined indirectly from volatile matter, ash and moisture content as shown in eqn (9.1).

$$FC = 100 - VM - M - Ash \qquad (9.1)$$

A typical ultimate (elemental) analysis is done to characterize biomass in terms of C, H, N and O although other elements such as S and Cl can be determined. ASTM Method E777 is used to determine carbon and hydrogen content in biomass, whereas E778 is used for nitrogen analysis. Oxygen is often calculated as 100 minus the summation of percentages of C, H and N. Ultimate analysis provides information about the quality of fuel. The higher the oxygen content, the lower the heating value of the biomass. The carbon content of biomass is lower than that of coal, whereas the oxygen content is higher. The heating value of biomass is relatively low compared to that of coal or liquid fuels such as gasoline and diesel. The heating value is defined as the amount of heat released by the unit mass of the fuel when it is combusted. When talking about the heating value, it is important to distinguish whether it is a higher heating value (HHV) or a lower heating value (LHV). The higher heating value includes the latent heat of vaporization of water, whereas the lower heating value does not include it. Table 9.1 compares proximate and ultimate analyses along with the heating value of different types of biomass with that of coal. Major differences between coal and biomass are in terms of FC and VM. Biomass has high VM but low FC compared to coal. Due to the high VM in biomass, it is highly reactive during gasification but creates challenges in gas clean-up. The FC conversion is the slowest reaction, and special attention is needed for sizing the gasifier for coal gasification.

9.1.2 Gasifier Types and Processes

Gasifiers can be classified in four major categories that are based on the fluid and/or solid movement inside the reactor.[18]

i. quasi nonmoving or self-moving feedstock;
ii. mechanically moved feedstock:
   a. downdraft gasifier;
   b. updraft gasifier;
   c. crossdraft gasifier;

Table 9.1 Proximate, ultimate and heating value analyses (dry weight basis) of selected biomass feedstocks.

	Switchgrass[a]	Hybrid poplar[a]	Pine woodchips[b]	Sugar cane bagasse[a]	Black Liquor[c]	Wyoming Elkol coal[d]
Proximate Analysis						
Fixed Carbon	14.34	12.49	18.01	11.95	11.80	51.4
Volatile Matter	76.69	84.81	81.71	85.61	40.98	44.4
Ash	8.97	2.70	0.28	2.44	47.22	4.2
Ultimate Analysis						
Carbon	46.68	50.18	49.33	48.64	23.47	71.5
Hydrogen	5.82	6.06	5.03	5.87	2.29	5.3
Nitrogen	0.77	0.60	0.53	0.16	<0.07	1.2
Oxygen[‡]	37.38	40.43	44.70	42.82	23.60	16.9
Sulfur	0.19	0.02	0.13	0.04	3.36	0.9
Chlorine	0.19	0.01	0.003	0.03	n/a	n/a
HHV, MJ/kg	18.06	19.02	19.40	18.99	13.1[¥]	29.50

[‡]calculated from difference. n/a = not available. a = Ref. 15. b = authors' own work. c = Ref. 16 d = Ref. 17. [¥] lower heating value.

   iii. fluidically moved feedstock:
     a. bubbling-bed (**BB**) gasifier;
     b. circulating-fluidized bed (**CFB**) gasifier;
     c. entrained-bed gasifier;
   iv. special reactors:
     a. spouted-bed gasifier;
     b. cyclone gasifier.

Among those listed above, downdraft, updraft, BB and CFB gasifiers are the most common.[19] Figures 9.1–9.4 show schematics of different gasifiers that are widely used in the commercial market. Commercially, about 75% of the gasifiers sold are downdraft gasifiers, 20% fluidized-bed, 2.5% updraft, and 2.5% are other types.[19] The updraft gasifier is popular for application choice when the primary purpose of gasification is heating only (below 10 MWt) due to its high thermal efficiency, and its ability to handle feedstock with wide variation in size and moisture content as high as 50%.[20] The movement of the feedstock and that of the gasifying agent are in opposite directions in this gasifier (also called a countercurrent gasifier). Since the producer gas formed is not forced to pass through a high-temperature (combustion) zone, tar content is high in this gas. Any compound with molecular weight greater than that of benzene is called tar.[21] On the other hand, the temperature of the producer gas exiting from this gasifier is lower (around 200–300 °C), and hence the thermal efficiency of this kind of gasifier is high. Due to the high tar content in the producer gas, a tar-cleaning system, which can become a major investment, is needed if the end-process requires tar-free producer gas.

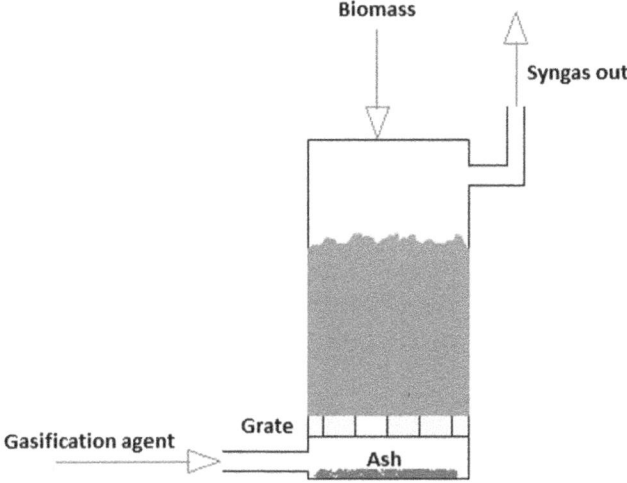

Figure 9.1 A schematic of an updraft gasifier.

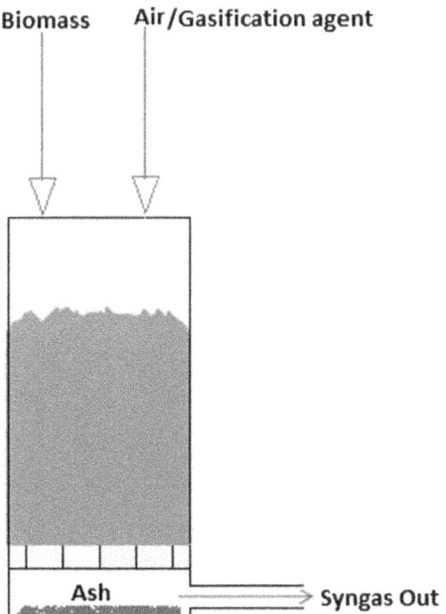

Figure 9.2 A schematic of a downdraft gasifier.

Downdraft gasifiers are preferred for small-scale power generation due to a low amount of tar content in the producer gas. In a downdraft gasifier, the feedstock and gasifying agent both move in the same direction. The gas has to pass through a high temperature so the amount of tar is significantly lower

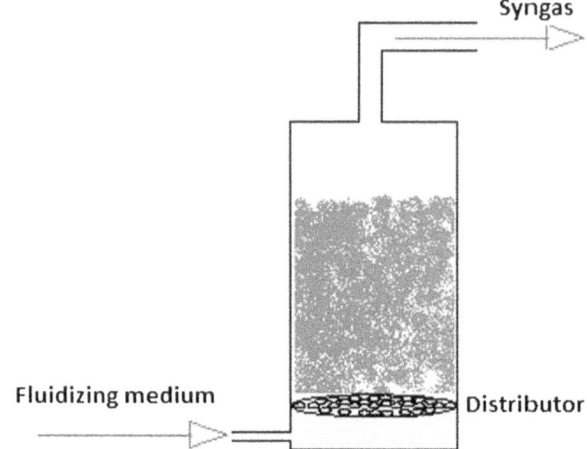

Figure 9.3   A schematic of a bubbling fluidized-bed gasifier.

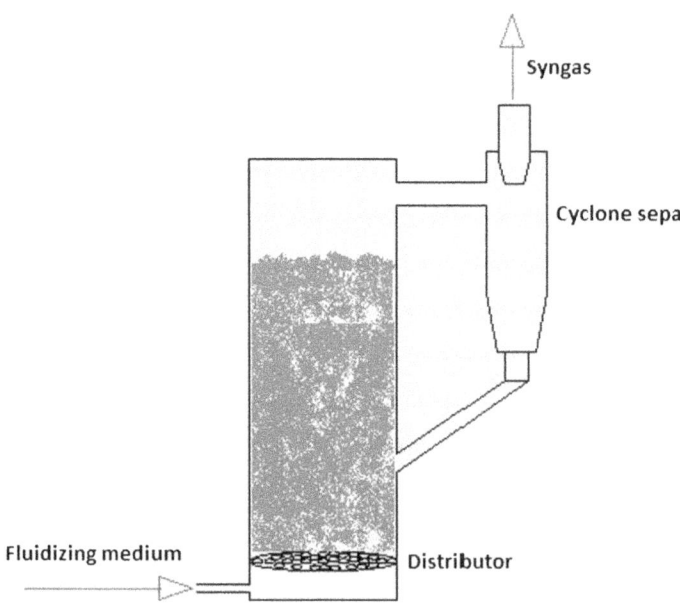

Figure 9.4   A schematic of a circulating fluidized-bed gasifier.

than in an updraft gasifier. Thermal efficiency is, however, lower because a high amount of CO_2 is being formed while the producer gas is passing through the high-temperature zone inside the gasifier. The problem with fixed-bed gasifiers is their inability to maintain uniform radial temperatures, which results in local slag, bridging, and clinkering problems. Lack of a uniform radial temperature is one of the reasons why this kind of gasifier cannot be scaled up, rendering them inflexible and of limited use.[20]

Fluidized-bed gasifiers provide higher throughput than those with a fixed bed. Fluidization enhances mass and heat transfer and can be operated at high pressures, making it an excellent choice for low-rank coal and biomass gasification. Entrained-bed gasification is similar to fluidized-bed gasification except for the operation range temperature – which is usually higher than 1900 °C – and a very high terminal velocity. This type can have higher throughput, but is limited to coal due to the particle size constraint for feedstock (less than 0.15 mm).[22] A significant amount of energy is required to ground biomass to the particular size required for entrained-bed gasification. In fluidized-bed gasifiers, feedstock is fluidized with some kind of bed material, like sand or silica, with a gasifying medium, which can be air or steam. Fluidized-bed gasifiers can further be classified into two types: bubbling and circulating. Circulating fluidized beds add one more feature to bubbling beds in which solid material is often trapped using cyclones, which is entrained with the gas phase, and recirculated back to the gasification bed. This provides significant advantages over the bubbling bed gasifier in terms of mass-conversion efficiency, and it reduces particulate content in the producer gas output.[22] A circulating fluidized-bed operates at a higher superficial velocity than a bubbling fluidized bed. Also, the gas–solid contact time is higher for a circulating fluidized bed.[23] Table 9.2 lists major characteristics of commonly used gasifiers in biomass gasification.

9.1.3 Chemical Reactions in the Gasification Process

Although gasification is a highly complex chemical process, gasification steps are modeled in series for simplicity. Gasification consists of four major steps: drying, pyrolysis, combustion (oxidation), and reduction, but there is no sharp boundary between them, and they overlap in many cases.[10] However, the overall process can be reasonably described by a set of reactions described below.[25–27] As mentioned previously, biomass consists of C, H, O and N with traces of other elements but, for simplicity, pure C is being used as a model compound.

$$Biomass \xrightarrow{heat(300-400\,°C)} gases\ (CO,\ CO_2,\ H_2, light\ HC's) \\ + pyrolyzed\ vapors + char \tag{9.2}$$

$$Pyrolyzed\ vapors \xrightarrow{(heat+high\ residence\ time)} gases\ (CO,\ CO_2, H_2,\ light\ HC's) \tag{9.3}$$

$$C + O_2 \leftrightarrow CO_2 \qquad \Delta H = -393.5\ kJ/mol \tag{9.4}$$

$$C + (1/2)O_2 \leftrightarrow CO \qquad \Delta H = -110.5\ kJ/mol \tag{9.5}$$

Table 9.2 Characteristics of common types of gasifier.

Characteristics	Gasifier Type			
	Downdraft	Updraft	BBG	CFBG
Gasifier size	High space requirement for higher throughput due to modular design of the gasifier and high residence time		Less space required due to enhanced heat transfer resulting in much faster gasification and lower residence time inside the gasifier	
Temperature Profile	Not uniform temperature distribution in the radial distribution		Uniform temperature distribution inside the gasifier	
Permissible particle size/Size sensitivity	<50 mm/good		<5 mm/more sensitive to feedstock size	
Reaction zone temperature	800–1100 °C		800–1000 °C	
Ability to handle fine particles	Limited		Good	
Moisture content	Very flexible	Flexible	Flexible	
Gas exit temperature	600–800 °C	250 °C	850 °C	
Tar concentration[a]	Low (0.01–6 g/Nm^3)	High (50 g/Nm^3)	6–12 g/Nm^3	

Carbon conversion efficiency	Very good		Fair
Thermal efficiency	Very good	Excellent	Good
LHV of producer gas	Poor	Poor	Poor
Cold gas efficiency	>80%		>90%
Gas clean-up	High cleaning required	relatively clean gas	Clean-up required for dust and tar
Dust content in producer gas	High	Low	Higher dust content
Energy requirement for operation	Low		High due to requirement of fans for fluidization
Investment	Higher investment for the energy generation compared to **BBG/CFBG** (for large scale output)		Lower investment
Process control	Cannot be controlled effectively as **BBG/CFBG**		Easy process control
Applications	Small to medium scales		Large scales
Residence time solids	Hours to days		Seconds to minutes
Residence time gas	Seconds		Seconds

^{a}Nm^3 = normal cubic meter.

$$C + CO_2 \leftrightarrow 2CO \quad \Delta H = 172.4 \ kJ/mol \tag{9.6}$$

$$C + H_2O \leftrightarrow H_2 + CO \quad \Delta H = 131.3 \ kJ/mol \tag{9.7}$$

$$C + 2H_2 \leftrightarrow CH_4 \quad \Delta H = -74.8 \ kJ/mol \tag{9.8}$$

$$CO + H_2O \leftrightarrow H_2 + CO_2 \quad \Delta H = -41.1 \ kJ/mol \tag{9.9}$$

$$CO + 2H_2 \leftrightarrow CH_4 + H_2O \quad \Delta H = 206.1 \ kJ/mol \tag{9.10}$$

$$CO + 0.5O_2 \leftrightarrow CO_2 \quad \Delta H = -284 \ kJ/mol \tag{9.11}$$

$$CH_4 + 2O_2 \leftrightarrow CO_2 + 2H_2O \quad \Delta H = -803 \ kJ/mol \tag{9.12}$$

$$H_2 + 0.5O_2 \leftrightarrow 2H_2O \quad \Delta H = -242 \ kJ/mol \tag{9.13}$$

$$2CO + 2H_2 \leftrightarrow CH_4 + CO_2 \quad \Delta H = -247 \ kJ/mol \tag{9.14}$$

$$CO_2 + 4H_2 \leftrightarrow CH_4 + 2H_2O \quad \Delta H = -165 \ kJ/mol \tag{9.15}$$

$$CH_4 + H_2O \leftrightarrow CO_2 + 3H_2 \quad \Delta H = 206 \ kJ/mol \tag{9.16}$$

$$CH_4 + 0.5O_2 \leftrightarrow CO + 2H_2 \quad \Delta H = -36 \ kJ/mol \tag{9.17}$$

Among the reactions described above, the char-oxidation (eqn (9.4)) and partial-oxidation (eqn (9.5)) reactions are the slowest and, consequently, the rate-controlling factors in the overall gasification process.[26] Pyrolysis of biomass also results in liquid formation, which is resistant to the cracking. This requires a subsequent cleaning setup for the tar, which can be a substantial investment in many cases.[10]

9.1.4 Effect of Various Parameters in the Gasification Process

Producer-gas composition varies widely and mostly depends upon the gasifier type, feedstock, feedstock pretreatment, and operating parameters like temperature, pressure, gasifying medium, and the nature of the interaction between reactants in the gasification process.[22,28] Effects of major parameters affecting the quality of producer gas are discussed in the sections below.

9.1.4.1 Moisture Content

Biomass contains moisture in two ways: intrinsically by its nature, and extrinsically, wherein moisture is absorbed from the surrounding atmosphere.[29,30] The moisture content of biomass is one of the major parameters that influence gas composition. The moisture content of biomass during gasification increases CO_2 concentration by the water-shift reaction (eqn (9.9)), that consumes CO and liberates H_2 and CO_2.[29,31,32] While the equilibrium constant for the water-shift reaction varies little over a wide range of temperatures, the direction tends to reverse at a higher temperature. More heat is required for moisture evaporation than the small amount of heat gained due to the exothermic behavior of the water-shift reaction and, therefore, the temperature inside the gasifier is reduced when gasifying biomass with higher moisture content.[26] The decrease in temperature further exacerbates the scenario and forms more CO_2 because the water-shift reaction is improved at the lower temperature. The overall effect is the reduction in calorific value of the producer gas because the small increase in H_2 is not sufficient to compensate for the significant loss of CO with an increase in moisture content.[29,31-35] However, the negative effect of moisture content on the calorific value of producer gas is lower with a lower equivalence ratio (ER). The ER is the ratio of actual air:fuel ratio to the stoichiometric air:fuel ratio, which provides the basis for evaluating the amount of air supplied for gasification with respect to the amount of air required for a complete combustion of the feedstock. Roy *et al.*[29] have observed that, in a downdraft gasifier, when the moisture content is increased from 0 to 40%, the heating value of the producer gas decreases by 8.72% at an ER of 0.45, while the decrease is 4.7% when the ER is 0.29. This result was reported from their equilibrium model and, thus, is applicable to any gasification process. Table 9.3 summarizes the effect of moisture content in three common gasifier types.

The autothermal limit is reported in the literature as 65% moisture content beyond which self-sustaining gasification is not possible due to an enthalpy deficiency for vaporization. In fact, supplemental fuel is required for most of the combustors when the moisture content is greater than 50% on a wet basis.[15,37] Moisture content up to 30% (wet basis) can be used for a downdraft gasifier.[24,36] When air is used as the gasification agent, the amount of methane

Table 9.3 Effect of moisture content upon major producer gas constituents.

Parameter	Gasifier type	CO	CO_2	H_2	CH_4	Maximum limit (% w.b)
Moisture Content (M.C)	Updraft	$-$[a]	$+$[b]	$+$	$\sim$[c]	<50 [20]
	Downdraft	$-$	$+$	$+$	$\sim$	<40 [36]
	Fluidized	$-$	$+$	$+$	$\sim$	<10 [22]

[a]Decreases with increase in M.C.
[b]Increases with increase in M.C.
[c]No significant change.

produced is small and stays almost constant with the change in moisture content.[31,38] Thus, the temperature decrease inside the gasifier due to moisture also results in lower mass conversion efficiency and increases tar content.[32,39–41] Sheth *et al.*[42] reported the decrease in biomass consumption rate with an increase in moisture content, which is due to the higher amount of heat necessary for drying those wood chips inside the reactor before the wood chips can be pyrolyzed. However, some moisture content is always desirable since it enhances steam reforming and helps to crack tar, and, at higher temperatures, also enhances other reactions such as char gasification.[43,44] In fact, steam injection is widely used in industrial applications to adjust producer gas composition in the gasification process.[45]

9.1.4.2 Equivalence Ratio

The equivalence ratio (ER) is the most influential parameter in any gasification process and often has significant impact on producer gas composition. An increase in ER increases the temperature inside the gasifier; while a decrease in ER increases char formation. While combustible products are reduced with an increase in ER, a higher amount of CO_2 is formed, which diminishes the heating value of the producer gas.[46–49] Zainal *et al.*[50] searched for an optimal value of ER for a downdraft gasifier using furniture wood and wood chips as a feedstock. The effect of ER for each producer gas component was analyzed with the conclusion that an optimal ER was 0.38 for the gasifier performance for that particular feedstock. At this ER, CO, CH_4 and calorific value each attain their maximum outputs while CO_2 reaches its minimum.

An optimum ER is necessary for accelerating pyrolysis and drying rate due to conduction and convection processes, which also increases the biomass consumption rate.[42] Both Skoulou *et al.*[51] and Sheth *et al.*[42] reported an optimal ER of 0.2 for gasification of olive kernels and olive tree cuttings, and furniture wood using a downdraft gasifier. Optimum ER varies for different types of biomass due to the amount of oxygen elementally present in the biomass as well as the ash content. For example, coal requires more oxygen than common biomass materials for gasification due to its lower oxygen content.[41] The existing literature shows that ER should be around 0.2 to 0.4 for successful gasification, but it is noteworthy that the ER depends on the operating temperature and moisture content of the biomass as well as the gasifier design. Therefore, ER data presented in Table 9.4 are for an illustration purpose only. Also in Table 9.4, the difference in optimal ER can also be observed for feedstock with the same elemental composition (pine wood chips and saw dust) in a fluidized bed. This is due to the difference in gasification temperature, which is lower for pine saw dust (780–830 °C) than pine wood chips (> 900 °C). The optimal ER for an updraft gasifier is not shown in the table due to the limitations of available literature for an updraft configuration.

Another benefit of increasing ER is to decrease the tar concentration in the producer gas. Tar concentration decreases with higher ER, primarily due to two reasons: (a) the higher temperature as a result of higher ER increases

Table 9.4 Optimal ER for selected feedstocks in downdraft and fluidized gasifiers.

Gasifier Type	Feedstock	Optimal ER	References
Downdraft	Furniture wood + charcoal	0.38	50
	Olive kernels and tree cutting	0.20	51
	Hazelnut shells	0.28	47
	Furniture waste	0.20	42
Fluidized-bed	Rice husk	0.2–0.55[a]	52
	Pine wood chips	0.30	53
	Pine saw dust	0.20	43

[a]The range is due to difference in operating conditions.

reaction rates of chemical products; and (b) the high ER supplies additional oxygen for cracking of tar into the lower hydrocarbons CO_2 and H_2O. Thus, it has been reported that at some point between the applicable ranges of ER (0.15–0.4), a shift between types of tar occurs. Light tar increases while heavy tar decreases.[47,54] Corella *et al.*[55] suggested using an ER above 0.36 for pine wood in a fluidized bed to reduce tar content below $2 \, g/Nm^3$.

9.1.4.3 Temperature

Biomass gasification typically occurs around 700–900 °C, and temperatures typically suggested for biomass gasification in a fluidized bed are around 800–900 °C.[56–58] The CO content increases with an increase in temperature due to the reverse water-gas shift reaction, which is more likely to occur at higher temperatures.[59] Mass-conversion efficiency decreases with a decrease in temperature.[60] An oxidation (combustion) zone below a temperature of 725 °C yields significantly lower mass conversion efficiency.[61] An increase in temperature reduces the tar content as well as char inside the gasifier,[51,62] but this could come at the expense of losing carbon in the form of CO_2. Drift *et al.*[54] suggest that the tar that is cracked due to temperature is mostly heavy tar, while light tar is not decomposed. Heavy tars are the product of the pyrolysis process and have not gone through cracking while light tars are the cracking products of heavy tar. In certain cases, light tar seems to increase with the increase in gasification temperature due to the subsequent breakdown of heavy tar into light tar and other compounds. Gas yield also increases due to higher tar cracking. One of the means of increasing temperature inside the gasifier is through internal recirculation of producer gas.[60] Tar-cracking temperatures are often reported to be around 1000–1100 °C with some dependency on gasifier design.[21,36]

Uniformity of temperatures in radial as well as in axial directions inside the reactor is very important for efficient mixing in a fluidized bed. Generally, less than 100 °C difference in total riser height is acceptable.[44] The heating value as well as producer gas yield is found to increase due to the increase in combustibles, particularly at temperatures above 800 °C with an increase in operating temperature driven by an external supply of heat in the gasifier for constant

ER.[44,46,48,56,63] However, this is different when the temperature is increased through burning carbon inside the reactor, which actually reduces the combustibles.[48] Temperature control cannot be independent in any gasification process and is an output variable, with the exception of small, lab-scale, or pilot plants that can be heated with an external heat source. The temperature of the reactor depends on different factors, such as moisture content of the fuel, ER, heat losses from the system, and the amount of steam added.[55,64,65] Thus, the temperature inside the gasifier should represent an optimal compromise with ER. The best approach includes proper insulation of the reactor and the use of waste heat. Although high temperatures increase the carbon conversion efficiency of the overall gasification system, consideration should be given to preventing the formation of ash-melt, especially while gasifying biomass materials such as rice husk which has a high ash content.[54]

9.1.4.4   Biomass Type

Biomass elemental composition has a significant effect on producer gas composition. Release of pyrolysis gas is highly dependent on hydrogen to carbon and oxygen to carbon ratios, and increases when these ratios increase, especially with an increase in the hydrogen/carbon ratio.[15] A higher oxygen concentration in biomass needs lower ER for gasification because of its inherent oxygen that will also be available for gasification.[41] Another important factor is the ash content of a feedstock. Table 9.5 provides the ash content and the elemental composition of common biomass feedstocks.[15,66] Although formation of clinkers can cause problems for the gasifier operation with biomass having ash content above 5%, successful gasification with ash contents up to 25% has been reported.[25,67] Higher ash content causes slagging and consequently ash agglomeration due to fusion, but the rate of this is dependent upon the ash content in biomass and ash composition.[15,68] Thus, high ash content biomass should be gasified at a temperature below the oxidation or reducing temperature of the mineral constituents in the ash, which is often not possible if the constituents have a relatively low ash-fusion temperature.[21,69] Common ash minerals in biomass are silica, potassium, calcium, aluminum, magnesium, iron, sodium, and chlorine. These minerals present in biomass can vaporize during the gasification process contaminating producer gas. Also, it is highly possible for these minerals to react

Table 9.5   Ash content (% dry basis) and its elemental composition (% in ash) for some common feedstocks.

Feedstock	Ash	CaO	K_2O	MgO	Na_2O	SiO_2
Pine	3.1	13	7.9	4.5	1.9	52
Poplar	3.4	33	18	3.7	0.14	2.8
Rice straw	13.1	8.9	16	3.5	2.8	51.0
Wheat straw	5.9	8.1	18	2.4	0.22	44.0
Switch-grass	8.97	2.03	11.6	3.0	0.58	65.18

with silicon in the presence of oxygen to create low-temperature melting silicates, which can create a severe deposition problem. Alkali metals such as potassium and calcium silicates have melting temperatures even below 700 °C.[15] One way to tackle the problem is to find some kind of metal-removal process, such as alkali-metal leaching. This has been reported to reduce mineral content by more than 80%. Removal of these alkali metals will increase the ash-fusion temperature, thus facilitating gasification.[15] However, it is unlikely that this process will be economical in a large-scale operation. The presence of ash in biomass requires careful control over the operating temperatures. They should neither be high enough to fuse minerals in the ash, forming a barrier to further gasification by the formation of clinkers, nor low enough to lead to unburnt carbon, resulting in lower carbon conversion efficiency.

9.1.4.5 Particle Size

Fixed-bed gasifiers have lower biomass feedstock size restrictions compared to fluidized-bed gasifiers. Usually, feed size less than 51 mm and 6 mm are recommended for fixed-bed and fluidized-bed, respectively.[22] Nonetheless, use of larger-size feedstock has been tried and reported by several authors.[70–72] Saravanakumar *et al.*[71] have successfully gasified long sticks with lengths of 680 mm and diameters of 60 mm in a top-lit, updraft gasifier. The maximum particle size suggested for a conventional downdraft gasifier with throated design is one-eighth of the reactor throat diameter.[73] The larger particles form bridges preventing an efficient flow of biomass inside a gasifier, while smaller particles interfere with the air or gasifying agent passage creating a high pressure drop, which can consequently result in a gasifier shut-down.[25]

Sharma[74] reported an increase in the temperature of oxidation and reduction zones with a decrease in the particle size of the biomass feedstock in a down-draft gasifier. A decrease in particle size reduces the heat loss due to radiation and enhances the thermal conductivity in the oxidation and reduction zones. On the other hand, a decrease in particle size increases the pressure drop inside the gasifier. The burning rate and, thus, the char oxidation period of fuel particles decrease with an increase in bulk density and particle size.[35,68] The biomass consumption rate is inversely related to particle size.[75,76] In other words, a higher residence time is recommended for larger biomass particle size. Ryu *et al.*[68] reported a decrease in CO from 18% to 13.5% when the size of wood cubes used in the experiments was increased from 10 mm to 35 mm. An increase in the size of biomass particles results in poor temperature distribution.

Carbon-conversion efficiency is not strongly affected by particle size except lower biomass size increases tar concentration because of high entrainment susceptibility during fluidization.[77] This is because particles can be easily transported to the upper part of the reactor, leaving little time for tar cracking. An axial temperature drop increases significantly with a decrease in size. This is due to the easy passage of feed particles from the feed point and, thus, little or no reaction takes place below the feed point and homogeneity of the bed material cannot be maintained throughout the reactor.[54,64] Valin *et al.*[78] reported an increase in gas-particle interactions with a decrease in particle size in a fluidized-bed reactor.

9.1.4.6 Pressure

High-pressure gasification reduces the size of a reactor for the same amount of throughput and can act to reduce the need for further compression when the gasification products are intended for subsequent use in the Fischer–Tropsch process or other chemical syntheses that require high pressure.[78] An increase in pressure in a fluidized bed increases turbulence, and thus an increase in gas–particle interaction is observed. An increase in pressure also results in bubble instability and bubble splitting in the fluidized bed. Valin *et al.*[78] have studied the effect of pressure on producer gas composition with pressure from 2 to 10 bar in a fluidized bed with wood sawdust as a feedstock. With an increase in pressure, increases in CO_2, CH_4 and H_2 were observed, while CO decreased. In their lab-scale reactor using steam and N_2 as gasification mediums, with an increase in pressure from 2 to 10 bar, H_2, CO_2 and CH_4 increase by 16%, 53% and 38%, respectively, and CO decreases by 33%. Overall, the increase in dry gas yield was reported to be 20% with an increase in pressure from 2 to 7 bar, after which the gas yield remains constant. The increase in different gases and total gas yield was due to the increase in char hold-up rate, which increases the catalytic activity of char as well as improved reaction kinetics due to high pressure.

9.1.4.7 Gasification Medium

Biomass gasification can be done with any of the following media: (i) air; (ii) oxygen; and (iii) steam. Gasification with air results in producer gas with a low higher heating value (HHV) due to the inherent dilution with N_2 present in the air. Conversely, gasification with oxygen yields producer gas with a heating value of 10–12 MJ/Nm^3, and steam gasification results in producer gas with a heating value that is even higher, 15–20 MJ/Nm^3.[10] Air gasification is widely used compared to oxygen and steam due to its economical and operational advantages.[79]

9.1.4.8 Bed Materials

Proper consideration of bed material in a fluidized bed is important for achieving proper homogenization of feed particles and efficient heat transfer so that a minimum temperature gradient is realized within the riser. In many cases, the bed material itself can act as a catalyst facilitating efficient tar cracking.[47,55] Skoulou *et al.*[47] compared the effect of olivine to that of silica sand, the latter of which was reported to have an adverse effect on the fluidization due to agglomeration and tar formation when operating at a temperature below 800 °C. Pfeifer *et al.*[80] studied an in-bed catalyst (Ni/ olivine) in a dual fluidized bed and observed significant tar reduction. The use of catalysts for tar cracking is in itself a vast subject, and further discussion is avoided here to remain within the scope of this chapter. Excellent reviews on tar cracking can be found in published documents elsewhere.[21,81,82]

9.1.5 Gasification of Black Liquor

Black liquor is the spent liquid that is extracted from wood pulp in the Kraft process. It is a mixture of hemicelluloses and lignin residues and some inorganic chemicals used for delignification. The general composition of black liquor is about 60% organic and 40% inorganic matter on a dry basis.[83] Despite the fact that black liquor is rich in lignin, the heating value is about 12 MJ/kg due to the high presence of inorganic materials.[16] Although most of the black liquor is burned in boilers to produce steam for electricity generation needed for the Kraft process, there are studies on gasification. Black liquor gasification technologies can be distinguished in two major classes: (i) low-temperature gasification, below melting point of inorganics, and (ii) high-temperature gasification, produces molten smelt.[84] The main advantage of black liquor gasification is that it separates sulfur from sodium that leaves the liquor in the form of H_2S. One main concern with black liquor gasification is formation of sodium formate due to the interaction of CO_2 with sodium at high temperatures greater than 1100 °C. Another concern is the low ash fusion temperature because of the high ash content.

9.1.6 Use of Producer Gas for Power and Fuels

Producer gas can be used for power generation using turbines or internal combustion engines or for the production of alternative fuels. Figure 9.5 depicts various chemicals and fuels that can be produced using producer gas. One of the major challenges using producer gas for producing fuels or chemicals is that the gas needs to be clean. Many of these processes require catalysts, and many of them are not tolerable to contaminants such as sulfur, chlorine, and tars. Also, in many processes, the ratio of H_2 to CO is governed

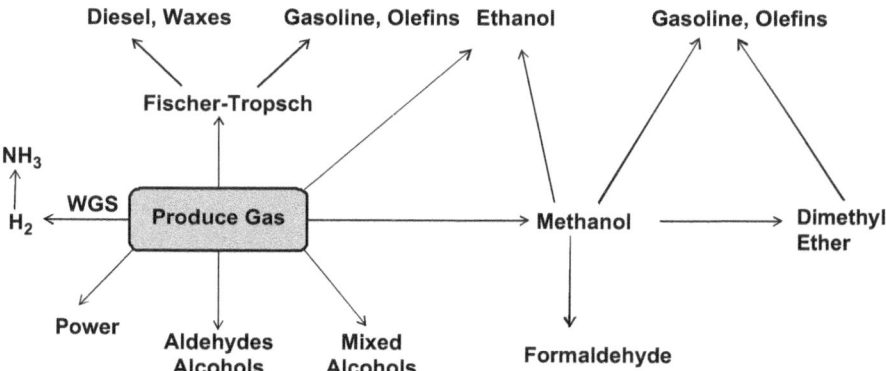

Figure 9.5 Various pathways for conversion of producer gas to fuels and chemicals.[85]

stoichiometrically. For example, in methanol synthesis, H_2/CO ratio is 2:1 but the producer gas has H_2/CO ratio of 1:1 in many cases. Despite these challenges, several companies – such as Range Fuels, Clear Fuels, and CHORAN – have created or are building demonstration plants.

9.2 Fast Pyrolysis

In this section, different reactor configurations for fast pyrolysis process, the reaction mechanism of biomass fast pyrolysis, bio-oil properties, and its applications will be discussed. In addition, bio-oil upgrading for transportation fuels and economic analysis of the fast pyrolysis process will be discussed. The history of pyrolysis started in ancient Egyptian times. However, research on pyrolysis for the production of liquid began in the 1980s.[86] Pyrolysis is a thermal degradation process that occurs in the absence of oxygen. The rate of thermal decomposition depends on biomass particle size and type, as well as the heating rate, final temperature, reactor configuration and presence of impurities. Major products of pyrolysis are char, liquid, and gas. The relative yield of these products can be changed by selecting appropriate heating rates and pyrolysis temperatures. Pyrolysis is broadly classified as slow pyrolysis and fast pyrolysis. Slow pyrolysis is a conventional method in which biomass is heated slowly to produce char as a major product. In fast pyrolysis, biomass is heated rapidly in the absence of oxygen, and it decomposes as vapors, aerosols, gases and char. The vapors and aerosols are quickly condensed to a liquid called bio-oil. The criteria for fast pyrolysis are: (i) moderate temperature (400–600 °C), (ii) high heating rate and heat-transfer rate, and (iii) short residence time (<2 s) for vapors.

The rate of pyrolysis reaction is limited by the rate of heat transfer and mass transfer. Finely ground and dried biomass is required to attain a high heating rate. In contrast to slow pyrolysis, fast pyrolysis requires careful control over the temperature in order to get maximum liquid yield. Rapid cooling is another requirement for fast pyrolysis to prevent further cracking of vapors as gases. As the residence time decreases, the liquid yield increases, and the yields of gas and char decrease. Since the major product from fast pyrolysis is in liquid form (bio-oil), it can be readily stored and transported. A mobile fast pyrolysis plant near the source of biomass and subsequent transportation of bio-oil to a biorefinery can reduce the cost of biomass harvesting and handling.[87]

Pyrolysis is an endothermic reaction. In order to gain the temperature gradient for heat transfer, the reactor temperature should be higher than the temperature of the reaction.[88] In fast pyrolysis, mainly conductive and convective modes of heat transfer occur, and their contribution towards complete heat transfer changes with the reactor configuration. In a fast pyrolysis process, all the steps occur in a small time scale. The major events that occur during the pyrolysis are described below:[89,90]

- heat transfer to increase the temperature of biomass;
- initiation of pyrolysis at higher temperatures that release volatiles and form char;

- heat flow from hot volatiles to cooler, unpyrolyzed biomass;
- condensation of some part of volatiles followed by secondary reactions to produce tar;
- autocatalytic secondary pyrolysis in competition with simultaneous primary pyrolysis; and
- further decomposition, reforming, water-gas shift reaction, radical combination and dehydration can also occur. The degree of these reactions depends on process parameters such as residence time and temperature.

Major steps in fast pyrolysis process are biomass pretreatment, fast pyrolysis, char and ash separation, and condensation and liquid collection[91] and these are summarized below.

Biomass Pretreatment In order to achieve high heat- and mass-transfer requirements for fast pyrolysis, biomass pretreatment is required before pyrolysis. The biomass needs to be dried below a moisture content of 10 wt.%, which can be done by utilizing the process heat of the gas produced during pyrolysis. A detailed review of drying technologies for an integrated gasification, bioenergy plant is presented elsewhere with these technologies being applicable to fast pyrolysis also.[92] Most of the reactor configurations for fast pyrolysis require very fine particles of biomass (approximately 2 mm or less) which makes the process expensive. Klass[93] reviewed a variety of size reduction equipment for biomass applications. Mani *et al.*[94] discussed the energy required for grinding feedstocks of different size and moisture content. For example, the energy required for grinding switchgrass (moisture content 8 wt.%) through a hammer mill (screen opening 1.6 mm) is 58.47 kWh$_e$ t^{-1} (210.5 kJ$_e$/kg), which accounts for 1% of the energy content of the switchgrass.

Fast Pyrolysis The fast pyrolysis reactor is the heart of the plant, and it requires only 10–15% of the capital cost of the plant.[9] Different types of reactor configurations for fast pyrolysis are briefly discussed in the next section.

Char and Ash Separation Ash in the biomass is retained with char, and successful char removal from bio-oil removes ash as well.[91] Char also contains different metals present in the biomass, and it acts as the catalyst for polymerization to increase the viscosity. Char, therefore, has to be removed from bio-oil before storing or further processing. A cyclone separator followed by a pyrolysis reactor is usually employed for char removal, but cyclones are not effective in removing char particles that are smaller than 3 μm. Another unit is required for removing fine particles of char, which are carried over from a cyclone separator. Hot-vapor filters are being developed for the removal of fine char. The char collected can be burned to produce heat necessary for biomass drying.

Condensation and Liquid Collection Pyrolytic vapors will be cracked continuously as they stay at a high temperature. Therefore, vapors have to be condensed quickly in order to prevent further cracking and loss in organic yield. A few milliseconds of vapor-phase residence time is necessary to optimize the yield of some chemicals; whereas a maximum vapor residence time of two seconds can be used for pyrolytic vapors if the bio-oil is being used as a fuel.[91,95] Collection of liquid is one of the most difficult operations in fast pyrolysis, and it needs careful design and control. Quenching and electrostatic precipitation used for liquid collection have been found to be effective.

9.2.1 Pyrolysis Reactor Configurations

Some basic, desirable criteria for designing a pyrolysis reactor are a simple design, ease of operation, scale-up potential, high thermal efficiency, and a compact design suitable for rural environments.[96] A wide variety of reactor configurations have been used for fast pyrolysis of biomass. Reactor configurations and the mode of heat transfer significantly influence the distribution of compounds in bio-oil. Major reactor configurations are bubbling and circulating fluidized bed, ablative, vacuum, rotating cone, and auger.[89] All of these reactors have some advantages and some drawbacks, with the latter acting as barriers for their commercial application to fast pyrolysis.[96]

9.2.1.1 Bubbling-Fluidized Bed

A bubbling-fluidized-bed reactor (Figure 9.6a) is a very popular configuration for fast pyrolysis, as it features good temperature control, an efficient heat-transfer rate, and short residence time.[9] This reactor configuration is simple in construction and operation and can, therefore, be scaled-up for large-scale pyrolysis. In addition, bubbling-fluidized-bed pyrolysis yields bio-oil up to 80 wt.% of dry biomass.[91] This configuration requires a carrier gas to fluidize the bed. The residence time of solids and vapors can be efficiently controlled by using a shallow bed and or by changing the flow rate of carrier gas in the bubbling-fluidized reactor. There are major heat-transfer limitations within the biomass particles. Therefore, small particle size of biomass gives high yield conversion.[91] Effective separation of char is required in order to prevent its catalytic effect for bio-oil polymerization.

9.2.1.2 Circulating-Fluidized Bed

A circulating-fluidized-bed reactor (Figure 9.6b) has many features similar to a bubbling-fluidized-bed reactor where the bed material from product vapors and gases are circulated back to the reactor by passing through a cyclone separator and a combustion chamber (to remove char from the bed material). If a catalyst is used in the fluidizing bed, the coke and char formed during pyrolysis can be burned out during recirculation.[98] In addition, the product gas can be circulated back to the bed as a carrier gas. Here, the residence time of vapors and char is almost the same. This configuration is suitable for higher yield of bio-oil. The circulating-fluidized bed reactor is more efficient than the

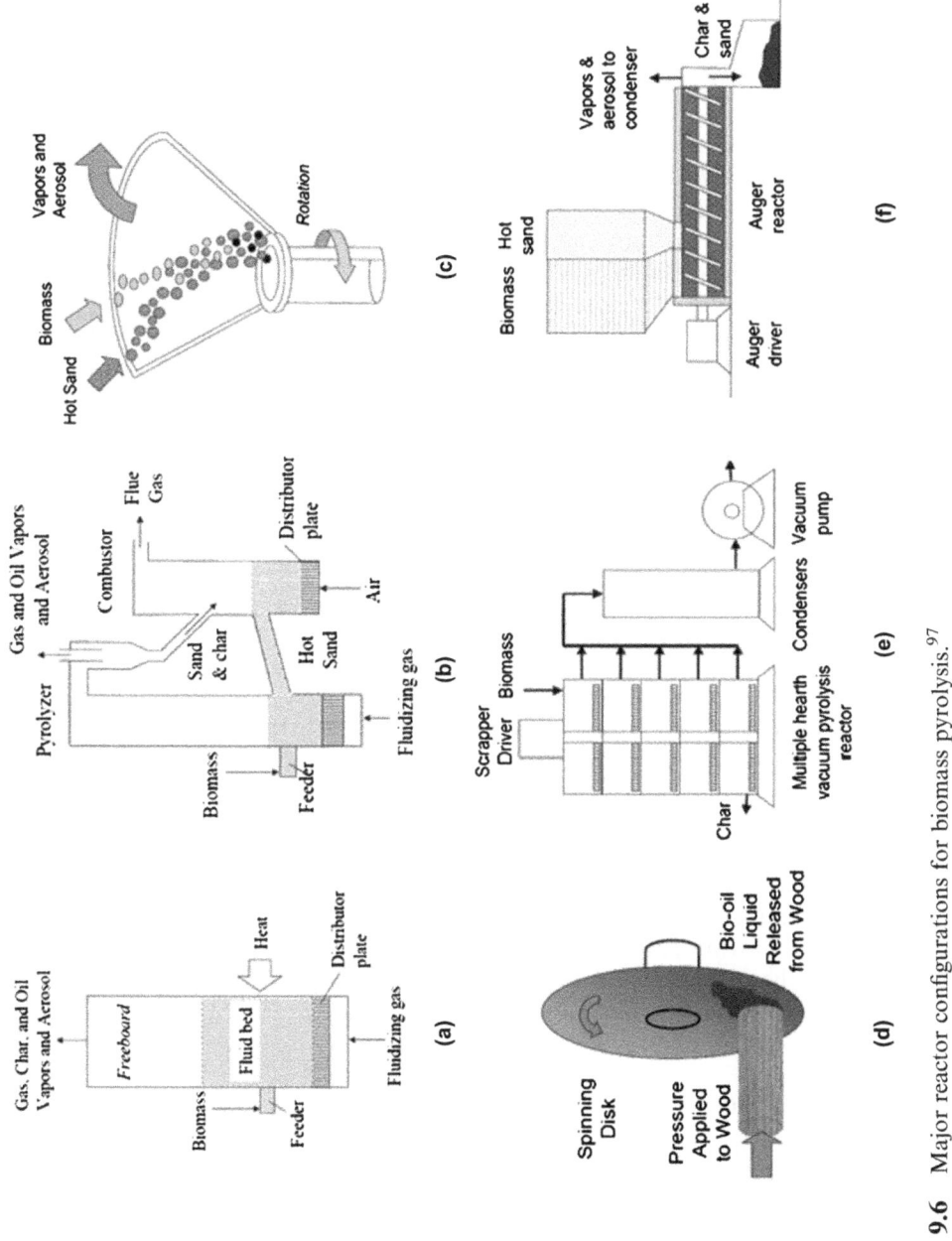

Figure 9.6 Major reactor configurations for biomass pyrolysis.[97] (Reproduced with permission from Robert C. Brown, Iowa State University.)

bubbling-fluidized-bed reactor in terms of temperature control and heat transfer. Post-treatment of bio-oil is required for removing char because of high char attrition and carryover with the vapors.[96] Accumulation of ash content in the circulating bed acts as a catalyst for cracking bio-oil components, which creates a loss of volatiles and improves some of the bio-oil properties.

9.2.1.3 Rotating-Cone Pyrolyzer

Twente University invented a new reactor design for fast pyrolysis called a rotating cone pyrolyzer (Figure 9.6c) that was further developed by the Biomass Technology Group (BTG) in the Netherlands. In a rotating-cone pyrolyzer, sand and biomass are transported by centrifugal forces.[99] Sand is added to avoid fouling the cone wall and to enhance heat transfer. Biomass and sand are fed to the bottom of a rotating cone, and the biomass is pyrolyzed while transporting upward through a spiral motion along the hot side wall of the cone. Very small particle size is required for this configuration. These reactors are very compact in design and can be used for high throughput. BTG and its daughter company, BTG BioLiquids (Trade name BTG-BTL), have optimized the design and developed a 200-kg/h rotating-cone pilot plant.[100,101] Lede et al.[102] describe a cyclone reactor, which is similar to a rotating-cone pyrolyzer, for fast pyrolysis.

9.2.1.4 Ablative Pyrolysis

When large particles of wood are mechanically pressed against a rapidly moving hot surface, they melt, evaporate or sublimate to produce vapors that are condensed as bio-oil. The residual oil on the hot surface provides lubrication and helps to enhance the evaporation of the successive biomass particles. The rate of reaction is influenced by the applied pressure and the relative velocity between biomass and the hot surface. Since the rate of reaction is not limited by the rate of heat transfer, an ablative reactor can utilize large wood pieces for pyrolysis. However, a high heat loss occurs since the hot surface stays at a higher temperature than the reaction temperature.[96] The reactor (Figure 9.6d) has hot moving parts, which make the operation mechanically complex. In addition, the process is controlled by surface area[90] and therefore, scaling-up an ablative reactor could be costly.

9.2.1.5 Vacuum Pyrolysis

Short residence time for vapors can be easily achieved by applying a vacuum in the reactor (Figure 9.6e). Here, secondary reactions are normally avoided by the rapid withdrawal of pyrolytic vapors from the pyrolysis zone. In this configuration, biomass is fed to the top of the reactor, which contains stacked hot circular plates. Biomass will be pyrolyzed when it contacts with hot plates while dropping from the top to bottom plate. The process occurs normally at a pressure of 15 kPa.[90] Although vapors have a very small residence time, the long residence time of biomass can be independently achieved. However, the reactor has poor heat- and mass-transfer rates, and it occupies a large amount of space.[96] In addition, the process is very expensive because of the need for a vacuum. Liquid yield will be typically lower than for previous configurations discussed.

9.2.1.6   Auger Reactor

An auger reactor (Figure 9.6f) is simple in operation and can be used for continuous processing. It does not require a carrier gas, and the mode of heat transfer is mainly due to conduction. Biomass particles are moved by an auger inside an externally heated, cylinder tube. In some designs, hot sand is used as a medium for heat transfer. The residence time of vapors can be controlled by the auger speed and the heated zone in the tube. The energy cost for the operation is very low compared to other designs.[103] Studies have reported the production of bio-oil using laboratory-scale auger reactors.[104–106] The major weakness of this configuration is the mechanical wear due to moving parts.

Entrained-flow reactors have also been tested for fast pyrolysis. Since it is difficult to attain the high heat-transfer rate from hot carrier gas to biomass, this configuration has not been developed further.[86] Some studies have also reported fast pyrolysis in fixed-bed reactors.[107,108] Fixed-bed reactors operate normally in batch mode, and a sweep gas is applied for the removal of gas and vapors from the reactor.[109] The major drawback of this configuration is the high residence time of vapors and therefore high char yield. Free-fall reactors comprise another class of reactors; they are simple in design and easy to operate. Moving parts and carrier gas are not required in this design.[110]

DynaMotive, KiOR, Ensyn, BTG-BTL, and Renewable Oil International are the major companies involved in bio-oil production through fast pyrolysis.[86] Ensyn Technologies Inc.'s fast pyrolysis process is patented as Rapid Thermal Processing (RTP). The reactor configuration for this technique is analogous to that of the circulating-fluidized-bed reactor. Their first commercial plant was built in 1984 and seven plants have been operated since then.[86,111] UOP, a Honeywell company, was awarded $25 million in 2010 by the U.S. Department of Energy (DOE) to build a demonstration unit in Hawaii to convert cellulosic biomass into green transportation fuels using fast pyrolysis technology. The plant will employ Ensyn's RTP process to produce bio-oil from biomass, and the plant is expected to be running in 2014 at the Tesoro Corp. refinery in Kapolei, Hawaii.[112] Ensyn Technologies Inc. and Tolko Industries Ltd. also announced in June 2010 that they have formed a partnership to build the world's largest, commercial, fast pyrolysis plant in High Level, Alberta, Canada. The plant will process 400 dry tons of biomass per day and will produce about 22.5 million gallons of bio-oil per year.[113] Dynamotive Technologies Corporation, another major company that conducts fast pyrolysis of biomass, was incorporated in 1991.[114] KiOR, another fast pyrolysis company, has demonstrated a catalytic pyrolysis plant, which is located just outside of Houston, Texas. The company has produced up to 15 barrels of bio-oil (named as renewable crude) per day from wood chips.[115] KiOR has a plan to build several plants in Mississippi and has already secured a loan from the Mississippi State Governor's Office. A spin-off company associated with the University of Massachusetts is named Anellotech. The goal of this company is to produce aromatic hydrocarbons from biomass using fast pyrolysis. The company has projected to produce about 50 gallons of chemicals per metric ton of biomass (wood), with a yield of 40 per cent.[116]

9.2.2 Pyrolysis Mechanism and Pathways

When biomass is heated, its constituents thermally degrade and vaporize. The thermal decomposition rates of biomass constituents differ. All three components (cellulose, hemicellulose, and lignin) of biomass are chemically stable until 150 °C. Numerous studies have been published on cellulose pyrolysis, and a number of reaction pathways and kinetic models have been proposed for cellulose decomposition.[117–124] One of the earlier works by Kilzer and Broido proposed a pathway for pyrolysis of cellulose that includes three distinct processes–dehydration, depolymerization, and decomposition.[117] Another major study on the cellulose pyrolysis mechanism has shown the formation of "active cellulose", which is further decomposed to volatiles, char, and gas.[123] The proposed Broido–Shafizadeh model was accepted for many years but has been subjected to further evaluation.[125] As the temperature and heating rate increased, the complexity of pyrolysis increased, and a variety of products were created.[119,120,126] One vital mechanism of fast pyrolysis of cellulose, proposed by Piskorz *et al.*,[120] is known as the Waterloo model (Figure 9.7). The Waterloo

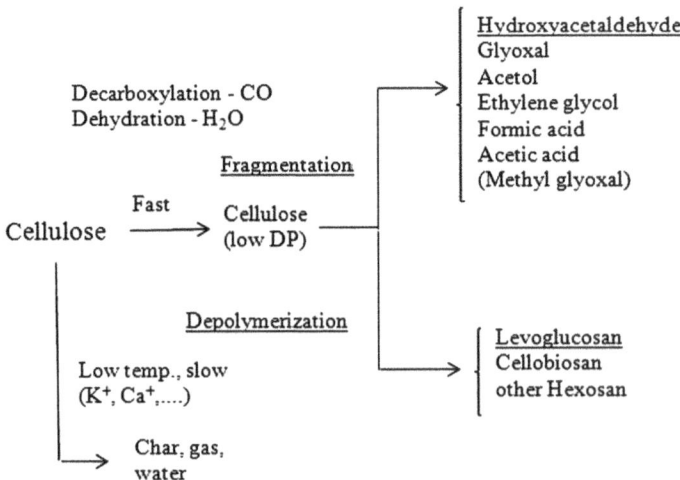

Figure 9.7   Waterloo model for cellulose pyrolysis.

model suggested two parallel pathways leading to a variety of products from fast pyrolysis of cellulose. A major intermediate product in cellulose pyrolysis is levoglucosan. However, these pyrolysis mechanisms will change with the change in heating rate, temperature and presence of impurities. A later study on mechanisms of cellulose pyrolysis suggested that the anhydro monosaccharide can be repolymerized to anhydro-oligomers or undergo further reactions like fragmentation or retro-aldol condensation, dehydration, decarbonylation, or decarboxylation to transform other volatile compounds. They proposed that the majority of char is formed not by the parallel pathway from "active cellulose" but by the repolymerization of volatile anhydro sugars and fragmented species.[127] Another study involving the Broido–Shafizadeh model showed that the char formation is exothermic, while its competing reaction (formation of anhydro sugars) is endothermic. Since char formation has a lower activation energy than the anhydro sugars formation, char yield decreases with the increase in temperature.[128]

Hemicellulose degrades at a lower temperature (150–300 °C) than cellulose. The degradation behavior of hemicellulose has been less studied because of the lower abundance and variety in structures.[129–132] Most of the thermal behavior studies of hemicellulose are limited to xylan (a major component in plant hemicellulose). Unlike the levoglucosan intermediate in cellulose pyrolysis, xylan produces furan derivatives as intermediate compounds before complete decomposition.[133] Antal *et al.*[134] proposed a mechanism for the degradation of xylan to furfural. Lignin is, thermally, the most stable component in biomass and is hydrophobic and aromatic in nature. There are three major monomers for lignin: p-coumaryl alcohol, coniferyl alcohol, and sinapyl alcohol. The two major techniques that can utilize lignin for biofuel production are gasification (discussed previously) and high-temperature fast pyrolysis. Even though there are some studies reported for lignin pyrolysis,[122,135–139] a well-defined reaction pathway for lignin has not been developed.

When holocellulose and lignin combine to form whole biomass, their pyrolytic behavior is influenced by other components. Therefore, it is difficult to propose a single reaction pathway for different types of lignocellulosic biomass. However, Evans and Milne[133] have proposed a reaction pathway for the thermal degradation of whole biomass. A detailed description of pyrolysis reaction mechanisms of biomass and its components can be found elsewhere.[140]

9.2.3 Bio-Oil Properties

Bio-oil is a dark brown liquid having a distinct odor. It has a higher heating value of around 17 MJ/kg and a water content of 25 wt.%. Compared to traditional biomass fuels such as black liquor or hog fuel, bio-oil has a higher energy density, which presents a much better opportunity for high-efficiency energy production. Typical properties of bio-oil in comparison with other conventional fuels are listed in Table 9.6. Bio-oil is a complex mixture of more than 300 compounds, containing alcohols, acids, aldehydes, ketones, esters, sugars, furans, phenols, guaiacols and other aromatics.[141,142] This complex

Table 9.6   Comparison of bio-oil, heavy fuel oil and black liquor.[146-150]

Properties		Bio-oil	Diesel	Heavy fuel oil	Black liquor
Specific gravity		1.2	0.85	0.96	*
Typical composition	% C	55–58	86.3	86.1	34–39
	% H	5.5–7	12.8	11.8	3–5
	% O	35–40	–	–	33–38
	% S	–	0.9	2.1	3–7
	% N	0–0.2	–	–	0.04–0.2
	Ash, wt%	0–0.2	0.01	<0.2	17.3–29
Viscosity	cP at 40 °C	40–100	1–3.5	340	*
Flash point	°C	66	60–80	>60	†
Pour point	°C	27	(–35)–(–15)	<30	†
Water	% wt	20–30	0.1	<1.0	>30
HHV	MJ/kg	16–19	45	40	14
Acidity	pH	2.5	–	–	12.5

'–' not significant; '*' changes with the composition; '†' not available.

Table 9.7   ASTM standard for bio-oil.

Property	Specification
Gross Heat of Combustion (MJ/kg)	15 min.
Water content (mass%)	30 max.
Solid content (mass%)	2.5 max.
Kinematic viscosity at 40 °C (mm^D/s)	125 max.
Density at 20 °C (kg/dm^3)	1.1–1.3
Sulfur content (mass%)	0.05 max.
Ash content (mass%)	0.25 max.
pH	Report
Flash point (°C)	45 min.
Pour point (°C)	–9 max.

nature of bio-oil provides potential and challenges for its applications. Since bio-oil is formed by rapid heating and cooling, the compounds in bio-oil are not in thermal equilibrium. Properties of bio-oil depend on the type of biomass and pyrolysis operating parameters such as temperature, heating rate, and residence time.[89,106,107,143–145]

Recently, an ASTM standard (ASTM D 7544–09) has been published for the commercial use of bio-oil.[151] The required physical properties of bio-oil, according to this ASTM standard, are listed in Table 9.7. A detailed discussion on the physical properties of bio-oil and their measurement techniques was presented elsewhere.[95]

9.2.4   Bio-Oil Applications

Bio-oil is a potential source for fuel and energy applications. In addition, many commodity chemicals can be extracted or produced from bio-oil. From the literature, final applications of bio-oil are summarized in Figure 9.8.[88,91] Piskorz and Radlein[152] conducted a biodegradability study of bio-oil by means

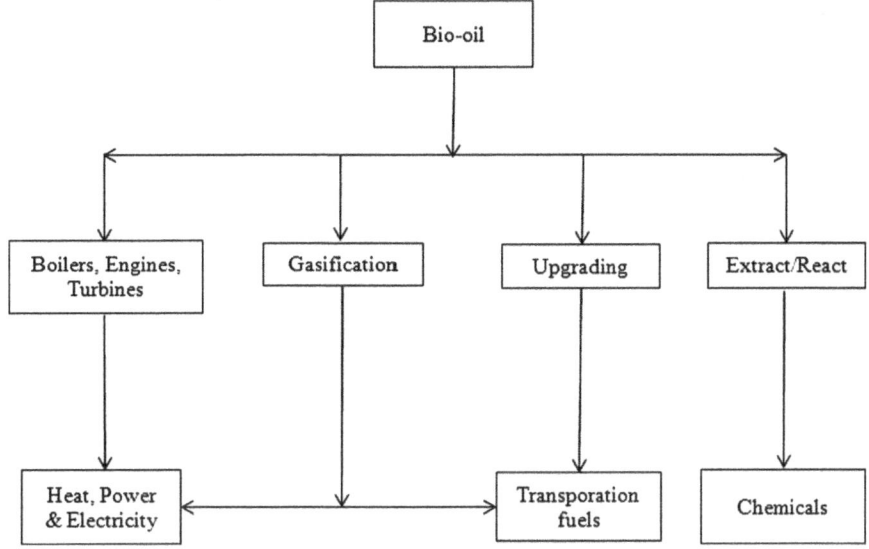

Figure 9.8 Applications of bio-oil.

of respirometry. The study showed that bio-oils degraded aerobically at a higher rate than hydrocarbon fuels, and this biodegradability rate was enhanced when bio-oil was neutralized for pH.

9.2.4.1 Combustion

Even though bio-oil has a high water content and low heating value, the combustion property of bio-oil has been tested for heat and power generations in engines, turbines, furnaces, and boilers. One advantage of bio-oil is the ease of handling liquid fuels, as opposed to solid or gas fuels, for combustion applications. Bio-oil is not easily combustible, and therefore the engine should be run first with conventional fuels before switching it over to bio-oil. One major problem of handling bio-oil is its high viscosity. Therefore, either the bio-oil delivery line has to be preheated before pumping or the viscosity has to be reduced by using some additives. Either way, bio-oil requires basic modifications to existing instruments and a supporting fuel for start up in combustion applications.[153] Emission of particulates are higher in bio-oil combustion than heavy fuel oil combustion; however, the emission of NO_x can be lowered in bio-oil combustion by increasing water content.[153] Even though CO emission from bio-oil combustion is high, it is acceptable (30–50 ppm).[154] Bio-oil has been tested for cofiring with coal at Manitowoc Public Utilities in Red Arrow, Wisconsin.[86,153] Ensyn[155] claims that their bio-oil can replace number 2 and 6 heating fuels for cofiring with natural gas and coal in a vast array of boiler applications. In addition, Orenda, a division of Magellan Aerospace, has developed a turbine with an efficiency range of 29% to 41% to generate 1 to 25 MW power from Ensyn bio-oil. Magellan's Orenda turbine can generate 2.5 MW power with Dynamotive's bio-oil.[114] The "silo" type combustion chamber in Orenda gas turbines can be easily modified and optimized to any fuel.[153] The major issues for bio-oil in turbines are acid corrosion and deposition on the combustion chamber and blades. Shihadeh and Hochgreb investigated fuel properties of bio-oil from two different sources and compared with No.2 diesel fuel in direct injection diesel engines. Thermal efficiencies of bio-oils were similar to those of the diesel fuel; however, bio-oils required greater ignition activation energies and had longer ignition delays. The longer ignition delay resulted from slow fuel–air mixing.[156] Chiaramonti *et al.*[157,158] conducted bioemulsions (bio-oil–diesel emulsions) for testing in diesel engines. These emulsions were more stable than normal bio-oil; however, they damaged the injectors and fuel pumps. Therefore, their study recommended special considerations in the design of injectors and fuel pumps.

9.2.4.2   Transportation Fuels

Bio-oil can be upgraded to a transportation fuel by removing its oxygen content. Gasoline-range liquid fuels are technically feasible from bio-oil. However, the economic feasibility of these techniques is under research.[159] The upgrading techniques are described in the next section.

9.2.4.3   Chemicals

Since bio-oil is a mixture of more than 300 chemicals, many commodity and specialty chemicals – such as pharmaceuticals and synthons, fertilizers, environmental chemicals and resins – can be prepared from bio-oil. Currently, the only commercial use of bio-oil reported is in the production of food flavors such as liquid smoke in the Red Arrows Products Co., Inc.[160] Chemicals that have been reported as recovered from bio-oil are phenol formaldehyde, calcium and/or magnesium acetate for biodegradable deicers, fertilizers, levoglucosan, levoglucosenone, hydroxyacetaldehyde (glycolaldehyde), and a range of flavorings and essences for the food industry.[88,153] Nonetheless, concentrations of all these compounds are fairly low, which makes the economics unfeasible. Bio-oil can be used as an insecticide or fungicide because of the presence of some phenolic compounds. Calcium salts produced by a reaction with carboxylic acids in the aqueous phase of bio-oil can be used as an environmentally friendly road de-icer.[153,161] Bio-oil can be converted to a nitrogen, slow-release fertilizer by a reaction with nitrogen sources like ammonia, urea, and proteinaceous materials like manure.[162] The carbonyl group in bio-oil reacts with any $-NH_2$ source and nitrogen will be converted to a stable, biodegradable organic form, which can be used as a slow, nitrogen-releasing fertilizer. An economical analysis of a plant using fertilizer (containing 10% N) produced from bio-oil showed that the capital cost for the plant of capacity 20 000 t/yr was $3.36 million.[163] The carboxylic acids and phenols in bio-oil can react with lime to form calcium salts and phenates (themolabilealkane earth compounds), which is called "BioLime". When this compound is introduced into flue gas, it decomposes at flue-gas temperatures and react with SO_2 in the flue gas, neutralizing the gas.[164] BioLime can remove more than 90% of sulfur dioxide from flue gas.[153]

Hydrogen, the primary energy carrier for the future, can be produced from bio-oil. Several studies have been conducted to produce hydrogen from the aqueous phase of bio-oil.[165,166] Hydrogen can be produced from bio-oil by steam reforming, partial oxidation, autothermal, aqueous phase, and supercritical water reforming. Most of the studies were based on steam reforming of the aqueous phase of bio-oil. The amount of hydrogen that can be produced from bio-oil depends on its hydrogen content. However, on average, two moles of hydrogen can be produced from one mole of carbon in bio-oil if water-gas shift reaction is also considered. Previous studies have shown the hydrogen yield was more than 80% of its theoretical yield.[167–169] The major challenge in hydrogen production from bio-oil is the catalyst deactivation due to coking. Therefore, most of these studies were focused on developing new catalysts for continuous reforming of hydrogen from an aqueous fraction of bio-oil.[167–171]

In addition to the above applications, bio-oil has the potential for other end-use markets. The sticky, resin-like quality of bio-oil means it can be utilized as an asphalt binder. Therefore, bio-oil can substitute for some petroleum products in asphalt emulsions for concrete paving.[86] Another potential application of bio-oil is in coal-dust suppression. The environmental and physical requirements of bio-oil – such as biodegradability, water immiscibility and the strength of its polymerization reactions – have to be analyzed in order to coat the coal piles.[86] Ensyn has produced a commercial resin called "MNRP" to replace 50% phenol in the manufacture of phenol formaldehyde and V-additive (plasticizer and emulsifier) from bio-oil.[155]

9.2.5 Bio-Oil Upgrading

Certain properties of bio-oil such as high density, high viscosity, high acidity, high water and oxygen content and low heating value negatively affect the fuel properties of bio-oil. In order to improve the fuel quality, bio-oil has to be completely deoxygenated. There are basically two different techniques to upgrade bio-oil as an end-use fuel: hydrotreating to produce alkane fuels characterized as CH_2, and zeolite-cracking to produce aromatic fuels characterized as $CH_{1.2}$.[140] Both hydrotreating and catalyst cracking were performed for pyrolytic vapors also. A new approach has been published for the production of commodity chemicals by upgrading bio-oil that involves the hydroprocessing of bio-oils over metal catalysts, followed by zeolite cracking.[159] Another technique that is being studied is catalytic pyrolysis, in which gasoline-range, hydrocarbon fuels can be recovered directly from biomass by introducing a catalyst during pyrolysis.[172] In catalytic pyrolysis, either the biomass is fed to a catalyst bed reactor for simultaneous pyrolysis and cracking, or a catalyst bed is inserted above the inert bed so that the vapor passes through the catalyst bed just after pyrolysis on the inert bed.[173,174] A detailed review of the catalytic hydroprocessing of bio-oil can be seen elsewhere.[9,175]

9.2.5.1 Hydrotreating

In hydrodeoxygenation (HDO), hydrogen reacts with oxygenated compounds in bio-oil under high pressure and moderate temperature (350–500 °C) in the presence of a heterogeneous catalyst to produce water and hydrocarbon. Normally this process is carried out in two steps. In the first step, bio-oil is treated at around 250 °C followed by conventional hydrodeoxygenation at a higher temperature and high pressure. The initial step is for stabilization in order to avoid polymerization reactions and coke formation. This step consumes a lower amount of hydrogen than the second step.[140] The second step involves HDO at higher temperature and complete deoxygenation. The catalysts normally used for the HDO process is sulfided NiMo or CoMo supported on Al_2O_3. The HDO process for a bio-oil of chemical formula $C_xH_yO_z$ can be simplified as in eqn (9.18).

$$C_xH_yO_z + (x + z - y/2)H_2 \rightarrow x\,CH_2 + z\,H_2O \tag{9.18}$$

Hydrodeoxygenation of bio-oil has been proved technically but not economically feasible. The conventional HDO catalysts are very expensive, and the high char content in bio-oil and coke formation shorten the life of these catalysts. Thermally unstable compounds in bio-oil lead to the coking of catalysts. Therefore, the regeneration of catalysts has to be done more frequently but this makes the process expensive. Another challenge with using conventional catalysts is the chance of stripping sulfur from the catalysts, increasing the sulfur content in the bio-oil. Nowadays, the research on hydrotreating of bio-oil has concentrated on the utilization of heterogeneous, noble-metal catalysts and transition-metal catalysts.[176,177] Another approach in HDO is the utilization of hydrogen-donor solvents, which also helps to decrease the viscosity of bio-oil. An upgrading of hemicellulose-derived compounds to produce jet and diesel fuels was also reported.[178,179] Establishment of new catalytic systems and optimization of process parameters is required for the economical development of HDO of bio-oil. Many studies have been carried out utilizing different or modified catalysts for hydrodeoxygenation of bio-oil or its model compounds.[160,180] However, none of this work has been successful in controlling the coke deposition to make the process a continuous operation.

9.2.5.2 Catalytic Cracking

In catalytic cracking, bio-oil compounds are deoxygenated in the presence of some shape-selective catalysts such as zeolites. Zeolite, ZSM-5, catalysts are acidic, shape selective, and highly active, converting oxygenated compounds to

C1–C10 hydrocarbons.[181] Deoxygenation with zeolite catalysts can be done at atmospheric pressure and does not require the supply of hydrogen. In zeolite cracking, the oxygen is removed as carbon dioxide and water. Cracking reactions involve the rupture of C–C bonds associated with dehydration, decarboxylation, and decarbonylation. The reaction can be simplified as shown in eqn (9.19).

$$\left(C_x H_y O_z\right)_n \rightarrow a.\,CH_{1.2} + b.\,H_2O + c.\,CO_2 \qquad (9.19)$$

Here a, b, c depends on x, y, and z. The aromatic yield is limited by the hydrogen available in the bio-oil. Bridgwater suggested the use of a multi-functional catalyst that can operate in a carbon-limited environment for cracking of bio-oil.[140] In a carbon-limited environment, the water-gas shift reaction is possible for the production of hydrogen that can be utilized during cracking. In some studies, catalytic cracking of bio-oil components has been carried out for pyrolytic vapors before condensing as bio-oil. When the catalysts were used for cracking, a high reduction in the oil yield was observed but the bio-oil contained less oxygen.[173]

9.2.5.3 Catalytic Pyrolysis

Catalytic pyrolysis is another approach to get deoxygenated, liquid fuel from biomass through pyrolysis. Most of the catalytic pyrolysis studies were conducted with zeolite catalysts;[111,182–185] in addition, some mesoporous catalysts such as Al-MCM-41, Al-MSU-F [186–189] were utilized. In the case of both types of catalysts, as the Si/Al increased, the quality of bio-oil also increased. Catalytic pyrolysis can produce bio-oil with very low oxygen content. However, a high reduction in liquid yield was noted because oxygen (40–50% in the elemental composition of biomass) was being removed. Both fluidized-bed and fixed-bed reactors can be utilized with catalyst particles as bed material. French and Czernik[182] studied catalytic pyrolysis with different types of zeolite catalysts and metal-impregnated zeolite catalysts; ZSM-5 group catalysts performed better than other zeolite catalysts. Microcatalytic pyrolysis studies by Carlson *et al.*[172,183,190] reported aromatics yield of 30% (on carbon basis) from glucose using ZSM-5. A study conducted by Lappas *et al.*[98] utilized different catalysts, such as a fluid-catalytic cracking (FCC) catalyst and a ZSM-5 additive, as bed materials for fluidization in a circulating-fluidized-bed reactor (CFB) where catalysts were regenerated and circulated continuously. Catalytic biomass pyrolysis led to the production of additional water, coke, and gases compared to conventional pyrolysis; however, the liquid was more stable in catalytic pyrolysis. A detailed description of catalytic pyrolysis in a pilot plant is provided elsewhere,[191] and that pilot study provided results similar to those in the study of Lappas *et al.* In most of the catalytic pyrolysis studies, catalysts particles are used as bed material instead of sand. However, in some cases, the catalyst particles and inert particles have been kept side-by-side so that biomass entered into the inert bed also passed through the catalyst

bed.[186–188] Iliopoulou *et al.*[186] achieved a similar yield of bio-oil for both catalytic and noncatalytic (Al-MCM-41 catalyst) pyrolysis, and the aromatic yield
was very high in the catalytic pyrolysis. Catalytic pyrolysis is a promising route
to produce aromatic hydrocarbons from biomass within a single-step process.
Currently, catalytic pyrolysis studies are mainly focused on modifying existing
catalysts or introducing new catalytic systems to reduce coke, water, and gas
formation, and increase aromatic yield.

9.2.6 Pyrolysis of Lignin

Pyrolysis studies on black liquor are rare compared to other biomass residues;
however, there are some studies on pyrolysis of black liquor.[150,192–195] Most of
these studies were performed at higher temperatures ($>700\,°C$) than typical
biomass pyrolysis (400–600 °C), and studies were mainly focused on the char
yield and composition of gas rather than bio-oil.[150,193,194] A high char yield
($>50\%$) was found for most of the studies because of high lignin and inorganic
content.[150,193] A study on pyrolysis (620–740 °C) of different fractions of black
liquor (recovered from vacuum distillation of black liquor) showed that
although black liquor solids had higher organic content (around 70%) than
that of the black liquor (around 39%), the char yield (around 50%) was relatively higher than the tar yield (around 35%).[193] A study on pyrolytic behavior
of black liquor showed that the thermal decomposition of organic matters
occurs below 550 °C, whereas decomposition of inorganic matters occurs above
550 °C.[195,196] A kinetic study on pyrolytic behavior of straw black liquor solids
showed the first-order decomposition for lower heating rates, but this first-
order relationship was not observed above a heating rate of 60 °C/min.[192]

9.2.7 Economical Analysis

Some economic studies are available for bio-oil production and its upgrading
as a surrogate of transportation fuels.[99,197–201] Ringer *et al.*[99] conducted a
detailed economic analysis of a pyrolysis plant with a capacity of 550 tons (50%
moisture content) per day. The total project investment was calculated as
$48.29 million with a total operating cost of $9.6 million, and the selling price of
the bio-oil was projected to be $7.62/GJ, LHV, and the bio-oil yield was
assumed to be 60 wt.% of dry biomass.[99]

Another economic study was conducted by the Pacific Northwest National
Laboratory (PNNL) for the production of 76 million gallons/year of gasoline
and diesel through fast pyrolysis.[197] In this study, the size of the plant was taken
as 2000 dry metric tons/day of hybrid poplar wood chips. The process included
six major steps, which were feed drying and size reduction, fast pyrolysis,
hydrotreating bio-oil to get hydrocarbon oil with $<2\%$ oxygen liquid, hydrocracking of a heavy portion of the hydrocarbon oil, distillation of hydrotreated
and hydrocracked oil to diesel and gasoline blends, and steam reforming of

off-gas to produce hydrogen. The minimum selling price for the fuels from this study was $2.04/gallon gasoline equivalent.

Recently, the National Renewable Energy Laboratory (NREL) conducted an economic analysis of the production of transportation fuels (naphtha range and diesel range) from corn stover biomass through fast pyrolysis.[198] In this study, two scenarios were used for bio-oil upgrading. One was a hydrogen production scenario where the hydrogen required for bio-oil upgrading would be produced from reforming a portion of the aqueous phase of bio-oil. The second scenario involved purchasing the hydrogen required for bio-oil upgrading. The study proposed to process 2000 MT/day of corn stover (25 wt.% moisture content and 10 to 25 mm in size). In the first scenario, the estimated production of fuel was 35.4 million gallons/yr. The capital expenditure for the combined pyrolysis and upgrading plant was estimated as $287 million, and the competitive product value was $3.09/gallon of gasoline equivalent. In the second scenario, the estimated production of fuel was 58.2 million gallons/yr. The capital expenditure for the combined pyrolysis and upgrading plant was estimated as $200 million, and the competitive product value was $2.11/gallon.[198,199] The cost of feedstock was estimated to be $ 83/ MT, the cost of electricity was projected as $0.054/kWh, and the catalyst replacement cost as $1.77 million/year.

References

1. H. L. Chum and R. P. Overend, Biomass and renewable fuels, *Fuel Processing Technology*, 2001, **71**(1–3), 187–195.
2. R. D. Perlack, L. L. Wright, A. F. Turhollow, R. L. Graham, B. J. Stokes, and D. C. Erbach, *Biomass as Feedstock for a Bioenergy and Bioproducts Industry: the Technical Feasibility of a Billion-ton Annual Supply*; Oak Ridge National Laboratory: Oak Ridge, 2005.
3. S. P. Babu, Thermal gasification of biomass technology developments: End of task report for 1992 to 1994. *Biomass and Bioenergy*, 1995, **9**, (1–5), 271–285.
4. A. Demirbas, Biomass resource facilities and biomass conversion processing for fuels and chemicals, *Energy Conversion and Management*, 2001, **42**(11), 1357–1378.
5. A. Demirbas, Biofuels sources, biofuel policy, biofuel economy and global biofuel projections, *Energy Conversion and Management*, 2008, **49**(8), 2106–2116.
6. K. B. Cantrell, T. Ducey, K. S. Ro and P. G. Hunt, Livestock waste-to-bioenergy generation opportunities, *Bioresource Technology*, 2008, **99**(17), 7941–7953.
7. H. Boerrigter and R. Rauch, *Syngas Production and Utilization; Review of Applications of Gases from Biomass gasification*. Biomass technology group (BTG): Netherland, 2005.

8. *New Renewable Energy Resources: A Guide to the Future.* Kogan Page: London, 1994.

9. G. W. Huber, S. Iborra and A. Corma, Synthesis of transportation fuels from biomass: chemistry, catalysts, and engineering, *Chemical Reviews*, 2006, **106**(9), 4044–4098.

10. A. V. Bridgwater, Renewable fuels and chemicals by thermal processing of biomass, *Chemical Engineering Journal*, 2003, **91**(2–3), 87–102.

11. A. Demirbas, Combustion characteristics of different biomass fuels, *Progress in Energy and Combustion Science*, 2004, **30**(2), 219–230.

12. G. J. Stiegel and R. C. Maxwell, Gasification technologies: the path to clean, affordable energy in the 21st century, *Fuel Processing Technology*, 2001, **71**(1–3), 79–97.

13. A. Demirbas, Biofuels securing the planet's future energy needs, *Energy Conversion and Management*, 2009, **50**(9), 2239–2249.

14. A. Demirbas, Progress and recent trends in biofuels, *Progress in Energy and Combustion Science*, 2007, **33**(1), 1–18.

15. B. M. Jenkins, L. L. Baxter, T. R. Miles Jr and T. R. Miles, Combustion properties of biomass, *Fuel Processing Technology*, 1998, **54**(1–3), 17–46.

16. P. Carlsson, H. Wiinikka, M. Marklund, C. Grönberg, E. Pettersson, M. Lidman and R. Gebart, Experimental investigation of an industrial scale black liquor gasifier. 1. The effect of reactor operation parameters on product gas composition, *Fuel*, 2010, **89**, 4025–4034.

17. R. L. Bain and K. Broer, Gasification. In *Thermochemical Processing of Biomass: Conversion into Fuels, Chemicals and Power*, R. C. Brown, ed. John Wiley & Sons, 2011, pp. 47–74.

18. R. Warnecke, Gasification of biomass: comparison of fixed bed and fluidized bed gasifier, *Biomass and Bioenergy*, 2000, **18**(6), 489–497.

19. H. A. M. Knoef, *Inventory of Biomass Gasifier Manufacturers and Installations, Final Report to European Commission, Contract DIS/1734/98-NL*, Biomass Technology Group B.V., University of Twente: Enschede, Netherland, 2000.

20. A. A. C. M. Beenackers, Biomass gasification in moving beds, a review of European technologies, *Renewable Energy*, **16**(1–4), 1180–1186.

21. T. A. Milne, and R. J. Evan, *Biomass Gasification 'Tars'; their Nature, Formation and Conversion*, NREL: Golden, 1998.

22. P. Basu, *Combustion and Gasification in Fluidized Beds.* Taylor and Francis Group, LLC: Boca Raton, FL, 2006.

23. A. Gomez-Barea and B. Leckner, Modeling of biomass gasification in fluidized bed, *Progress in Energy and Combustion Science*, 2010, **36**, 444–509.

24. M. Dogru, A. Midilli and C. R. Howarth, Gasification of sewage sludge using a throated downdraft gasifier and uncertainty analysis, *Fuel Processing Technology*, 2002, **75**(1), 55–82.

25. P. McKendry, Energy production from biomass (part 3): gasification technologies. *Bioresource Technology* 2002, **83**, (1), 55–63.

26. S. Lee, *Alternative Fuels.* Taylor & Francis: Washington, 1996.

27. R. C. Brown, *Biorenewable Resources-Engineering New Products from Agriculture*. 1st edn, Iowa State Press: Ames, Iowa, 2003.
28. J. M. Prins, Thermodynamic analysis of biomass gasification and torrefaction. PhD thesis, Eindhoven University of Technology, Netherlands, 2005.
29. P. C. Roy, A. Datta and N. Chakraborty, Modelling of a downdraft biomass gasifier with finite rate kinetics in the reduction zone, *International Journal of Energy Research*, 2009, **33**(9), 833–851.
30. P. McKendry, Energy production from biomass (part 1): overview of biomass. *Bioresource Technology*, 2002, **83**, (1), 37–46.
31. Z. A. Zainal, R. Ali, C. H. Lean and K. N. Seetharamu, Prediction of performance of a downdraft gasifier using equilibrium modeling for different biomass materials, *Energy Conversion and Management*, 2001, **42**, 1499–1515.
32. A. Melger, J. F. Perez, H. Laget and A. Horillo, Thermochemical equilibrium modeling of a gasifying process, *Energy Conversion and Management*, 2007, **48**, 59–67.
33. A. K. Sharma, Equilibrium modeling of global reduction reactions for a downdraft (biomass) gasifier, *Energy Conversion and Management*, 2008, **49**(4), 832–842.
34. C. R. Altafini, P. R. Wander, and R. M. Barretoa, Prediction of the working parameters of a wood waste gasifier through an equilibrium model *Energy Conversion and Management*, 2003, **44**(17), 2763–2777.
35. D. Shin and S. Choi, The combustion of simulated waste particles in a fixed bed, *Combustion and Flame*, 2000, **121**(1–2), 167–180.
36. M. Dogru, C. R. Howarth, G. Akay, B. Keskinler and A. A. Malik, Gasification of hazelnut shells in a downdraft gasifier, *Energy*, 2002, **27**(5), 415–427.
37. T. B. Reed and A. Das, *Handbook of Biomass Downdraft Gasifier Engine Systems*. SERI: Golden, CO, 1988.
38. G. Gautam, S. Adhikari and S. Bhavnani, Estimation of biomass synthesis gas using equilibrium modeling, *Energy & Fuels*, 2010, **24**, 2692–2698.
39. T. H. Jayah, L. Aye, R. J. Fuller and D. F. Stewart, Computer simulation of a downdraft wood gasifier for tea drying, *Biomass and Bioenergy*, 2003, **25**(4), 459–469.
40. J. K. Ratnadhariya and S. A. Channiwala, Three zone equilibrium and kinetic free modeling of biomass gasifier - a novel approach, *Renewable Energy*, 2009, **34**(4), 1050–1058.
41. K. J. Ptasinski, M. J. Prins and A. Pierik, Exergetic evaluation of biomass gasification, *Energy*, 2007, **32**(4), 568–574.
42. P. N. Sheth and B. V. Babu, Experimental studies on producer gas generation from wood waste in a downdraft biomass gasifier, *Bioresource Technology*, 2009, **100**(12), 3127–3133.
43. I. Narvaez, A. Orio, M. P. Aznar and J. Corella, Biomass gasification with air in an atmospheric bubbling fluidized bed. Effect of six operational

variables on the quality of the produced raw gas, *Industrial & Engineering Chemistry Research*, 1996, **35**(7), 2110–2120.

44. X. T. Li, J. R. Grace, C. J. Lim, A. P. Watkinson, H. P. Chen and J. R. Kim, Biomass gasification in a circulating fluidized bed, *Biomass and Bioenergy*, 2004, **26**(2), 171–193.

45. X. Li, J. R. Grace, A. P. Watkinson, C. J. Lim and A. Ergüdenler, Equilibrium modeling of gasification: a free energy minimization approach and its application to a circulating fluidized bed coal gasifier, *Fuel*, 2001, **80**(2), 195–207.

46. C. Hanping, L. Bin, Y. Haiping, Y. Guolai and Z. Shihong, Experimental Investigation of Biomass Gasification in a Fluidized Bed Reactor, *Energy & Fuels*, 2008, **22**(5), 3493–3498.

47. V. Skoulou, G. Koufodimos, Z. Samaras and A. Zabaniotou, Low temperature gasification of olive kernels in a 5-kW fluidized bed reactor for H_2-rich producer gas, *International Journal of Hydrogen Energy*, 2008, **33**(22), 6515–6524.

48. P. J. van den Enden and E. S. Lora, Design approach for a biomass fed fluidized bed gasifier using the simulation software CSFB, *Biomass and Bioenergy*, 2004, **26**(3), 281–287.

49. A. van der Drift, J. van Doorn and J. W. Vermeulen, Ten residual biomass fuels for circulating fluidized-bed gasification, *Biomass and Bioenergy*, 2001, **20**(1), 45–56.

50. Z. A. Zainal, A. Rifau, G. A. Quadir and K. N. Seetharamu, Experimental investigation of a downdraft gasifier, *Biomass and Bioenergy*, 2002, **23**(4), 283–289.

51. V. Skoulou, A. Zabaniotou, G. Stavropoulos and G. Sakelaropoulos, Syngas production from olive tree cuttings and olive kernels in a downdraft fixed-bed gasifier, *International Journal of Hydrogen Energy*, 2008, **33**(4), 1185–1194.

52. E. Natarajan, A. Nordin and A. N. Rao, Overview of combustion and gasification of rice husk in fluidized bed reactors, *Biomass and Bioenergy*, 1998, **14**(5–6), 533–546.

53. J. Gil, J. Corella, M. P. Aznar and M. A. Caballero, Biomass gasification in atmospheric and bubbling fluidized bed: effect of the type of gasifying agent on the product distribution, *Biomass and Bioenergy*, 1999, **17**, 389–403.

54. A. van der Drift and J. van Doom, Effect of Fuel Size and Process Temperature on Fuel Gas Quality from CFB Gasification of Biomass. In *Progress in Thermochemical Biomass Conversion*, A. V. Bridgwater, ed. 2001, Blackwell Science Ltd, Oxford, pp. 265–271.

55. J. Corella, J. M. Toledo and G. Molina, Calculation of the conditions to get less than 2 g tar/mn3 in a fluidized bed biomass gasifier, *Fuel Processing Technology*, 2006, **87**(9), 841–846.

56. C. Wu, X. Yin, L. Ma, Z. Zhou and H. Chen, Design and Operation of A 5.5 MWe Biomass integrated gasification combined cycle demonstration plant, *Energy & Fuels*, 2008, **22**(6), 4259–4264.

57. Z. Wu, C. Wu, H. Huang, S. Zheng and X. Dai, Test results and operation performance analysis of a 1-MW biomass gasification electric power generation system, *Energy & Fuels*, 2003, **17**(3), 619–624.
58. R. C. Saxena, D. Seal, S. Kumar and H. B. Goyal, Thermo-chemical routes for hydrogen rich gas from biomass: A review, *Renewable and Sustainable Energy Reviews*, 2008, **12**(7), 1909–1927.
59. A. Zabaniotou, O. Ioannidou and V. Skoulou, Rapeseed residues utilization for energy and 2nd generation biofuels, *Fuel*, 2008, **87**(8–9), 1492–1502.
60. P. R. Wander, C. R. Altafini and R. M. Barreto, Assessment of a small sawdust gasification unit, *Biomass and Bioenergy*, 2004, **27**(5), 467–476.
61. A. Rogel and J. Aguillón, The 2D Eulerian approach of entrained flow and temperature in a biomass stratified downdraft gasifier, *American Journal of Applied Sciences*, 2006, **3**(10), 2068–2075.
62. M. M. Küçük and A. Demirbas, Biomass conversion processes, *Energy Conversion and Management*, 1997, **38**(2), 151–165.
63. P. Weerachanchai, M. Horio and C. Tangsathitkulchai, Effects of gasifying conditions and bed materials on fluidized bed steam gasification of wood biomass, *Bioresource Technology*, 2009, **100**(3), 1419–1427.
64. S. R. A. Kersten, W. Prins, A. van der Drift and W. P. M. van Swaaij, Experimental fact-finding in CFB biomass gasification for ECN's 500 kWth pilot plant, *Industrial & Engineering Chemistry Research*, 2003, **42**(26), 6755–6764.
65. J. Corella and A. Sanz, Modeling circulating fluidized bed biomass gasifiers. A pseudo-rigorous model for stationary state, *Fuel Processing Technology*, 2005, **86**(9), 1021–1053.
66. M. J. F. Llorente and J. E. C. García, Comparing methods for predicting the sintering of biomass ash in combustion, *Fuel*, 2005, **84**, 1893–1900.
67. X. L. Yin, C. Z. Wu, S. P. Zheng and Y. Chen, Design and operation of a CFB gasification and power generation system for rice husk, *Biomass and Bioenergy*, 2002, **23**(3), 181–187.
68. C. Ryu, Y. B. Yang, A. Khor, N. E. Yates, V. N. Sharifi and J. Swithenbank, Effect of fuel properties on biomass combustion: Part I. Experiments--fuel type, equivalence ratio and particle size, *Fuel*, 2006, **85**(7–8), 1039–1046.
69. L. Gerun, M. Paraschiv, R. Vîjeu, J. Bellettre, M. Tazerout, B. Gøbel and U. Henriksen, Numerical investigation of the partial oxidation in a two-stage downdraft gasifier, *Fuel*, 2008, **87**(7), 1383–1393.
70. K. M. Bryden and K. W. Ragland, Numerical modeling of a deep, fixed bed combustor, *Energy & Fuels*, 1996, **10**(2), 269–275.
71. A. Saravanakumar, T. M. Haridasan, T. B. Reed and R. K. Bai, Experimental investigation and modelling study of long stick wood gasification in a top lit updraft fixed bed gasifier, *Fuel*, 2007, **86**(17–18), 2846–2856.
72. R. Bilbao, A. Millera and M. B. Murillo, Temperature profiles and weight loss in the thermal decomposition of large spherical wood particles. *Industrial & Engineering Chemistry Research* 1993, **32**, (9), 1811–1817.

73. D. M. Earp, The gasification of biomass in a downdraft reactor. Aston University, UK, 1988.
74. A. K. Sharma, Modeling fluid and heat transport in the reactive, porous bed of downdraft (biomass) gasifier, *International Journal of Heat and Fluid Flow*, 2007, **28**(6), 1518–1530.
75. F. V. Tinaut, A. Melgar, J. F. Pérez and A. Horrillo, Effect of biomass particle size and air superficial velocity on the gasification process in a downdraft fixed bed gasifier. An experimental and modelling study, *Fuel Processing Technology*, 2008, **89**(11), 1076–1089.
76. Y. B. Yang, V. N. Sharifi and J. Swithenbank, Effect of air flow rate and fuel moisture on the burning behaviours of biomass and simulated municipal solid wastes in packed beds, *Fuel*, 2004, **83**(11–12), 1553–1562.
77. D. Y. C. Leung and C. L. Wang, Fluidized-bed gasification of waste tire powders, *Fuel Processing Technology*, 2003, **84**(1–3), 175–196.
78. S. Valin, S. Ravel, J. Guillaudeau, and S.Thiery, Comprehensive study of the influence of total pressure on products yields in fluidized bed gasification of wood sawdust. *Fuel Processing Technology*, 2010, **91**(10), 1222–1228.
79. A. V. Bridgwater, The technical and economic feasibility of biomass gasification for power generation, *Fuel*, 1995, **74**, 631–653.
80. C. Pfeifer, R. Rauch and H. Hofbauer, In-bed catalytic tar reduction in a dual fluidized bed biomass steam gasifier, *Industrial & Engineering Chemistry Research*, 2004, **43**(7), 1634–1640.
81. Z. Abu El-Rub, E. A. Bramer and G. Brem, Review of catalysts for tar elimination in biomass gasification processes, *Industrial & Engineering Chemistry Research*, 2004, **43**(22), 6911–6919.
82. L. Devi, K. J. Ptasinski and F. J. J. G. Janssen, A review of the primary measures for tar elimination in biomass gasification processes, *Biomass and Bioenergy*, 2003, **24**(2), 125–140.
83. V. Sricharoenchaikul, W. J. FrederickJr and P. Agrawal, Carbon distribution in char residue from gasification of kraft black liquor, *Biomass and Bioenergy*, 2003, **25**, 209–220.
84. M. Naqvi, J. Yan and E. Dahlquist, Black liquor gasification integrated in pulp and paper mills: A critical review, *Bioresource Technology*, 2010, **101**, 8001–8015.
85. V. Subramani and S. K. Gangwal, A review of recent literature to search for an efficient catalytic process for the conversion of syngas to ethanol, *Energy & Fuels*, 2008, **22**, 814–839.
86. I. H. Farag, C. E. LaClair, and C. J. Barrett, *Technical, Environmental and Economic Feasibility of Bio-Oil in New Hampshire's North Country*; New Hampshire Industrial Research Center (NHIRC): Durham, NH, 2002, pp. 1–95.
87. P. C. Badger and P. Fransham, Use of mobile fast pyrolysis plants to densify biomass and reduce biomass handling costs—a preliminary assessment, *Biomass and Bioenergy*, 2006, **30**, 321–325.

88. A. Bridgwater, D. Meier and D. Radlein, An overview of fast pyrolysis of biomass, *Organic Geochemistry*, 1999, **30**, 1479–1493.
89. D. Mohan Jr. and C. U. P. P. H. Steele, Pyrolysis of wood/biomass for bio-oil: a critical review, *Energy & Fuels*, 2006, **20**, 848–889.
90. R. B. Gupta and A. Demirbas, Gasoline, *Diesel and Ethanol Biofuels from Grasses and Plants*. In Cambridge University Press: New York, USA, 2010, pp. 140–157.
91. A. V. Bridgwater and G. V. C. Peacocke, Fast pyrolysis processes for biomass, *Renewable and Sustainable Energy Reviews*, 2000, **4**, 1–73.
92. J. G. Brammer and A. V. Bridgwater, Drying technologies for an integrated gasification bio-energy plant, *Renewable and Sustainable Energy Reviews*, 1999, **3**, 243–289.
93. D. L. Klass, *Biomass for Renewable Energy, Fuels and Chemicals*. Academic Press: San Diego, USA, 1998, pp. 230–231.
94. S. Mani, L. G. Tabil and S. Sokhansanj, Grinding performance and physical properties of wheat and barley straws,corn stover and switchgrass, *Biomass and Bioenergy*, 2004, **27**, 339–352.
95. A. Oasmaa and C. Peacocke, *A Guide to Physical Property Characterisation of Biomass - Derived fast pyrolysis liquids*; Technical Research Centre of Finland: 2001.
96. D. S. Scott, P. Majerski, J. Piskorz and D. Radlein, A second look at fast pyrolysis of biomass—the RTI process, *Journal of Analytical and Applied Pyrolysis*, 1999, **51**, 23–37.
97. R. C. Brown, and J. Holmgren, Fast Pyrolysis and Bio-Oil Upgrading. In www.ars.usda.gov/.../biomasstoDiesel/RobertBrown&JenniferHolmgren presentationslides.pdf, Iowa State University and UOP.
98. A. A. Lappas, M. C. Samolada, D. K. Iatridis, S. S. Voutetakis and I. A. Vasalos, Biomass pyrolysis in a circulating fluid bed reactor for the production of fuels and chemicals, *Fuel*, 2002, **81**, 2087–2095.
99. M. Ringer, V. Putsche, and J. Scahill, Large-Scale Pyrolysis Oil Production: A Technology Assessment and Economic Analysis, 2006, National Renewable Energy Laboratory, CO (Technical Report-NREL/TP-510-37779).
100. Available from www.btgworld.com. Access on 2011. 01.12.
101. Available from www.btg-btl.com. Access on 2011. 01.12.
102. J. Lede, F. Broust, F.-T. Ndiaye and M. Ferrer, Properties of bio-oils produced by biomass fast pyrolysis in a cyclone reactor, *Fuel*, 2007, **86**, 1800–1810.
103. J. D. Adjaye, and N. N. Bakhshi, Production of hydrocarbons by catalytic upgrading of a fast pyrolysis bio-oil. Part I: Conversion over various catalysts. *Fuel Processing Technology*. 1995, **45**, 161–183.
104. M. Garcia-Perez, T. T. Adams, J. W. Goodrum, D. P. Geller and K. C. Das, Production and fuel properties of pine chip bio-oil/biodiesel blends, *Energy & Fuels*, 2007, **21**, 2363–2372.
105. L. Ingram, D. Mohan, M. Bricka, P. Steele, D. Strobel, D. Crocker, B. Mitchell, J. Mohammad, K. Cantrell and C. U. Pittman Jr., Pyrolysis of

wood and bark in an auger reactor: physical properties and chemical analysis of the produced bio-oils, *Energy & Fuels*, 2008, **22**, 614–625.

106. S. Thangalazhy-Gopakumar, S. Adhikari, H. Ravindran, R. B. Gupta, O. Fasina, M. Tu and S. D. Fernando, Physiochemical properties of bio-oil produced at various temperatures from pine wood using an auger reactor, *Bioresource Technology*, 2010, **101**, 8389–8395.

107. M. Asadullah, M. A. Rahman, M. M. Ali, M. S. Rahman and M. A. Motin, M. B. Sultan, and M. R. Alam, Production of bio-oil from fixed bed pyrolysis of bagasse, *Fuel*, 2007, **86**, 2514–2520.

108. N. Ozbay, A. E. Putun, B. B. Uzun and E. Putun, Biocrude from biomass: pyrolysis of cottonseed cake, *Renewable Energy*, 2004, **24**, 615–625.

109. P. Basu, Pyrolysis In *Biomass Gasification and Pyrolysis*, Academic Press: Burlington, MA, USA, 2010, pp. 65–92.

110. O. Onay and O. M. Kockar, Pyrolysis of rapeseed in a free fall reactor for production of bio-oil, *Fuel*, 2006, **85**, 1921–1928.

111. A. Aho, N. Kumar, K. Eranen, T. Salmi, M. Hupa and D. Y. Murzin, Catalytic pyrolysis of biomass in a fluidized bed reactor: influence of the acidity of h-beta zeolite, *Trans IChemE, Process Safety and Environmental Protection*, 2007, **85**(B5), 473–480.

112. Available from http://www.ensyn.com/news/DOEBiomassToGreenFuels Grant.pdf. Access on 2010.18.09.

113. Available from http://www.ensyn.com/news/Ensyn-Press%20Release-HighNorth.pdf. Access on 2010.18.09.

114. Available from http://www.dynamotive.com. Access on 2010.18.09.

115. Available from http://www.kior.com/index.shtml. Access on 2010.18.09.

116. Available from http://anellotech.com/index.html. Access on 2010.18.09.

117. F. J. Kilzer and A. Broido, Speculations on the nature of cellulose pyrolysis, *Pyrodynamics*, 1965, **2**, 151–163.

118. F. Shafizadeh, Introduction to pyrolysis of biomass, *Journal of Analytical and Applied Pyrolysis*, 1982, **3**, 283–305.

119. D. Radlein, J. Piskorz and D. S. Scott, Fast pyrolysis of natural polysaccharides as a potential industrial process, *Journal of Analytical and Applied Pyrolysis*, 1991, **19**, 41–63.

120. J. Piskorz, D. S. A. G. Radlein, D. S. Scott and S. Czernik, Pretreatment of wood and cellulose for production of sugars by fast pyrolysis, *Journal of Analytical and Applied Pyrolysis*, 1989, **16**, 127–142.

121. F. Shafizadeh, Industrial pyrolysis of cellulosic materials, *Applied Polymer Symposia*, 1975, **28**, 153–174.

122. A. Demirbas, Mechanisms of liquefaction and pyrolysis reactions of biomass, *Energy Conversion & Management*, 2000, **41**, 633–646.

123. A. G. Bradbury, Y. Sakai and F. Shafizadeh, A kinetic model for pyrolysis of cellulose, *Journal of Applied Polymer Science*, 1979, **23**, 3271–3280.

124. G. Varhegyi, M. J. Antal Jr., E. Jakab and P. Szabo, Kinetic modeling of biomass pyrolysis, *Journal of Analytical and Applied Pyrolysis*, 1997, **42**, 73–87.

125. G. Varhegyi, M. J. Antal Jr, P. Szabo, E. Jakab and F. Till, Application of complex reaction kinetic models in thermal analysis, *Journal of Thermal Analysis*, 1996, **47**, 535–542.
126. J. P. Diebold, A unified, global model for the pyrolysis of cellulose, *Biomass and Bioenergy*, 1994, **7**(1–6), 75–85.
127. Y.-C. Lin, J. Cho, G. A. Tompsett, P. R. Westmoreland, and G. W. Huber, Kinetics and mechanism of cellulose pyrolysis. *The Journal of Physical Chemistry C*, 2009, **113**, 20097–20107.
128. J. Cho, J. M. Davis and G. W. Huber, The intrinsic kinetics and heats of reactions for cellulose pyrolysis and char formation, *ChemSusChem*, 2010, **3**, 1162–1165.
129. T. Hosoya, H. Kawamoto and S. Saka, Pyrolysis behaviors of wood and its constituent polymers at gasification temperature, *Journal of Analytical and Applied Pyrolysis*, 2007, **78**(2), 328–336.
130. G. R. Ponder and G. N. Richards, Thermal synthesis and pyrolysis of a xylan, *Carbohydrate Research*, 1991, **218**, 143–155.
131. A. D. Pouwels, A. Tom, G. B. Eijkel and J. J. Boon, Characterization of beech wood and its holocellulose and xylan fractions by pyrolysis-gas chromatography-mass spectroscopy, *Journal of Analytical and Applied Pyrolysis*, 1987, **11**, 417–436.
132. F. Shafizadeh, G. D. Mcginnis and C. W. Philpot, Thermal degradation of xylan and related model compounds, *Carbohydrate Research*, 1972, **25**, 23–33.
133. R. J. Evans and T. A. Milne, Molecular characterization of the pyrolysis of biomass. 1. Fundamentals, *Energy & Fuels*, 1987, **1**(2), 123–137.
134. M. J. Antal Jr., T. Leesomboon and W. S. Mok, Mechanism of formation of 2-furaldehyde from D-xylose, *Carbohydrate Research*, 1991, **217**, 71–85.
135. P. F. Britt, A. C. BuchananIII, and D. R. Matineau, *Flash Vacuum Pyrolysis of Lignin Model Compounds: Reaction Pathways of Aromatic Methoxy Groups*, 1999, Oak Ridge National Laboratory (ORNL), TN (Technical Report-ORNL/CP-101275).
136. M. Windt, D. Meier, J. H. Marsman, H. J. Heeres and S. D. Koning, Micro-pyrolysis of technical lignins in a new modular rig and product analysis by GC–MS/FID and GCxGC–TOFMS/FID, *Journal of Analytical and Applied Pyrolysis*, 2009, **85**, 38–46.
137. F. P. Petrocelli and M. T. Klein, Model reaction pathways in kraft lignin pyrolysis, *Macromolecules*, 1984, **17**(2), 161–169.
138. K.-I. Kuroda and Y. Inoue, Analysis of lignin by pyrolysis-gas chromatography i effect of inorganic substances on guaiacol-derivative yield from softwoods and twir lignins, *Journal of Analytical and Applied Pyrolysis*, 1990, **18**, 59–69.
139. D. J. Nowakowski, A. V. Bridgwater, D. C. Elliott, D. Meier and P. Wild, d., Lignin fast pyrolysis: Results from an International Collaboration, *Journal of Analytical and Applied Pyrolysis*, 2010, **88**, 53–72.
140. A. Bridgwater, *Thermal Biomass Conversion and Utilization - Biomass Information System*; Luxembourg, 1996.

141. J. P. Diebold, A Review of the Chemical and Physical Mechanisms of the Storage Stability of Fast Pyrolysis Bio-Oils *NREL/SR-570-27613*, 2000.

142. T. A. Milne, F. Agblevor, M. Davis, S. Deutch and D. Johnson, *A Review of the Chemical Composition of Fast Pyrolysis Oils*. Blackie Academic & Professional: London, New York, 1997.

143. R. He, X. P. Ye, B. C. English and J. A. Satrio, Influence of pyrolysis condition on switchgrass bio-oil yield and physicochemical properties, *Bioresource Technology*, 2009, **100**, 5305–5311.

144. P. A. Horne and P. T. Williams, Influence of temperature on the products from the flash pyrolysis of biomass, *Fuel*, 1996, **75**, 1051–1059.

145. S. Thangalazhy-Gopakumar, S. Adhikari, R. B. Gupta, and S. D. Fernando, Influence of pyrolysis operating conditions on bio-oil components: a microscale study in a pyroprobe. *Energy & Fuels*, 2011, **25**, 1191–1199.

146. A. V. Bridgwater, A. J. Toft and J. G. Brammer, A techno-economic comparison of power production by biomass fast pyrolysis with gasification and combustion, *Renewable and Sustainable Energy Reviews*, 2002, **6**, 181–248.

147. IEA-Bioenergy Task 34 - Pyrolysis, http://www.pyne.co.uk/?_id = 73. (2011 - 01 - 13).

148. NREL, Biodiesel Handling and Use Guide. 2009, NREL/TP-540-43672.

149. CONCAWE, Heavy Fuel Oils. *product dossier no. 98/109* 1998, http://www.accede.org/prestige/documentos/Tox_fuel_pesado.pdf.

150. V. Sricharoenchaikul, A. L. Hicks and W. J. Frederick, Carbon and char residue yields from rapid pyrolysis of kraft black liquor, *Bioresource Technology*, 2001, **77**, 131–138.

151. ASTMD7544, Standard Specification for Pyrolysis Liquid Biofuel. In *Petroleum Standards (D02.E0), Book of Standards Volume: 05.04*, ASTM International: West Conshohocken, PA, USA, 2009.

152. J. Piskorz and D. Radlein, *Determination of Biodegradation Rates of Bio-oil by Respirometry*. Cpl Press: Newbury, HW, 2008, p. 119–134.

153. S. Czernik and A. V. Bridgwater, Overview of applications of biomass fast pyrolysis oil, *Energy & Fuels*, 2004, **18**(2), 590–598.

154. C. R. Shaddix and D. R. Hardesty, *Combustion Properties of Biomass Flash Pyrolysis Oils: Final Project Report*; Sandia National Laboratories: 1999.

155. Ensyn http://www.ensyn.com. (2010 - 08 - 18).

156. A. Shihadeh and S. Hochgreb, Diesel engine combustion of biomass pyrolysis oils, *Energy & Fuels*, 2000, **14**, 260–274.

157. D. Chiaramonti, M. Bonini, E. Fratini, G. Tondi, K. Gartner, A. Bridgwater, H. P. Grimm, I. Soldaini, A. Webster and P. Baglioni, Development of emulsions from biomass pyrolysis liquid and diesel and their use in engines—Part 1: Emulsion production, *Biomass and Bioenergy*, 2003, **25**, 85–99.

158. D. Chiaramonti, M. Bonini, E. Fratini, G. Tondi, K. Gartner, A. Bridgwater, H. P. Grimm, I. Soldaini, A. Webster and P. Baglioni,

Development of emulsions from biomass pyrolysis liquid and diesel and their use in engines—Part 2: Tests in diesel engines, *Biomass and Bioenergy*, 2003, **25**, 101–111.

159. T. P. Vispute, H. Zhang, A. Sanna, R. Xiao and G. W. Huber, Renewable chemical commodity feedstocks from integrated catalytic processing of pyrolysis oils, *Science*, 2010, **330**, 1222–1227.

160. O. I. Senol, E. M. Ryymin, T. R. Viljava and A. O. I. Krause, Effect of hydrogen sulphide on the hydrodeoxygenation of aromatic and aliphatic oxygenates on sulphided catalysts, *Journal of Molecular Catalysis A: Chemical*, 2007, **277**, 107–112.

161. K. H. Oehr, D. S. Scott, and S. Czernik, Method of Producing Calcium Salts from Biomass, 1993.

162. D. Radlein, J. Piskorz, and P. Majerski, Method of Producing Slow-Release Nitrogenous Organic Fertilizer from Biomass, 1997. U.S patent number: 5676727.

163. D. Beckman, and D. Radlein, *Slow Release Fertilizer Production Plant from Bio-Oil Technical-Economic Assessment*; VTT Technical Research Centre Of Finland, 2000.

164. Oehr, K. Acid Emission Reduction, 1995. U.S patent number: 5458803.

165. J. R. Galdmez, L. Garca and R. Bilbao, Hydrogen production by steam reforming of bio-oil using coprecipitated Ni − Al catalysts. Acetic acid as a model compound, *Energy & Fuels*, 2005, **19**(3), 1133–1142.

166. S. Czernik and R. J. French, Production of hydrogen from plastics by pyrolysis and catalytic steam reform, *Energy & Fuels*, 2006, **20**(2), 754–758.

167. S. Czernik, R. Evans and R. French, Hydrogen from biomass-production by steam reforming of biomass pyrolysis oil, *Catalysis Today*, 2007, **129**, 265–268.

168. S. Czernik, R. French, C. Feik and E. Chornet, Hydrogen by catalytic steam reforming of liquid byproducts from biomass thermoconversion processes, *Industrial and Engineering Chemistry Research*, 2002, **41**(17), 4209–4215.

169. D. Wang, S. Czernik and E. Chornet, Production of hydrogen from biomass by catalytic steam reforming of fast pyrolysis oils, *Energy & Fuels*, 1998, **12**(1), 19–24.

170. A. C. Basagiannis and X. E. Verykios, Steam reforming of the aqueous fraction of bio-oil over structured $Ru/MgO/Al_2O_3$ catalysts, *Catalysis Today*, 2007, **127**, 256–264.

171. L. Garcia, R. French, S. Czernik and E. Chornet, Catalytic steam reforming of bio-oils for the production of hydrogen: effects of catalyst composition, *Applied Catalysis A: General*, 2000, **201**, 225–239.

172. T. R. Carlson, T. P. Vispute and G. W. Huber, Green gasoline by catalytic fast pyrolysis of solid biomass derived compounds, *ChemSusChem*, 2008, **1**, 397–400.

173. P. T. Williams and N. Nugranad, Comparison of products from the pyrolysis and catalytic pyrolysis of rice husks, *Energy*, 2000, **25**, 493–513.

174. P. A. Horne, N. Nugranad and P. T. Williams, Catalytic coprocessing of biomass-derived pyrolysis vapours and methanol, *Journal of Analytical and Applied Pyrolysis*, 1995, **34**, 87–108.
175. D. C. Elliott, Historical developments in hydroprocessing bio-oils, *Energy & Fuels*, 2007, **21**, 1792–1815.
176. J. Wildschut, F. H. Mahfud, R. H. Venderbosch and H. J. Heeres, Hydrotreatment of fast pyrolysis oil using heterogeneous noble-metal catalysts, *Industrial and Engineering Chemistry Research*, 2009, **48**, 10324–10334.
177. S. Ramanathan and S. T. Oyama, New catalysts for hydroprocessing: transition metal carbides and nitrides, *Journal of Physical Chemistry*, 1995, **99**, 16365–16372.
178. R. Xing, A. V. Subrahmanyam, H. Olcay, W. Qi, G. P. v. Walsum, H. Pendse and G. W. Huber, Production of jet and diesel fuel range alkanes from waste hemicellulose-derived aqueous solutions, *Green Chemistry*, 2010, **12**, 1933–1946.
179. A. V. Subrahmanyam, S. Thayumanavan and G. W. Huber, C-C bond formation reactions for biomass-derived molecules, *ChemSusChem*, 2010, **3**, 1158–1161.
180. A. Centeno, E. Laurent and B. Delmon, Influence of the support of como sulfided catalysts and of the addition of potassium and platinum on the catalytic performances for the hydrodeoxygenation of carbonyl, carboxyl, and guaiacol -type molecules, *Journal of Catalysis*, 1995, **154**, 288–298.
181. E. Costa, J. Aguado, G. Ovejero and P. Cafiizares, Conversion of n-butanol-acetone mixtures to C1-Cl0 hydrocarbons on HZSM-5 type zeolites, *Industrial and Engineering Chemistry Research*, 1992, **31**(4), 1021–1025.
182. R. French and S. Czernik, Catalytic pyrolysis of biomass for biofuels production, *Fuel Processing Technology*, 2010, **91**, 25–32.
183. T. R. Carlson, G. A. Tompsett, W. C. Conner and G. W. Huber, Aromatic production from catalytic fast pyrolysis of biomass-derived feedstocks, *Topics in Catalysis*, 2009, **52**, 241–252.
184. A. Aho, N. Kumar, K. Eranen, T. Salmi, M. Hupa and D. Y. Murzin, Catalytic pyrolysis of woody biomass in a fluidized bed reactor: influence of the zeolite structure, *Fuel*, 2008, **87**, 2493–2501.
185. A. A. Boateng, C. A. Mullen, C. M. McMahan, M. C. Whalen and K. Cornish, Guayule (Parthenium argentatum) pyrolysis and analysis by PY–GC/MS, *Journal of Analytical and Applied Pyrolysis*, 2010, **87**, 14–23.
186. E. F. Iliopoulou, E. V. Antonakou, S. A. Karakoulia, I. A. Vasalos, A. A. Lappas and K. S. Triantafyllidis, Catalytic conversion of biomass pyrolysis products by mesoporous materials: effect of steam stability and acidity of Al-MCM-41 catalysts, *Chemical Engineering Journal*, 2007, **134**, 51–57.
187. A. Pattiya, J. O. Titiloye and A. V. Bridgwater, Evaluation of catalytic pyrolysis of cassava rhizome by principal component analysis, *Fuel*, 2010, **89**, 244–253.

188. A. Pattiya, J. O. Titiloye and A. V. Bridgwater, Fast pyrolysis of cassava rhizome in the presence of catalysts, *Journal of Analytical and Applied Pyrolysis*, 2008, **81**, 72–79.

189. J. Adam, E. Antonakou, A. Lappas, M. Stocker, M. H. Nilsen, A. Bouzga, J. E. Hustad and G. Øye, In situ catalytic upgrading of biomass derived fast pyrolysis vapours in a fixed bed reactor using mesoporous materials, *Microporous and Mesoporous Materials*, 2006, **96**, 93–101.

190. T. R. Carlson, J. Jae, Y.-C. Lin, G. A. Tompsett and G. W. Huber, Catalytic fast pyrolysis of glucose with HZSM-5: The combined homogeneous and heterogeneous reactions, *Journal of Catalysis*, 2010, **270**, 110–124.

191. E. V. Antonakou, V. S. Dimitropoulos and A. A. Lappas, Production and characterisation of bio-oil from catalytic biomass pyrolysis, *Thermal Science*, 2006, **10**(3), 151–160.

192. J. L. Sánchez, G. Gea, A. Gonzalo, R. Bilbao and J. Arauzo, Kinetic study of the thermal degradation of alkaline straw black liquor in nitrogen atmosphere, *Chemical Engineering Journal*, 2004, **104**, 1–6.

193. P. K. Bhattacharya, V. Parthiban and D. Kunzru, Pyrolysis of black liquor solids, *Industrial and Engineering Chemistry Process Design and Development*, 1986, **25**, 420–426.

194. J. Wintoko, H. Theliander and T. Richards, Experimental investigation of black liquor pyrolysis using single droplet TGA, *TAPPI Journal*, 2007, **6**(5), 9–15.

195. G. Gea, M. B. Murillo and J. Arauzo, Thermal degradation of alkaline black liquor from straw. thermogravimetric study, *Industrial and Engineering Chemistry Research*, 2002, **41**, 4714–4721.

196. Y. Zhao, R. Bie, J. Lu and T. Xiu, Kinetic Study on pyrolysis of NSSC black liquor in a nitrogen atmosphere, *Chemical Engineering Communications*, 2010, **197**, 1033–1047.

197. S. B. Jones, C. Valkenburg, C. Walton, D. C. Elliott, J. E. Holladay, D. J. Stevens, C. Kinchin, and S. Czernik, Production of gasoline and diesel from biomass via fast pyrolysis, hydrotreating and hydrocracking: A design case. *DE-AC05-76RL01830*, 2009, PNNL-18284 Rev. 1.

198. M. M. Wright, J. A. Satrio, R. C. Brown, D. E. Daugaard and D. D. Hsu *Techno-Economic Analysis of Biomass Fast Pyrolysis to Transportation Fuels*, National Renewable Energy Laboratory: Golden, Colorado, USA, 2010.

199. M. M. Wright, D. E. Daugaard, J. A. Satrio and R. C. Brown, Techno-economic analysis of biomass fast pyrolysis to transportation fuels, *Fuel*, 2010, **89**, S2–S10.

200. M. N. Islam and F. N. Ani, Techno-economics of rice husk pyrolysis, conversion with catalytic treatment to produce liquid fuel, *Bioresource Technology*, 2000, **73**, 67–75.

201. Mullaney, H. *Technical, Environmental and Economic Feasibility of Bio-Oil in New Hampshire's North Country*, New Hampshire Industrial Research Center (NHIRC): Durham, NH, 2002.

CHAPTER 10

Biohydrogen Production from Cellulosic Biomass

DAVID B LEVIN,[*a] JI HYE JO[b] AND PIN-CHING MANESS[b]

[a] Department of Biosystems Engineering, University of Manitoba, Winnipeg, Manitoba, R3T 5V6, Canada; [b] National Renewable Energy Laboratory, 1617 Cole Blvd., Golden, Colorado, 80401 USA
*Email: levindb@ad.umanitoba.ca

10.1 Biohydrogen

Biological hydrogen (biohydrogen, $BioH_2$) is an attractive alternative to chemical and electrochemical methods because it is a potentially carbon-neutral process that is carried out at lower temperatures and pressures, and is therefore less energy intensive than chemical and electrochemical processes.[1] The processes of biological hydrogen production can be broadly classified into two distinct groups, Light-dependent biohydrogen processes and dark fermentation processes. Light-dependent processes include direct or indirect biophotolysis and photofermentation, and are based on photosynthesis (oxygenic and anoxygenic). Hydrogen production by direct or indirect biophotolysis is carried out by photoautotrophic organisms, such as algae and cyanobacteria, while photofermentation is carried out by photoheterotrophic organisms, such as green sulfur and purple nonsulfur bacteria. Dark fermentation, in which carbohydrates are converted to H_2, CO_2, and organic acids, provides a promising alternative to light-dependent processes, particularly when waste biomass is

RSC Green Chemistry No. 18
Integrated Forest Biorefineries
Edited by Lew Christopher

used as a feedstock for the generation of H_2. Since fermentation does not require a constant light supply, it can be run continuously using inexpensive and commercially used systems. Furthermore, hydrogen production rates are much higher using fermentations when compared to photosynthesis-based systems,[2] thus reducing bioreactor running costs.

A number of species, including *Bacillus*, *Escherichia*, *Enterobacter*, *Ruminococcoi* and *Clostridium*, are capable of producing hydrogen *via* fermentation.[2-5] Some of these organisms not only operate at thermophillic (40–60 °C) conditions, at which the solubility of H_2 is lower preventing product inhibition, but are also capable of degrading lignocellulose, a primary component of biomass. Although fermentative hydrogen production occurs at high rates and can take place continuously in the absence of light, the issue of poor hydrogen yields must be addressed to make this technology more economically feasible. Thermodynamic limitations, product inhibition, the presence of branched catabolic pathways, media composition, and the nature of substrate all have an impact on hydrogen yields.

10.1.1 Dark Fermentative Hydrogen Production

Fermentation is an anaerobic process during which an organic substrate such as glucose is oxidized to provide both building blocks and metabolic energy in support of cell growth. The reducing equivalent must be disposed of in order to regenerate its own oxidant and when the latter is linked to proton reduction, H_2 production was observed. Hydrogen production is associated with the conversion of glucose to pyruvate and NADH during glycolysis. From this branchpoint two major pathways are responsible for H_2 production. In facultative anaerobes, such as *E. coli*[6-9] and *Enterobacter* species,[10,11] pyruvate-formate lyase (PFL) catalyzes the conversion of pyruvate to acetyl-CoA and formate, with the latter converted to H_2 and CO_2 *via* the action of formate-hydrogen lyase (FHL) (Figure 10.1, Pathway A). The production of two moles of formate limits the maximal yield of H_2 per mole of glucose (H_2 molar yield) to two in enterobacteriaceae. In strict anaerobes such as the *Clostridium* species, pyruvate conversion proceeds to acetyl-CoA and ferredoxin catalyzed by pyruvate-ferredoxin oxidoreductase (PFO). The reduced ferredoxin serves as the electron donor for H_2 production mediated by a ferredoxin-linked hydrogenase (Figure 10.1, Pathway B). The NADH generated during glycolysis could yield additional H_2 *via* two routes. In *Clostridium* microbes NADH could reduce ferredoxin *via* the NADH, ferredoxin oxidoreductase (NFO), with the reduced ferredoxin drives H_2 production *via* a hydrogenase.[12,13] Alternatively, in a more direct route, NADH-linked hydrogenase has been reported in thermophiles such as *Thermoanaerobacter tengcongensis*[14] and in *Clostridium thermocellum*,[15] the latter based on bioinformatic study. Using both ferredoxin and NADH as reductants could collectively increase the H_2 molar yield to more than two, at the cost of accumulating less reduced end products such as lactic acid and ethanol.

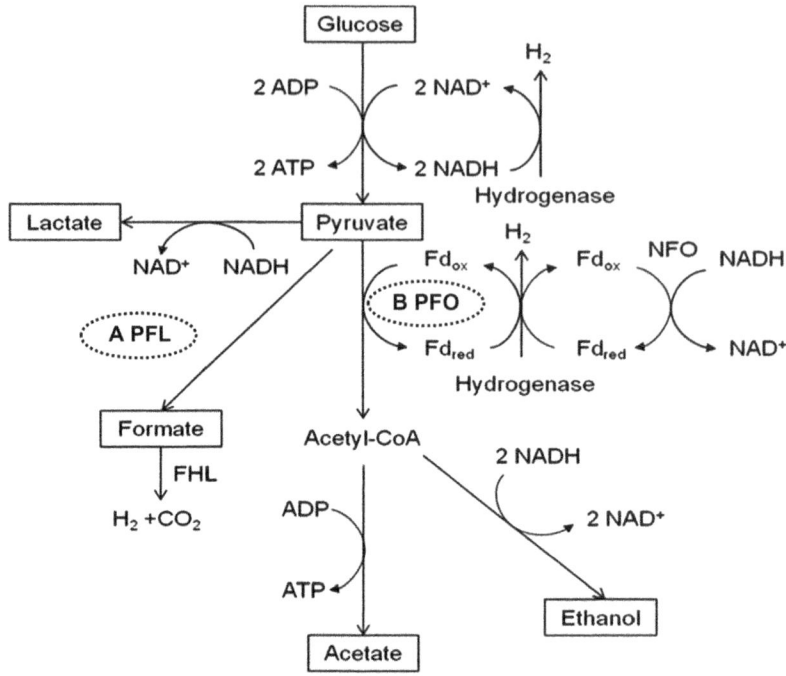

Figure 10.1   Fermentative hydrogen production pathway. PFL, Pyruvate-formate lyase, a key enzyme in Pathway A. FHL, Formate-hydrogen lyase, PFO, Pyruvate-ferredoxin oxidoreductase, a key enzyme in Pathway B. NFO, NADH, ferredoxin oxidoreductase, ADP, Adenosine diphosphate, ATP, Adenosine triphosphate, NADH, Nicotine adenine dinucleotide, reduced, NAD^+, Nicotine adenine dinucleotide, oxidized.

10.1.2   Hydrogenase Enzymes

Central to biological H_2 production is the hydrogenase enzyme. Hydrogenase catalyzes the oxidation or evolution of molecular hydrogen according to eqn (10.1),

$$2H^+ + 2e^- \leftrightarrow H_2 \qquad (10.1)$$

Based on the structure and chemical composition of the metallo-cluster within the active site, hydrogenase can be classified into two main groups, [FeFe]-hydrogenase and [NiFe]-hydrogenase. Albeit phylogenetically unrelated, both [FeFe]- and [NiFe]-hydrogenases contain the unusual CN^- and CO ligands attached to the iron atoms of the binuclear metal center to efficiently catalyze the interconversion between H^+ and H_2.[16]

10.1.2.1   *[FeFe]-Hydrogenases*

[FeFe]-hydrogenases are primarily found in anaerobic bacteria and eukarya. The modular structure and domain organization of these hydrogenases are diverse from simple monomer to more complex trimer or tetramer. The model [FeFe]-hydrogenases from *Clostridium pasteurianum* and *C. acetobutylicum* are monomeric enzymes containing up to four [Fe-S] accessory clusters assisting in electron transfer to the active site, the H-cluster. The H-cluster is composed of a [4Fe-4S]-subcluster coordinated to a unique binuclear Fe–Fe center *via* a conserved cysteine residue. [FeFe]-hydrogenases are usually involved in H_2 production, but the periplasmic [FeFe]-hydrogenase of *Desulfovibrio vulgaris* Hildenborough can function as an uptake hydrogenase. Maturation proteins, HydE, HydF, and HydG, are required for H-cluster assembly. The HydE and HydG maturases contain the CxxxCxxC motif of three conserved cysteines which coordinate a [4Fe-4S] cluster, and belong to the class of radical *S*-adenosylmethionine (SAM) proteins.[17] Recent research shows that HydG catalyzes the synthesis of CO using tyrosine as the substrate.[18] The HydF is also a metalloprotein with GTPase activity at its N-terminal domain and binds a [Fe-S] cluster at its C-terminus. HydF is considered to be the scaffold protein where assembly of the binuclear part of the H-cluster occurs.[19] Chemical additives like tyrosine, ferrous ammonium sulfate and SAM can increase *in vitro* catalytic activities by facilitating the reconstitution of the H-cluster.

10.1.2.2 [NiFe]-Hydrogenases

[NiFe]-hydrogenases are found in many bacteria, cyanobacteria, and archaea, but not in eucaryotes. The model enzyme from *D. gigas* is a heterodimer with the catalytic large subunit (~ 60 kDa) harboring the bimetallic NiFe-active site, and the small subunit (~ 30 kDa) containing the [Fe–S] clusters that assist in electron transfer from/to the catalytic site.[20] The NiFe-active site is deeply buried within the large subunit, and is coordinated to the protein *via* four conserved cysteines. The maturation pathway for the [NiFe] catalytic site is based on the model developed for the three hydrogenases in *E. coli.*[21] HypE and HypF generate the CN^- ligand from carbamoyl phosphate (CP) in an ATP-dependent process. The origin of CO ligand is still under debate.[16] HypC and HypD bind the Fe atom containing the CN^- and CO ligands and donate it to the catalytic subunit. HypA, HypB, and SlyD are responsible for nickel delivery in a GTP-dependent reaction.[22] The activation of the catalytic subunit requires a final proteolytic cleavage of the C-terminal residues, catalyzed by HycI, allowing the protein to fold properly around the [NiFe] active site.

A common type of the [NiFe] hydrogenases reported in the *Clostridium* species is the membrane-localized energy-converting hydrogenase (Ech). Analysis of the genome sequences of four cellulolytic *C. thermocellum*, *C. cellulolyticum*, *C. papyrosolvens* and *C. phytofermentans* reveals the presence of a hexameric Ech hydrogenase. The Ech hydrogenase appears to be of archeal origin and shares homology with subunits of the NADH-quinone oxido-reductase (complex I) in respiring organisms.[23] Its role in H_2 production or H_2 oxidation during fermentation remains unresolved, nor the mechanism governing energy generation.

10.2 Thermodynamic Considerations

Microbial fermentation has been optimized during evolution toward maximizing cell biomass accumulation in lieu of H_2 production. Energy conservation is an important aspect of bacterial fermentation during which electron mediators such as NAD^+ or oxidized ferredoxin must be regenerated quickly to allow continuous breakdown of substrate to maximize cell growth. To maintain homeostasis, numerous metabolic pathways must compete for the same pool of the intracellular equivalents including the H_2 production reaction, which lowers the yield of H_2. Theoretically up to 12 mol of H_2 can be produced from one mol of glucose if solely based on enthalpy of the substrate input and the product output (eqn (10.2)). Yet the standard Gibbs free energy of eqn (10.2) suggests this reaction is thermodynamically unfavorable and that no energy is conserved from this endothermic reaction to afford microbial cell growth. Woodward et al.[24] demonstrated the production of stoichiometric amount of H_2 (near 11.6 mol) from one mol of glucose in an *in vitro* system with enzymes of the pentose phosphate pathway. Yet the reaction occurred under near-equilibrium conditions with a low rate and under very low H_2 partial pressure. Thauer et al.[25] predicted that 4 mol of H_2 per mol glucose is the biological maximum if acetate is the sole byproduct (eqn (10.3)). This refers to as the "Thauer limit". Under this condition only approximately 33% of the chemical energy in glucose is conserved in H_2. In theory this route would also yield the highest amounts of ATP (4 mol ATP/mol glucose) based on Figure 10.1. In reality, the actual energy yield in eqn (10.3) is insufficient to support such a high yield of ATP (–70 kJ for 1 mol ATP synthesis).[25–27] If butyrate is the sole byproduct, the reaction is thermodynamically more favorable to drive the synthesis of three mol of ATP albeit with only two mol of H_2 produced (eqn (10.4)).

$$C_6H_{12}O_6 + 12H_2O \rightarrow 6CO_2 + 12H_2 \quad \Delta G^{0\prime} = 241 \, \text{kJ mol}^{-1} \quad (10.2)$$

$$C_6H_{12}O_6 + 4H_2O \rightarrow 2CH_3COOH + 2CO_2 + 4H_2 \quad \Delta G^{0\prime} = -182.4 \, \text{KJ mol}^{-1} \quad (10.3)$$

$$C_6H_{12}O_6 + 2H_2O \rightarrow CH_3CH_2CH_2COOH + 2CO_2 + 2H_2$$
$$\Delta G^{0\prime} = -257.1 \, \text{KJ mol}^{-1} \quad (10.4)$$

It is observed that bacteria undergo mixed acid fermentation producing an array of waste byproducts (acetic, formic, lactic, butyric acids, ethanol, butanol, *etc.*). This metabolic diversity affords a quick regeneration of NAD^+ yet at the expense of lower H_2 molar yield, ranging from one to three as reported in most laboratories.[28–34] Thermodynamic constraints ultimately dictates the practical yield of H_2 that cannot be addressed by genetic engineering of the metabolic pathway alone. Hydrogen production becomes more exergonic with increasing temperature.[35] This accounts for the higher rate observed in

Enterobacter aerogenes and higher rate and molar yield in the hyperthermophiles *Caldicellulosiruptor saccharolyticus* and *Thermotoga elfii*, the latter cultured at 70 °C.[36–38] Similarly, a 70% improvement in molar yield (from 0.97 to 1.67) was reported when temperature was raised from 33 °C to 41 °C in a mixed anaerobic culture.[39] Under standard condition, NADH-mediated H_2 production is an up-hill reaction, only proceeding at very low H_2 partial pressure (pH_2 less than 10^{-3} atm) to achieve a more negative free-energy change of the overall reaction.[2,30,40,41] Recently, the trimeric FeFe-hydrogenase in *Thermotoga maritima* was reported to utilize both NADH and ferredoxin (1,1 ratio) as electron donors synergistically for H_2 production.[42] This bifurcating nature employs the exergonic oxidation of ferredoxin (midpoint potential, −453 mV) to drive the unfavorable oxidation of NADH ($E_0' = -320$ mV) to produce H_2 ($E_0' = -420$ mV). The use of bifurcating enzyme overcomes the thermodynamic barrier and warrants energy conservation. This finding provides a new perspective on bioenergetics and further underscores the importance of H_2 production in fermentative microbes.

10.3 Hydrogen Yields from Lignocellulosic Biomass

A wide variety of organic materials have been used in different processes for the hydrogen generation, using mixed, undefined microbial communities or defined pure cultures. Substrates used for biohydrogen production *via* dark fermentation include sugars, such as glucose, fructose, galactose and arabinose,[43–45] sucrose,[46–52] xylose,[53] starch,[54–59] and cellulose.[33,60–62] Various industrial waste streams that contain heterogeneous mixtures of sugars, starch, and in some cases cellulose, have also been investigated [see reviews [63,64]].

Substrate costs play a significant role for the overall economics of biohydrogen production. Abundant, inexpensive feedstocks are essential if biohydrogen production is to become an economically viable process of generating usable H_2. Lignocellulosic biomass is a complex of biopolymers that makes up the structural components of plant material. The approximate composition of lignocellulose found in most biomass feedstocks is roughly 45-60% cellulose, 20–40% hemicellulose, 25% lignin, and 1–5% pectin.[65–67] Cellulose consists of linear, insoluble polymers consisting of up to 25 000 repeating β-1,4 linked β-D-glucopyranose units. Cellulose is a highly ordered molecule consisting of 15–45 crystalline microfibril chains, which in turn associate to form cellulose fibers. In nature, cellulose is found primarily in plant cell walls and is associated with varying degrees of other bipolymers, including, (i) hemicellulose, a random, amorphous heteropolysaccharide composed of typically β-1,3 linked xylans, arabinoxylan, gluco-mannan, and galactomannan; (ii) lignin, a complex hydrophobic network of phenylpropaniod units; (iii) pectins, composed of α-(1-4)-linked D-galacturonic acid; and (iv) proteins.[64–66]

Lignocellulosic biomass is renewable, inexpensive, constitutes a large fraction of waste biomass from municipal, agricultural, and forestry sectors, and thus offers excellent potential as a feedstock for renewable biofuels.[68–70] Cellulose is, however, difficult to hydrolyze due to its crystalline structure. Current strategies that produce fuel ethanol from lignocellulosic biomass (or "second-generation" biofuels) use simultaneous saccharification and fermentation (SSF) or simultaneous saccharification and cofermentation (SSCF).[66,71] Both SSF and SSCF require extensive pretreatment of the cellulosic feedstock by steam-explosion and/or acid treatment, followed by addition of exogenously produced cocktails of cellulolytic enzymes to hydrolyse cellulose chains and release the glucose monomers required for fermentation. These pretreatments are costly, and some of the byproducts generated, for example furfurals, can inhibit downstream processes.

10.3.1 Biohydrogen from Hydrolyzed Cellulose

Steam explosion of corn stover, with or without and acid treatment step, was a suitable substrate for H_2 production.[32,72] A mixed microbial community, derived from the heated sludge of a local wastewater treatment plant, was capable of efficiently fermenting the aqueous hydrolyzate derived from the hemicellulose fraction of the steam-pretreated corn stover. Biogas contained equal amounts of H_2 and CO_2. Acetic and butyric acids were the major soluble fermentation byproducts. Molar yields of 2.84 and 3.0 mol H_2/mol glucose equivalent were obtained using the mixed sugars present in the hydrolyzate derived from neutral and acidic steam explosion, respectively.[32] *Clostridium acetobutylicum* strain X9 produced H_2 from corn stalks pretreated with H_2SO_4, NaOH, or NH_3 or by H_2SO_4 pretreatment followed by steam explosion.[72] *C. acetobutylicum* X9 produced H_2 at a rate of 664 mL H_2 per liter of culture with acetate and butyrate as the major soluble fermentation byproducts.

Hydrogen production from dilute acid treated corn stover by the *Thermoanaerobacterium thermosaccharolyticum* W16 was investigated by Cao *et al.*[73] The effects of sulfuric acid concentration and reaction time in the hydrolysis stage of the process were determined based on a 2×2 central composite experimental design with respect to maximum H_2 productivity. The optimal hydrolysis conditions to yield the maximum quantity of H_2 by *T. thermosaccharolyticum* W16 were 1.69% sulfuric acid and 117 min reaction time. At these conditions, the H_2 yield was shown to be 3.3 L H_2 L^{-1} medium, which corresponds to 2.24 mol H_2 mol^{-1} sugar. The present results indicate the potential of using acid-hydrolyzed corn stover as a feedstock for bioH_2 production using *T. thermosaccharolyticum* W16.

Lo *et al.*[74] employed a "temperature-shift" strategy to improve reducing sugar production from bacterial hydrolysis of cellulosic materials. In this strategy, production of cellulolytic enzymes by *Cellulomonas uda* E3-01 was carried out at 35 °C, while hydrolysis of cellulosic substrates was achieved at 45 °C, at which cell growth was inhibited to avoid consumption of reducing sugar. This temperature-shift strategy was shown to markedly increase the reducing sugar (especially, monosaccharide and disaccharide) concentration in the hydrolysate while hydrolyzing pure (carboxymethyl-cellulose, xylan, avicel and cellobiose) and natural (rice husk, rice straw, bagasse and Napier-grass) cellulosic materials. The cellulosic hydrolysates from CMC and xylan were successfully converted to H_2 *via* dark fermentation with *Clostridium butyricum* CGS5, attaining a maximum hydrogen yield of 4.8 mmol H_2/g reducing sugar.

Hydrogen was produced by simultaneous saccharification and fermentation from steam-exploded corn straw (SECS) using *Clostridium butyricum* AS1.209.[75] In these experiments, cellulolytic enzymes were added to the bioreactor containing the SECS substrate, during the fermentation reaction. The effects of various process parameters, such as solid to liquid ratio, enzyme loading and initial pH, *etc.*, were examined with respect to maximum H_2 productivity, which was obtained by fitting the cumulative H_2 production data to a modified Gompertz equation. The maximum specific H_2 production rate and maximal hydrogen yield were 126 ml/g VSS d and 68 ml/g SECS, respectively. The yield of soluble metabolites was 197.7 mg/g SECS. Acetic acid accounted for 46% of the total and was the most abundant product and this shows that hydrogen production from SECS was essentially acetate-type fermentation. Hydrogen production by simultaneous saccharification and fermentation of SECS has the predominance of a short lag stage and high maximum specific H_2 production rate and it was a promising method for H_2 production and straw biomass conversion.

Kaparaju *et al.*[76] have investigated biohydrogen wheat straw within a biorefinery framework. Initially, wheat straw was hydrothermally liberated to a cellulose-rich fiber fraction and a hemicellulose-rich liquid fraction (hydrolysate). Enzymatic hydrolysis and subsequent fermentation of cellulose yielded 0.41 g-ethanol/g-glucose, while dark fermentation of hydrolysate produced 178.0 ml-H_2/g-sugars. Multiple biofuels production from wheat straw can increase the efficiency for material and energy and can presumably be more economical process for biomass utilization.

Hydrogen was produced by the extreme thermophile *Caldicellulosiruptor saccharolyticus* using H_2SO_4 treated paper sludge hydrolysate as the sole carbon source.[77,78] Acetate and lactate were the major soluble fermentation byproducts. The H_2 yield was dependent on lactate formation and varied between 50 and 94% of the theoretical maximum. The carbon balance in the medium with glucose and xylose was virtually 100%. The carbon balance was not complete in the paper sludge medium because the measurement of biomass was impaired owing to interfering components in the paper sludge hydrolysate. Nevertheless, >85% of the carbon could be accounted for in the products acetate and lactate. The maximal volumetric H_2 production rate was 5 to 6 mmol/(L h), which was lower than the production rate in media with glucose, xylose, or a combination of these sugars (9–11 mmol/ [L h]). The reduced H_2 production rate suggests the presence of inhibiting components in paper sludge hydrolysate.

10.3.2 Biohydrogen from Direct Cellulose Fermentation

Fermentative H_2 production using cellulose as the sole carbon source is under extensive investigation. The mesophilic, cellulolytic bacterium, *C. termitidis* strain CT1112 displayed a cell generation time of 18.9 h when grown on 2 g/L α-cellulose.[79] The major soluble fermentation byproducts were acetate and ethanol. Maximum yields of acetate, ethanol, H_2, and formate on α-cellulose, the yields were 7.2, 3.1, 7.7 and 2.9 mmol/L of culture, respectively. Although, the generation time was longer when cultured on α-cellulose than on the soluble cellulodextrin cellobiose, acetate and H_2 synthesis were favored over ethanol synthesis, indicating that carbon flow to ethanol and formate was restricted. During the log phase, H_2 was produced at a specific rate of 2.79 mmol/h/g dry weight⁻ of cells on α-cellulose.

The thermophilic, cellulolytic bacterium *Clostridium thermocellum* strain 27405 produced greater amounts of H_2 when cultured (in Balch tubes) on cellulosic substrates compared with the soluble cellulodextran cellobiose, with an average yield of 1.6 mol H_2/mol glucose equivalent.[33,60,80] The major soluble fermentation byproducts include ethanol, acetate, and formate, with lactate being produced when the pH drops below 6.3.[33,81] Hydrogen production by *C. thermocellum* 27405 was also investigated using dried distillers grain (DDGS), barley hulls (BH), or fusarium head blight contaminated barley hulls (CBH) as the carbon source in batch fermentation experiments.[61] Overall, DDGS produced the highest concentration of H_2 gas at 1.27 mmol H_2/glucose equivalent, while CBH and BH produced 1.18 and 1.24 mmol H_2/glucose equivalent, respectively.

10.4   Process Engineering for Fermentation

Although industrial-scale biological H_2 production faces many challenges, incremental progress through small and pilot-scale studies is underway. The use of cellulosic biomass feedstocks, either hydrolyzed by acid hydrolysis and/or enzymatic pretreatments to sugar monomers, or converted *via* consolidated bioprocessing, is a novel approach to biohydrogen that has only recently begun to be investigated.

Hydrogen production in a continuous culture of *C. thermocellum* 27405 was investigated by Magnusson *et al.*[62] A 5 L working volume fermentor was established and growth experiments were maintained for over 3000 h. Substrate concentrations were varied from 1 to 4 g/L and the feed was introduced with continuous N_2 gas sparging to prevent clogging of the feed line. The pH and temperature of the reactor were maintained at 7.0 and 60 °C, respectively, throughout the study. At concentrations above 4 g/L, the delivery of α-cellulose was impaired due to feed-line clogging and it became difficult to maintain a homogenous suspension. The highest total gas (H_2 plus CO_2) production rate, 56.6 mL/L/h, was observed at a dilution rate of 0.042/h and substrate concentration of 4 g/L. Under these conditions, the H_2 production rate was 5.06 mmol/L/h. Acetate and ethanol were the major soluble end-products, while lactate and formate were greatly reduced compared to production in batch cultures. The concentrations of all metabolites increased with increasing substrate concentration, with the exception of lactate. Despite a number of short-term electrical and mechanical failures during the testing period, the system recovered quickly, exhibiting substantial robustness. A carbon balance indicated near 100% carbon recovery. This study shows that long-term, stable H_2 production can be achieved during direct fermentation of an insoluble cellulosic substrate under continuous culture conditions.

10.4.1 Single-Phase Fermentation Reactions

Two primary factors control the thermodynamic feasibility of H_2 production, the internal redox balance and the partial pressure of H_2. The pH of the culture medium has also been shown to affect the free energy of proton reduction to H_2.[82] Furthermore, pH also has an impact on the regulation of carbon and electron flux through the various branches of the fermentation pathways. Hydrogen synthesis by *C. thermocellum*, for example, decreased as the pH of the medium drops below 6.8[33,80] and is completely absent at pH 6.3.[83] Thus, maintaining the pH of H_2 producing cultures in the optimal range of the microorganism is extremely important for continuous H_2 production.

In bacteria with branched fermentative pathways, H_2 synthesis decreases and metabolic pathways shift to production of alternate reduced products such as lactate, ethanol, acetone, butanol, or alanine as H_2 concentrations increase. H_2 concentrations of greater than 2000 Pa can completely inhibit growth of some organisms.[30,84]

Veit *et al.*[85] demonstrated that *in vivo* H_2 partial pressures are associated with $NADH/NAD^+$ and $NADPH/NADP^+$ ratios observed during growth. They constructed formate hydrogen lyase (FHL) mutant strains of *E. coli* that expressed either the NADP,Ferredoxin oxidoreductase of *B. subtilis*, or the NAD(P),ferredoxin oxidoreductase of *Chlorobium tepidum*. In these strains, fermentation would yield H_2 only from NADH or NADPH through ferredoxin and HydA. They confirmed that when H_2 was produced from NADH, the equilibrium H_2 pressure was less than 50 Pa. Moreover, these strains rapidly consumed added H_2 until the same H_2 partial pressures were re-established. When H_2 was produced from NADPH, they measured H_2 equilibrium concentrations of up to 400 Pa. Nath *et al.*[86] also demonstrated that significant increases in H_2 yields may be obtained by decreasing H_2 partial pressure. By lowering the total gas pressure in the headspace of a batch fermentation bioreactor containing *Enterobacter cloacae* DM11 from 760 mm Hg to 380 mm Hg, the molar yield of H_2 increased from 1.9 mol to 3.9 mol H_2/mol glucose.

The concentration of oxidized intermediates and end-products such as CO_2, which compete for electrons from NADH, also affect the rate of synthesis and final yield of H_2. In some organisms, cells synthesize succinate and formate using CO_2, phosphoenolpyruvate, and reduced nicotinamide adenine dinucleotide (NADH) *via* a truncated reductive TCA cycle.[87] As with ethanol formation, this pathway competes with reactions in which H_2 is synthesized by NADH-dependent hydrogenases (which oxidize NADH to NAD^+). Efficient removal of CO_2 from the fermentation system would reduce competition for NADH and could result in increased H_2 synthesis.

Various methods for maintaining subnanomolar concentrations of H_2 fermentors have been developed for the pure culture growth of syntrophic organisms involving palladium catalyzed oxidation,[88] or rapid sparging.[89] Sparging a bioreactor with nitrogen (N_2) gas was demonstrated as an effective method of maintaining a low concentration of dissolved H_2 in the fermentation broth.[90,91] These studies indicated that H_2 production does in fact increase with lower concentration of dissolved H_2, with shifts in metabolism and thermo-dynamics presented as the main possible drivers.[91]

Removal and selective purification of H_2 has also been demonstrated using membrane technologies. A hollow fiber/silicone rubber membrane effectively reduced biogas partial pressure in a dark-fermentation system, resulting in a 10% improvement in the rate of H_2 production and a 15% increase in H_2 yield,[92] and a nonporous, synthetic polyvinyltrimethylsilane (PVTMS) membrane was used for production of high-purity H_2 from three different H_2 producing bioreactor systems.[93] In a more recent study, membrane contactors were used to effectively produce fuel gases such as methane and H_2 from organic wastes, or remove CO_2 from microbial gas mixtures.[94] Belafi-Bako *et al.*[95] demonstrated the integration of a combination of porous and nonporous membranes into a biohydrogen fermentation system, which was based on a themophilic, heterotrophic archaebacterium, to achieve simultaneous gas separation and concentration. Substantial gains in H_2 production can also be achieved through optimization of bioreactor designs. Rates of H_2 sythesis far greater than other studies were obtained using activated carbon as a support matrix that allowed retention of an undefined consortium of mesophilic, H_2 producing bacteria within fixed-bed bioreactors.[96]

10.4.2   Two-Phase Systems

Although the theoretical maximum yield of H_2 from a single dark-fermentation reaction is limited to 4 mol H_2/mol glucose, yields of >4 mol H_2/mol glucose can be achieved through two-phased systems in which a second reaction system is used to liberate H_2 contained within fermentation end-products. Two-phase system strategies include dark fermentation followed by photofermentation and dark fermentation followed by use of a microbial electrolysis cell (MEC), which is also referred to as "electrohydrogenesis".

10.4.2.1   Dark Fermentation Followed by Photofermentation

Lo *et al.*[97] investigated a two-phase dark-fermentation process followed by a photofermentation process using acid-hydrolyzed wheat starch. Hydrolyzed starch was continuously introduced to a dark fermentation bioreactor containing *Clostridium butyricum* CGS2, where the hydrolysate was converted to H_2 at a rate of 220 mL/L of culture/h ($= 9.8$ mmol/L/h). The resulting effluent from dark fermentation became the influent of continuous photo H_2 production process inoculated with *Rhodopseudomonas palustris* WP3-5 at 35 °C, pH 7.0, and with 100 W/m^2 irradiation, and 48 h HRT. Combining enzymatic hydrolysis, dark fermentation and photofermentation led to a marked improvement of overall H_2 yield, up to 16.1 mmol H_2/g COD or 3.09 mol H_2/mol glucose, and COD removal efficiency (ca. 54.3%), suggesting the potential of using the proposed integrated process for efficient and high-yield biohydrogen production from starch feedstock.

Nath *et al.*[98] conducted similar experiments in which glucose was fermented to acetate, CO_2 and H_2 in an anaerobic dark fermentation by *Enterobacter cloacae* DM11. This was followed by a second-phase reaction whereby the acetate was converted to H_2 and CO_2 in a photobioreactor by the photo-synthetic bacterium, *Rhodobacter sphaeroides* strain O.U. 001. The yield of H_2 in the first stage was approx. 3.3 mol H_2/mol glucose (approx. 82% of the-oretical). The yield of H_2 in the second stage was between 1.5–1.7 mol H_2/mol acetic acid (37–43% of theoretical). The combined yield of H_2 in the two-stage process was 4.8–5.0 mol H_2/mol substrate, significantly higher than the 3.3 mol H_2/mol glucose obtained in the dark fermentation alone.

10.4.2.2 Dark Fermentation Followed by Electrohydrogenesis

Electrohydrogenesis is a novel process in which H_2 gas may be released from soluble organic molecules, in a device called a microbial electrolysis cell (MEC).[99] In an MEC, bacteria on the anode oxidize organic matter, releasing electrons through the circuit to the cathode where H_2 is formed from protons in the water. The H_2 producing reaction is endothermic, and therefore additional electrons must be provided by a power source. The MEC efficiency relative to the electrical input has reached over 400%.[100]

Hydrogen has been produced in MECs using a variety of substrates, including acetic acid, butyric acid, lactic acid, glucose, cellulose, and waste-water.[101–104] Lalaurette *et al.*[105] used a two-stage dark-fermentation and elec-trohydrogenesis process to convert the recalcitrant lignocellulosic materials into H_2 gas at high yields and rates. Fermentation using *Clostridium thermo-cellum* produced 1.67 mol H_2/mol-glucose at a rate of 0.25 L H_2/L-d with a corn stover lignocellulose feed, and 1.64 mol H_2/mol-glucose and 1.65 L H_2/L-d with a cellobiose feed. The major soluble fermentation byproducts from both lignocelluose and cellobiose substrates were primarily acetic acid, lactic acid, succinic acid, formic acid and ethanol. An additional 800 ± 290 mL H_2/g-COD was produced from a synthetic effluent with a wastewater inoculum (fermentation effluent inoculum; FEI) by electrolysis using MECs. Hydrogen yields were increased to 980 ± 110 mL H_2/g-COD with the synthetic effluent by combining in the inoculum samples from multiple microbial fuel cells (MFCs), each preacclimated to a single substrate (single substrate inocula; SSI).

Hydrogen yields and production rates with SSI and the actual fermentation effluents were 980 ± 110 mL/g-COD and 1.11 ± 0.13 L/L-d (synthetic); 900 ± 140 mL/g-COD and 0.96 ± 0.16 L/L-d (cellobiose); and 750 ± 180 mL/g-COD and 1.00 ± 0.19 L/L-d (lignocellulose). A maximum hydrogen pro-duction rate of 1.11 ± 0.13 L H_2/L reactor/d was produced with synthetic effluent. Energy efficiencies based on electricity needed for the MEC using SSI were $270 \pm 20\%$ for the synthetic effluent, $230 \pm 50\%$ for lignocellulose effluent and $220 \pm 30\%$ for the cellobiose effluent. COD removals were $\sim 90\%$ for the synthetic effluents, and 70–85% based on VFA removal (65% COD removal) with the cellobiose and lignocellulose effluent. The overall hydrogen yield was 9.95 mol-H_2/mol-glucose for the cellobiose. These results show that pre-acclimation of MFCs to single substrates improves performance with a complex mixture of substrates, and that high H_2 yields and gas production rates can be achieved using a two-stage fermentation and MEC process.

References

1. D. B. Levin and R. Chahine, Challenges for renewable hydrogen production from biomass., *Intl. J. Hydrogen Energy*, 2010, **35**, 4962–4969.
2. D. B. Levin, L. Packer and M. Love, Biohydrogen production, prospects and limitations to practical application., *Intl. J. Hydrogen Energy*, 2004, **29**, 173–185.
3. R. Nandi and S. Sengupta, Microbial production of hydrogen, an overview., *Crit. Rev. Microbiol.*, 1998, **24**, 61–84.
4. P. A. M. Claassen, J. B. van Lier, A. M. Lopez Contreras, E. W. J. van Niel, L. Sijtsma, A. J. M. Stams, S. S. de Vries and R. A. Weusthuls, Utilisation of biomass for the supply of energy carriers., *Appl. Microbiol. Biotechnol.*, 1999, **52**, 741–755.
5. K. Nath and D. Das, Improvement of fermentative hydrogen production, various approaches., *Appl. Microbiol. Biotechnol.*, 2004, **65**, 520–529.
6. A. Pecher, F. Zinoni, C. Jatisatienr, R. Wirth, H. Hennecke and A. Bock, On the redox control of synthesis of anaerobically induced enzymes in enterobacteriaceae., *Arch. Microbiol.*, 1983, **136**, 131–136.
7. A. Yoshida, T. Nishimura, H. Kawaguchi, M. Inui and H. Yukawa, Enhanced hydrogen production from formic acid by formate hydrogen lyase-overexpressing *Escherichia coli* strains., *Appl. Environ. Microbiol.*, 2005, **71**, 6762–6768.
8. G. Vardar-Schara, T. Maeda and T. K. Wood, Metabolically engineered bacteria for producing hydrogen via fermentation., *Microbiol. Biotechnol.*, 2008, **1**, 107–125.
9. T. Maeda, V. Sanchez-Torres and T. K. Wood, Metabolic engineering to enhance hydrogen production., *Microbiol. Biotechnol.*, 2008, **1**, 30–39.
10. S. Tanisho and Y. Ishiwatam, Continuous hydrogen production from molasses by the bacterium *Enterobacter aerogenes*, *Intl. J. Hydrogen Energy*, 1994, **19**, 807–812.
11. A. Converti and P. Perego, Use of carbon and energy balances in the study of the anaerobic metabolism of *Enterobacter aerogenes* at variable starting glucose concentrations., *Appl. Microbiol. Biotechnol.*, 2002, **59**, 303–309.
12. E. Guedon, S. Payot, M. Desvaux and H. Petitdemange, Carbon and electron flow in *Clostridium cellulolyticum* grown in chemostat culture on synthetic medium., *J. Bacteriol.*, 1999, **181**, 3262–3269.

13. S. Saint-Amans, L. Girbal, J. Andrade, K. Ahrens and P. Soucaille, Regulation of carbon and electron flow in *Clostridium butyricum* VPI 3266 grown in glucose-glycerol mixtures., *J. Bacteriol.*, 2001, **183**, 1748–1754.

14. B. Soboh, D. Linder and R. Hedderich, A multisubunit membrane-bound [NiFe] hydrogenase and an NADH-dependent Fe-only hydrogenase in the fermenting bacterium *Thermoanaerobacter tengcongensis.*, *Microbiol.*, 2004, **150**, 2451–2463.

15. C. R. Carere, V. Kalia, R. Sparling, N. Cicek and D. B. Levin, Pyruvate catabolism and hydrogen synthesis pathway genes of *Clostridium thermocellum* ATCC 27405. *Indian J.*, *Microbiol.*, 2008, **48**, 252–266.

16. J. C. Fontecilla-Camps, A. Volbeda, C. Cavazza and Y. Nicolet, Structure/function relationships of [NiFe]- and [FeFe]-hydrogenases, *Chem. Rev.*, 2007, **107**, 4273–4303.

17. M. C. Posewitz, P. King, S. L. Smolinski, L. Zhang, M. Seibert and M. L. Ghirardi, Discovery of two novel radical S-adenosylmethionine proteins required for the assembly of an active [Fe]-hydrogenase., *J. Biol. Chem.*, 2004, **279**, 25711–25720.

18. E. M. Shepard, B. R. Duffus, S. J. George, S. E. McGlynn, M. R. Challand, K. D. Swanson, P. L. Roach, S. P. Cramer, J. W. Peters and J. B. Broderick, [FeFe]-hydrogenase maturation, HydG-catalyzed synthesis of carbon monoxide., *J. Am. Chem. Soc.*, 2010, **132**, 9247–9249.

19. I. Czech, A. Silakov, W. Lubitz and T. Happe, The [FeFe] hydrogenase maturase HydF from Clostridium acetobutylicum contains a CO and CN-ligated iron cofactor., *FEBS Lett.*, 2010, **584**, 638–642.

20. A. Volbeda, M. H. Charon, C. Piras, E. C. Hatchikian, M. Frey and J. C. Fontecilla-Camps, Crystal structure of the nickel-iron hydrogenase from Desulfovibrio gigas., *Nature*, 1995, **373**, 580–587.

21. S. Reissmann, E. Hochleitner, H. Wang, A. Paschos, F. Lottspeich, R. S. Glass and A. Böck, Taming of a poison, biosynthesis of the NiFe-hydrogenase cyanide ligands., *Science*, 2003, **299**, 1067–1070.

22. L. Casalot and M. Rousset, Maturation of the [NiFe] hydrogenases., *Trends Microbiol.*, 2001, **9**, 228–237.

23. A. Joe Shaw, D. A. Hogsett and L. R. Lynd, Identification of the [FeFe]-hydrogenase responsible for hydrogen generation in *Thermoanaerobacterium saccharolyticum* and demonstration of increased ethanol yield via hydrogenase knockout., *J. Bacteriol.*, 2009, **191**, 6457–6464.

24. J. Woodward, M. Orr, K. Cordray and E. Greenbaum, Enzymatic production of biohydrogen., *Nature*, 2000, **405**, 1014–1015.

25. R. K. Thauer, K. Jungermann and K. Decker, Energy conservation in chemotrophic anaerobic bacteria., *Bacteriol. Rev.*, 1977, **41**, 100–180.

26. D. Noguera, R. Araki and B. E. Rittmann, Soluble microbial products (SMP) in anaerobic chemostats., *Biotechnol. Bioeng.*, 1994, **44**, 1040–1047.

27. N. S. Lee, M. B. Salerno and B. E. Rittmann, Thermodynamic evaluation of H_2 production in glucose fermentation, *Environ. Sci. Technol.*, 2008, **42**, 2401–2407.

28. S. V. Ginkel and S. Sung, Biohydrogen production as a function of pH and substrate concentration., *Environ. Sci. Technol.*, 2001, **35**, 4726–4730.
29. N. Kumar, A. Ghosh and D. Das, Redirection of biochemical pathways for the enhancement of H$_2$ production by *Enterobacter cloacae*., *Biotechnol. Lett.*, 2001, **23**, 537–541.
30. P. C. Hallenbeck, Fundamentals of the fermentative production of hydrogen., *Water Sci. Technol.*, 2005, **52**, 21–29.
31. W. Park, S. H. Hyun, S. E. Oh, B. E. Logan and I. S. Kim, Removal of headspace CO$_2$ increases biological H$_2$ production., *Environ. Sci. Technol.*, 2005, **39**, 4416–4420.
32. R. Datar, J. Huang, P.-C. Maness, A. Mahagheghi, S. Czernik and E. Chounet, Hydrogen production from the fermentation of corn stover biomass pretreated with a steam explosion process., *Intl. J. Hydrogen Energy*, 2007, **32**, 932–939.
33. R. Islam, N. Cicek, R. Sparling and D. B. Levin, Influence of initial cellulose concentration on the carbon flow distribution during batch fermentation by *Clostridium thermocellum* ATCC 27405., *Appl. Microbiol. Biotechnol.*, 2009, **82**, 141–148.
34. D. B. Levin, C. R. Carere, N. Cicek and R. Sparling, Challenges for biohydrogen production via direct lignocellulose fermentation., *Intl. J. Hydrogen Energy*, 2009, **34**, 7390–7403.
35. R. Conrad and B. Wetter, Influence of temperature on energetic of hydrogen metabolism in homoacetogenic, methanogenic, and other anaerobic bacteria., *Arch. Microbiol.*, 1990, **155**, 94–98.
36. B. Fabiano and P. Perego, Thermodynamic study and optimization of hydrogen production by *Enterobacter aerogenes*., *Intl. J. Hydrogen Energy*, 2002, **27**, 149–156.
37. E. W. J. van Niel, M. A. W. Budde, G. G. de Haas, F. J. van der Wal, P. A. M. Claassen and A. J. M. Stams, Distinctive properties of high hydrogen producing extreme thermophiles, *Caldicellulosiruptor* saccharolyticus and *Thermotoga elfii*., *Intl. J. Hydrogen Energy*, 2002, **27**, 1391–1398.
38. T. de Vrije, A. E. Mars, M. A. Budden, M. H. Lai, C. Dijkema and P. de Waard, Glycolytic pathway and hydrogen yield studies of the extreme thermophile *Caldicellulosiruptor saccharolyticus*., *Appl. Microbiol. Biotechnol.*, 2007, **74**, 1358–1367.
39. Y. Mu, X. J. Zheng, H. Q. Yu and R. F. Zhu, Biological hydrogen production by anaerobic sludge at various temperatures., *Intl. J. Hydrogen Energy*, 2006, **31**, 780–785.
40. B. Schink, Energetics of syntrophic cooperation in methanogenic degradation., *Microbiol. Mol. Biol. Rev.*, 1977, **61**, 262–280.
41. P. C. Hallenbeck and D. Ghosh, Advances in fermentative biohydrogen production, the way forward., *Trends Biotechnol.*, 2009, **27**, 287–297.
42. G. J. Schut and M. W. W. Adams, The iron-hydrogenase of *Thermotoga maritima* utilizes ferredoxin and NADH synergistically, a new perspective on anaerobic hydrogen production., *J. Bacteriol.*, 2009, **191**, 4451–4457.

43. K.-J. Wu, C.-F. Chang and J.-S. Chang, Simultaneous production of biohydrogen and bioethanol with fluidized-bed and packed-bed bio-reactors containing immobilized anaerobic sludge., *Proc. Biochem.*, 2007, **42**, 1165–1171.

44. A. S. Danko, A. A. Abreu and M. M. Alves, Effect of arabinose concentration on dark fermentation hydrogen production using different mixed cultures., *Intl. J. Hydrogen Energy*, 2008, **33**, 4527–4532.

45. J. Li, N. Ren, B. Li, Z. Qin and J. He, Anaerobic biohydrogen production from monosaccharides by a mixed microbial community culture., *Biores. Technol.*, 2008, **99**, 6528–6537.

46. F. Y. Chang and C. Y. Lin, Calcium effect on fermentative hydrogen production in an anaerobic up-flow sludge blanket system., *Water Sci. Technol.*, 2006, **54**, 105–112.

47. G. Kyazze, N. Martinez-Perez, R. Dinsdale, G. C. Premier, F. R. Hawkes, A. J. Guwy and D. L. Hawkes, Influence of substrate concentration on the stability and yield of continuous biohydrogen production., *Biotechnol. Bioeng.*, 2006, **93**, 971–979.

48. C.-N. Lin, S.-Y. Wu and J.-S. Chang, Fermentative hydrogen production with a draft tube fluidized bed reactor containing silicone-gel-immobilized anaerobic sludge., *Intl. J. Hydrogen Energy*, 2006, **31**, 2200–2210.

49. C.-N. Lin, S.-Y. Wu, J.-S. Chang and J.-S. Chang, Biohydrogen production in a three-phase fluidized bed bioreactor using sewage sludge immobilized by ethylene-vinyl acetate copolymer., *Biores. Technol.*, 2009, **100**, 3298–3301.

50. S. K. Khanal, W. H. Chen, L. Li and S. Sung, Biohydrogen production in continuous-flow reactor using mixed microbial culture., *Water Environ Res.*, 2006, **78**, 110–117.

51. M. Fritsch, W. Hatmeier and J.-S. Chang, Enhancing hydrogen production of *Clostridium butyricum* using a column reactor with square-structured ceramic fittings., *Intl J. Hydrogen Energy*, 2008, **33**, 6549–6555.

52. L. J. Thompson, V. M. Gray, B. Kalala, D. Lindsay, K. Reynolds and A. von Holy, Biohydrogen production by *Enterobacter cloacae* and *Citrobacter freundii* in carrier induced granules., *Biotechnol. Lett.*, 2008, **30**, 27127–27134.

53. C.-Y. Lin and W.-C. Hung, Enhancement of fermentative hydrogen/ethanol production from cellulose using mixed anaerobic cultures., *Intl. J. Hydrogen Energy*, 2008, **33**, 3660–3667.

54. H. Yu, Z. Zhu, W. Hu and H. Zhang, Hydrogen production rice winery wastewater in an upflow anaerobic reactor by using mixed anaerobic cultures., *Intl. J. Hydrogen Energy*, 2002, **27**, 1359–1365.

55. H. Yokoi, R. Maki, J. Hirose and S. Hayashi, Microbial production of hydrogen from starch-manufactering wastes., *Biomass Bioenergy*, 2002, **22**, 389–395.

56. M. F. Arooj, S. K. Han, S. H. Kim, D. H. Kim and H. S. Shin, Sludge characteristics in anaerobic SBR system producing hydrogen gas., *Water Res.*, 2007, **41**, 1177–1184.

57. H. Argun, F. Kargi, I. K. Kapdan and R. Qztekin, Batch dark fermentation of powdered wheat starch to hydrogen gas, Effects of the initial substrate and biomass concentrations., *Intl. J. Hydrogen Energy*, 2008, **33**, 6109–6115.

58. C.-H. Cheng, C.-H. Hung, K.-S. Lee, P.-Y. Liau, C.-M. Liang, L.-H. Yang, P.-J. Lin and C.-Y. Lin, Microbial community structure of a starch-feeding fermentative hydrogen production reactor operated under different incubation conditions., *Intl. J. Hydrogen Energy*, 2008, **33**, 5242–5249.

59. K.-J. Wang and J.-S. Chang, Continuous biohydrogen production from starch with granulated mixed bacterial microflora., *Energy Fuels*, 2008, **22**, 93–97.

60. D. B. Levin, R. Sparling, R. Islam and N. Cicek, Hydrogen production by *Clostridium thermocellum* 27405 from cellulosic biomass substrates., *Intl. J. Hydrogen Energy*, 2006, **31**, 1496–1503.

61. L. Magnusson, R. Islam, R. Sparling, D. Levin and N. Cicek, Direct hydrogen production from cellulosic waste materials with a single-step dark fermentation process., *Intl. J. Hydrogen Energy*, 2008, **33**, 5398–5403.

62. L. Magnusson, N. Cicek, R. Sparling and D. B. Levin, Continuous hydrogen production during fermentation of a-cellulose by the thermophillic bacterium *Clostridium thermocellum*., *Biotechnol. Bioeng.*, 2009, **102**, 759–766.

63. D. Das and T. N. Vizeroglu, Advances in biological hydrogen production processes., *Intl. J. Hydrogen Energy*, 2008, **33**, 6046–6057.

64. D. Das, Advances in biohydrogen production processes, An approach towards commercialization, *Intl. J. Hydrogen Energy*, 2009, **34**, 7349–7357.

65. A. L. Demain, M. Newcomb and Wu. J. H. D. Cellulase, clostridia, and ethanol., *Microbiol. Mol. Biol. Rev.*, 2005, **69**, 124–154.

66. M. Desvaux, *Clostridium cellulolyticum*, model organism of mesophilic cellulolytic clostridia., *FEMS Microbiol. Rev.*, 2005, **29**, 741–764.

67. L. R. Lynd, P. J. Weimer, W. H. van Zyl and I. S. Pretorius, Microbial cellulose utilization, Fundamentals and biotechnology., *Microbiol. Mol. Biol. Rev.*, 2002, **66**, 506–577.

68. C. R. Carere, R. Sparling, N. Cicek and D. B. Levin, Third generation biofuels via direct cellulose fermentation., *Intl. J. Mol. Sci.*, 2008, **9**, 1342–1360.

69. N. Ren, A. Wang, G. Cao, J. Xu and L. Gao, Bioconversion of lignocellulosic biomass to hydrogen, Potential and challenges., *Biotechnol. Adv.*, 2009, **27**, 1051–1060.

70. G. D. Saratale, S.-D. Chen, Y.-C. Lo, R. G. Satatale and J.-S. Chang, Outlook of biohydrogen production from lignocellulosic feedstock using dark fermentation - A review., *J. Sci. Indust. Res.*, 2009, **67**, 962–979.

71. L. R. Lynd, W. H. van Zyl, J. E. McBride and M. Laser, Consolidated bioprocessing of cellulosic biomass, an update., *Curr. Opin. Biotechnol.*, 2005, **16**, 577–583.

72. N. Ren, A. Wang, L. Gao, L. Xin, D.-J. Lee and A. Su, Bioaugmented hydrogen production from carboxymethyl cellulose and partially delignified corn stalks using isolated cultures., *Intl. J. Hydrogen Energy*, 2008, **33**, 5250–5255.
73. G. Cao, N. Ren, A. Wang, D.-J. Lee, W. Guo, B. Liu, Y. Feng and Q. Zhao, Acid hydrolysis of corn stover for biohydrogen production using *Thermoanaerobacterium thermosaccharolyticum* W16., *Intl. J. Hydrogen Energy*, 2009, **34**, 7182–7188.
74. Y.-C. Lo, Y.-C. Su, C.-Y. Chen, W.-M. Chen, K.-S. Lee and J.-S. Chang, Biohydrogen production from cellulosic hydrolysate produced via temperature-shift-enhanced bacterial cellulose hydrolysis., *BioRes. Technol.*, 2009, **100**, 5802–5807.
75. D. Li and H. Chen, Biological hydrogen production from steam-exploded straw by simultaneous saccharification and fermentation., *Intl. J. Hydrogen Energy*, 2007, **32**, 1742–1748.
76. P. Kaparaju, M. Serrano, A. B. Thomsen, P. Kongjan and I. Angelidaki, Bioethanol, biohydrogen, and biogas production from wheat straw in a biorefinery concept., *BioRes. Technol.*, 2009, **100**, 2562–2568.
77. Z. Kadar, T. de Vrije, M. A. W. Budde, Z. Szengyel, K. Reczey and P. A. M. Claassen, Hydrogen production from paper sludge hydrolysate., *Appl. Biochem. Biotechnol.*, 2003, **107**, 557–566.
78. Z. Kadar, T. de Vrije, G. E. van Noorden, M. A. W. Budde, Z. Szengyel, K. Reczey and P. A. M. Claassen, Yields from glucose, xylose, and paper sludge hydrolysate during hydrogen production by the extreme thermophile *Caldicellulosiruptor saccharolyticus.*, *Appl. Biochem. Biotechnol.*, 2004, **114**, 497–508.
79. U. Ramachandran, N. Wrana, N. Cicek, R. Sparling and D. B. Levin, Hydrogen production and end-product synthesis patterns by *Clostridium termitidis* strain CT1112 in batch fermentation cultures with cellobiose or α-cellulose., *Intl. J. Hydrogen Energy*, 2008, **33**, 7006–7012.
80. R. Islam, N. Cicek, R. Sparling and D. B. Levin, Effect of substrate loading on hydrogen production during anaerobic fermentation by *Clostridium thermocellum* 27405., *Appl. Microbiol. Biotechnol.*, 2006, **72**, 576–583.
81. R. Sparling, R. Islam, N. Cicek, C. Carere, H. Chow and D. B. Levin, Formate synthesis by *Clostridium thermocellum* during anaerobic fermentation., *Can. J. Microbiol.*, 2006, **52**, 681–688.
82. D. L. Valentine, W. S. Reeburgh and D. C. Blanton, A culture apparatus for maintaining H_2 at sub-nanomolar concentrations., *J. Microbiol. Methods*, 2000, **39**, 243–251.
83. T. Rydzak, D. B. Levin, N. Cicek and R. Sparling, Growth phase dependant enzyme profile of pyruvate catabolism and end-product formation in *Clostridium thermocellum* ATCC 27405., *J. Biotechnol.*, 2009, **140**, 169–175.
84. P. C. Hallenbeck and J. R. Benemann, Biological hydrogen production, fundamentals and limiting processes., *Intl. J. Hydrogen Energy*, 2002, **27**, 1185–1193.

85. A. Veit, M. K. Akhtar, T. Mizutani and P. R. Jones, Constructing and testing the thermodynamic limits of synthetic NAD(P)H,H$_2$ pathways., *Microbiol. Biotechnol.*, 2008, **1**, 382–394.

86. K. Nath, A. Kumar and D. Das, Effect of some environmental parameters on fermentative hydrogen production by *Enterobacter cloacae* DM11., *Can. J. Microbiol.*, 2006, **52**, 525–532.

87. J. Sridhar, M. A. Eiteman and J. W. Wiegel, Elucidation of Enzymes in Fermentation Pathways Used by *Clostridium thermosuccinogenes* Growing on Inulin., *Appl. Environ. Microbiol.*, 2000, **66**, 246–251.

88. D. O. Mountfort and H. F. Kaspar, Palladium-mediated hydrogenation of unsaturated hydrocarbons with hydrogen gas released during anaerobic cellulose degradation., *Appl. Environ. Microbiol.*, 1986, **52**, 744–750.

89. O. Mizuno, R. Dinsdale, F. R. Hawkes, D. L. Hawkes and T. Noike, Enhancement of hydrogen production from glucose by nitrogen gas sparging., *Biores. Technol.*, 2000, **73**, 59–65.

90. I. Hussy, F. R. Hawkes, R. Dinsdale and D. L. Hawkes, Continuous fermentative hydrogen production from a wheat starch co-product by mixed microflora., *Biotechnol. Bioeng.*, 2003, **84**, 619–626.

91. J. T. Kraemer and D. M. Bagley, Supersaturation of dissolved H$_2$ and CO$_2$ during fermentative hydrogen production with N$_2$ sparging, *Biotechnol. Lett.*, 2006, **28**, 1485–1491.

92. T.-M. Liang, S.-S. Cheng and K.-L. Wu, Behavioral study on hydrogen fermentation reactor installed with silicone rubber membrane., *Int. J. Hydrogen Energy*, 2002, **27**, 1157–1165.

93. V. V. Teplyakov, L. G. Gassanova, E. G. Sostina, E. V. Slepova, M. Modigell and A. I. Netrusov, Lab scale bioreactor integration with active membrane system for hydrogen production, Experience and prospects., *Intl. J. Hydrogen Energy*, 2002, **27**, 1149–1155.

94. L. G. Gassanova, A. I. Netrusov, V. V. Teplyakov and M. Modigell, Fuel gases from organic wastes using membrane bioreactors., *Desalination*, 2006, **198**, 56–66.

95. K. Belafi-Bako, D. Bucsu, Z. Pientka, B. Balint, K. L. Kovacs and M. Wessling, Integration of biohydrogen fermentation and gas separation process to recover and enrich hydrogen., *Intl. J. Hydrogen Energy*, 2006, **31**, 1490–1495.

96. J.-S. Chang, K.-S. Lee and P.-J. Lin, Biohydrogen production with fixed-bed bioreactors., *Intl. J. Hydrogen Energy*, 2002, **27**, 1167–1174.

97. Y.-C. Lo, S.-D. Chen, C.-Y. Chen, T.-I. Huang, C.-Y. Lin and J.-S. Chang, Combining enzymatic hydrolysis and dark-photo fermentation processes for hydrogen production from starch feedstock, A feasibility study., *Intl. J. Hydrogen Energy*, 2008, **33**, 5224–5233.

98. K. Nath, M. Muthukumar, A. Kumar and D. Das, Kinetics of two-stage fermentation process for the production of hydrogen., *Intl. J. Hydrogen Energy*, 2008, **33**, 1195–1203.

99. H. Liu, S. Grot and B. E. Logan, Electrochemically assisted microbial production of hydrogen from acetate., *Environ. Sci. Technol.*, 2005, **39**, 4317–4320.
100. D. Call and B. E. Logan, Hydrogen production in a single chamber microbial electrolysis cell lacking a membrane., *Environ. Sci. Technol.*, 2008, **42**, 3401–3406.
101. S. Cheng and B. E. Logan, Sustainable and efficient biohydrogen production via electrohydrogenesis., *Proc. Natl. Acad. Sci. USA*, 2007, **104**, 18871–18873.
102. J. Ditzig, H. Liu and B. E. Logan, Production of hydrogen from domestic wastewater using a bioelectrochemically assisted microbial reactor (BEAMR)., *Intl J. Hydrogen Energy*, 2007, **32**, 2296–2304.
103. R. Rozendal, H. V. M. Hamelers, R. J. Molenkamp and C. J. N. Buisman, Performance of single chamber biocatalyzed electrolysis with different types of ion exchange membranes., *Water Res.*, 2007, **41**, 1984–1994.
104. R. A. Rozendal, T. H. J. A. Sleutels, H. V. M. Hamelers and C. J. N. Buisman, Effect of the type of ion exchange membrane on performance, ion transport, and pH in biocatalyzed electrolysis of wastewater., *Water Sci. Technol.*, 2008, **57**, 1757–1762.
105. E. Lalaurette, S. Thammannagowda, A. Mohagheghi, P.-C. Maness and B. E. Logan, Hydrogen production from cellulose in a two-stage process combining fermentation and electrohydrogenesis., *Intl J. Hydrogen Energy*, 2009, **34**, 6201–6210.

CHAPTER 11

Integrated Technology for Biobased Composites

ZHIYONG CAI,* ALAN W. RUDIE AND
THEODORE H. WEGNER

Forest Product Laboratory, USDA Forest Service, One Gifford Pinchot
Drive, Madison, WI 53726-2398, USA
*Email: zcai@fs.fed.us

11.1   Introduction

Forests play a major role in the ecosystem sustainability and general health of
our planet. The biomass contained in our forests and other green vegetations
affects the carbon cycle, climate change, habitat protection, clean water supplies, and sustainable economy. Exciting new opportunities are emerging for
sustainably meeting global energy needs and simultaneously creating high-
value biobased products from wood, forest, agricultural residues, and other
biobased materials.

Biomass is commonly referred to as lignocellulosics because the two dominant chemical components of plants are cellulose, the structural polymer that
represents about 50% of the plant material, and lignin, a crosslinked phenolic
polymer that performs the role of adhesive bonding the components of the cell
wall together. Globally, the vast lignocellulosic resource provides about half of
all major industrial raw materials for renewable energy, chemical feedstock,
and biocomposites. Conversion of woody biomass to biofuels is technically
feasible, but this conversion process is marginally economical with the current
technology and price of crude petroleum. An integrated utilization of biomass

RSC Green Chemistry No. 18
Integrated Forest Biorefineries
Edited by Lew Christopher
© The Royal Society of Chemistry 2013
Published by the Royal Society of Chemistry, www.rsc.org

has been proposed as means to overcome economic shortcomings by optimizing biomass use and value for a wider array of products.[1] Engineered biobased composites provide a tool for resource management because they can add value to low-value fiber resources and thereby promote demand for diverse biofiber feedstocks, including small-diameter timber, fast-growing plantation trees, exotic-invasive species, hazardous forest fuels, and agricultural crop residues. At the same time, engineering biobased composites serve as a means for economic development of rural communities and provide value-added commodity products from recycled or undervalued materials or problematic natural resources.[2]

The term biobased composite is being used to describe any woody material adhesively bonded together. This product mix ranges from fiberboard to laminated beams and components. Composites are used for a number of nonstructural and structural applications in product lines ranging from panels for interior covering purposes, panels for exterior uses, in furniture and support structures (Figure 11.1). This chapter describes the general composition, materials and processes used to manufacture biobased composite materials. This chapter also describes wood–nonwood composites.

Conventional biobased composites are primarily made from wood and other woody materials with only a few per cent resin and other additives. Product types can be subcategorized based on the physical configuration of the woody elements used to make these products. The morphology of the woody elements influences the properties of composite materials, and can be controlled by

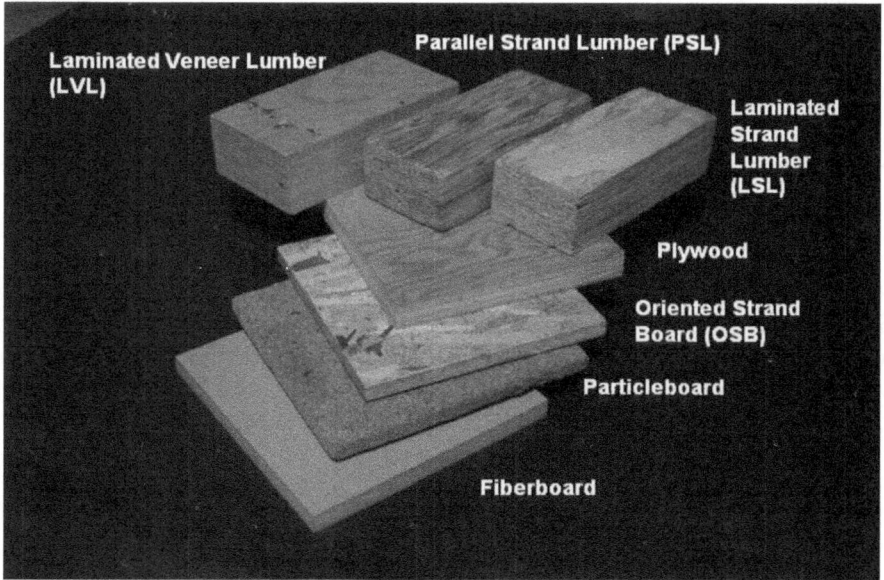

Figure 11.1 Examples of various composite products. From clockwise from top left: laminated veneer lumber, parallel strand lumber, laminated strand lumber, plywood, oriented strand board, particleboard, and fiberboard.

Table 11.1 Static bending properties of different wood and wood-based composites.

Material	Specific gravity	Static Bending Properties			
		Modulus of Elasticity		Modulus of Rupture	
		GPa	*(×10⁶ psi)*	*MPa*	*(psi)*
Clear wood					
– White Oak	0.68	12.27	(1.78)	104.80	(15 200)
– Red Maple	0.54	11.31	(1.64)	92.39	(13 400)
– Douglas-fir (Coastal)	0.48	13.44	(1.95)	85.49	(12 400)
– Western white pine	0.38	10.07	(1.46)	66.88	(9700)
– Longleaf pine	0.59	13.65	(1.98)	99.97	(14 500)
Panel products					
– Hardboard	0.9–1.0	3.10–5.52	(0.45–0.80)	31.02–56.54	(4500–8200)
– Medium densityfiberboard	0.7–0.9	3.59	(0.52)	35.85	(5200)
– Particleboard	0.6–0.8	2.76–4.14	(0.40–0.60)	15.17–24.13	(2200–3500)
– Orientedstrand board	0.5–0.8	4.41–6.28	(0.64–0.91)	21.80–34.70	(3161–5027)
– Plywood	0.4–0.6	6.96–8.55	(1.01–1.24)	33.72–42.61	(4890–6180)
Structural timber Products					
– Glued laminated timber	0.4–0.6	9.00–14.50	(1.30–2.10)	28.61–62.62	(4150–9080)
– Laminated veneer lumber	0.4–0.7	8.96–19.24	(1.30–2.79)	33.78–86.18	(4900–12500)
Wood–nonwood composites					
– Wood plastic	0.8–1.1	1.53–4.23	(0.22–0.61)	25.41–52.32	(3684–7585)

selection of the raw material and by the processing techniques used to generate the wood elements. Composite properties can also be controlled by segregation and stratification of wood elements having different morphologies in different layers of the composite material (Table 11.1). In conventional wood-based composites, properties can also be controlled. The physical configuration of the wood element, adjusting the density of the composite, adjusting adhesive resin, or adding additives are just a few of the many ways to influence properties. Performance standards are in place for many conventional wood-based composite products (Table 11.2).

11.2 Conventional Biobased Composite Materials

11.2.1 Composite Elements

Biobased composites are composed primarily of woody elements (often 90% or more by mass) bound together with a resin and other additives. The vast

Table 11.2 Commercial product or performance standards for wood-based composites.

Product Category	Applicable standard	Name of Standard	Source
Plywood	PS 1-09	Voluntary product standard PS 1-07 construction and industrial plywood	4
	PS 2-04	Voluntary product standard PS 2-04 performance standard for wood-based structural-use panels	11
	HP-1-04	Voluntary product standard HP-1-04 Hardwood and decorative plywood	5
Oriented strandboard (OSB)	PS 2-04	Voluntary product standard PS 2-04 performance standard for wood-based structural-use panels	11
Particleboard	ANSI A208.1-1999	Particleboard standard	12
Fiberboard	ANSI A208.2-2002	MDF standard	7
	ANSI A135.4-2004	Basic Hardboard	13
	ANSI A135.5-2004	Prefinished hardboard paneling	14
	ANSI A135.6-2006	Hardboard siding	15
	ANSI A194.1	Cellulosic fiberboard	16
Glue laminated timber (Glulam)	ANSI/AITC 190.1	American National Standard for Wood Products – Structural Glue-laminated timber	17
Structural composite lumber (including laminated veneer lumber (LVL), laminated strand lumber (LSL), and parallel strand lumber (PSL))	ASTM D 5456-08	Standard specification for evaluation of structural composite lumber products	6

lignocellulosic resource provides the basic element for composite products such as the fiber, as it is in paper, much larger particles varying in size and geometry. Raw materials providing these elements include shavings, straws, hemp, sugarcane, bamboo, sawdust, fiber, particles, wafers, strands, and veneer.

11.2.2 Adhesives

Commonly used resin or binder systems in biobased composites include phenol–formaldehyde, urea–formaldehyde, melamine–formaldehyde, and iso-cyanate. The selection of the resin system is dependent upon the process, cost, product standards, and applications.

Phenol–formaldehyde (PF) resins, commonly referred to as phenolic resins, are typically used in the manufacture of construction plywood and oriented strandboard in structural applications where exposure to weather during construction is a concern. Phenolic resins are relatively slow-curing compared with other thermosetting resins. Cured phenolic resins remain chemically stable at elevated temperatures, even under wet conditions. The PF resin bonds are sometimes referred to as being "boil-proof" because of their ability to maintain structural integrity and adequate bonding after boiling water test. The inherently darker color of PF resin compared with other resins may make them aesthetically unsuitable for product applications such as interior paneling and furniture.

Urea–formaldehyde (UF) resins are typically used in the manufacture of products used in interior applications, for example, particleboard and medium-density fiberboard (MDF). They cure at lower temperatures than PF resins. Urea–formaldehyde resins are the lowest-cost thermosetting adhesive resins. They offer light color, which often is a requirement in the manufacture of decorative products. However, the release of formaldehyde from products bonded with UF is a growing health and environmental concern. Recently enacted EPA regulations (http://www.epa.gov/opptintr/chemtest/formaldehyde/) are forcing changes in UF chemistry, driving some manufacturers to switch away from UF.

Melamine–formaldehyde (MF) resins are used primarily for decorative laminates, paper treating, and paper coating. They are typically more expensive than PF resins. MF resins may, despite their high cost, be used in bonding conventional wood-based composites. When used in this application, they typically are blended with UF resins. Melamine–UF resins are used to pass formaldehyde emission standards, where an inconspicuous (light color) adhesive is needed, and greater water resistance than can be attained with UF resin is required.

Isocyanate as diphenylmethane di-isocyanate (MDI) resin is commonly used, as an alternative to PF resin, primarily in composite products fabricated from strands. Polymeric MDI (pMDI) resin, which is closely related to MDI resin, is also commonly used in this application. Isocyanate resins are typically more costly than PF resins, but require much lower addition rates, (for example 2% compared to 6–7% w/w UF in particleboard), have higher cure rates, and will tolerate higher moisture contents in the wood source. Facilities that use MDI are required to take special precautionary protective measures, as the uncured resin can result in chemical sensitization of persons exposed to it. Cured isocyanate resin poses no recognized health concerns.

Biobased adhesives, primarily protein glues, were widely used prior to the early 1970s in construction plywood. In the mid 1970s, they were supplanted by PF adhesives, on the basis of the superior bond durability provided by phenolics. Several soy-protein-based resin systems for interior application have recently been developed and commercialized. Durable adhesive systems may also be derived from tannins, lignin, or soy blended with PF.

11.2.3 Additives

A number of additives are used in the production of conventional composite products. One of the most notable additives is wax, which is used to provide finished products with some resistance to liquid-water absorption. In particle- and fiberboard products, wax emulsions provide limited-term water resistance and dimensional stability when the board is wetted. Even small amounts (0.5% to 1%) act to retard the rate of liquid water pickup for limited time periods. Other additives used for specialty products include preservatives, moldicides, fire retardants, and impregnating resins.

11.2.4 Products

Conventional wood-based composites made from various constituent materials can be subcategorized based on the physical configuration of the wood elements used to make these products: veneer, particle, strand or fiber.

11.2.4.1 Oriented Strandboard

Oriented strandboard (OSB) is an engineered structural-use panel manufactured from thin wood strands bonded together with waterproof resin, typically PF or MDI. Since its debut in 1978, OSB has been rapidly accepted in new residential construction in many areas of North America. It is used extensively for roof, wall, and floor sheathing in residential and commercial construction. The wood strands typically have an aspect ratio (strand length divided by width) of at least 3. OSB panels are usually made with three layers of strands, the outer faces having longer strands aligned in the long direction of the panel and a core layer that is crossaligned or laid randomly using the smaller strands or fines. The orientation of different layers of aligned strands gives OSB its unique characteristics, including greater bending strength and stiffness in the oriented or aligned direction. Control of strand size, orientation, and layered construction allows OSB to be engineered to suit different uses.

In North America, aspen is the predominant wood used for OSB. Other species, such as southern pine, spruce, birch, yellow-poplar, sweetgum, sassafrass, and beech are also suitable raw materials for OSB production. High-density species such as beech and birch are often mixed with low-density species such as aspen to maintain panel properties.[3]

Figure 11.2 shows an OSB manufacturing process. Logs are debarked and then sent to a soaking pond or directly to the stranding process. Long log disk or ring stranders are commonly used to produce wood strands typically measuring 114 to 152 mm (4.5 to 6 in.) long, 12.7 mm (0.5 in.) wide, and 0.6 to 0.7 mm (0.023 to 0.027 in.) thick. Green strands are stored in wet bins and dried prior to panel assembly. Dried strands are blended with adhesive and wax in a highly controlled operation, with separate rotating blenders used for face and core strands.

The strands with adhesive applied are sent to mat formers. Mat formers take on a number of configurations, ranging from electrostatic equipment to mechanical devices containing spinning disks to align strands along the panel's length and star-type crossorienters to position strands across the panel's width. All formers use the long and narrow characteristic of the strands to place it

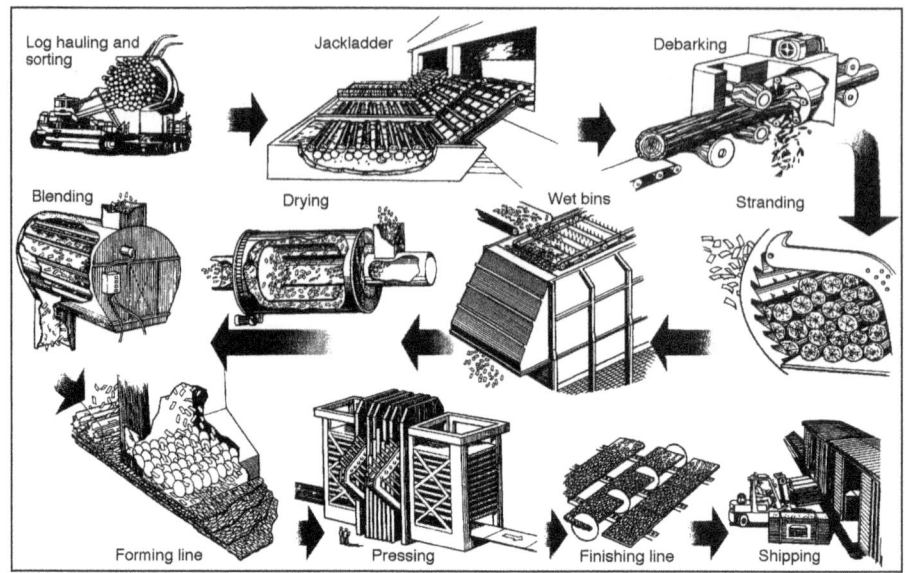

Figure 11.2 Schematic of OSB manufacturing process.
(Courtesy of TECO, Sun Prairie, Wisconsin.)

between the spinning disks or troughs before it is ejected onto a moving screen or conveyor belt below the forming heads. Oriented layers of strands within the mat are dropped sequentially onto a moving conveyor. The conveyor carries the mat into the press.

Once the mat is formed, it is hot pressed. In hot pressing, the loose layered mat of oriented strands is compressed under heat and pressure to cure the resin. As many as sixteen 3.7- by 7.3-m (12- by 24-ft) panels may be formed simultaneously in a multiple-opening press. A more recent development is the continuous press for OSB. The press compacts and consolidates the oriented and layered mat of strands and heats it to 177–204 °C (350 °F to 400 °F) to cure the resin.

OSB is produced to comply with voluntary industry product performance standards. These inspection or certification programs also generally require that the quality control system of a production plant meet specified criteria. OSB panels conforming to these product performance standards are marked with grade stamps.

11.2.4.2 *Plywood*

Plywood generally requires larger-diameter, high-value trees and has limited value for many forest management activities. But in integrated utilization, it is a high-value use for large-diameter straight grain trees. Plywood is a flat panel built up wholly or primarily of sheets of veneer called plies. It is constructed with an odd number of layers with the grain direction of adjacent layers oriented perpendicular to one another. The outside plies are called faces, or face and back plies. Inner plies are plies other than the face or back plies. Inner plies whose grain direction runs parallel to that of the faces are termed "centers", whereas inner plies whose grain direction runs perpendicular to that of the faces are termed "crossbands" or "cores". The outer layers and all odd-numbered layers have their grain direction oriented parallel to the long dimension of the panel. The grain in even-numbered layers is perpendicular to the length of the panel. The center layer may be composed of veneer, lumber, particleboard, or fiberboard; however all-veneer construction is most common in structural plywood.

The properties of plywood depend on the quality of the veneer plies used, the order of layer placement, the adhesive used, and the degree to which bonding conditions are controlled during production. The durability of the adhesive-to-wood bond depends largely on the adhesive used, but also on control of bonding conditions and on veneer quality.

Two classes of plywood are commonly available, covered by separate standards: (a) structural plywood, and (b) hardwood and decorative plywood. The bulk of construction and industrial plywood is used where strength, stiffness, durability, and construction utility are more important than appearance. Structural plywood has traditionally been made from softwoods such as Douglas fir and Southern yellow pine. Construction and industrial plywood is categorized by exposure capability and grade using Voluntary Product Standard PS 1–09.[4]

Hardwood and decorative plywood is made of many different species. Hardwood plywood is normally used in applications including decorative wall panels and furniture and cabinet panels where appearance is more important than strength. Most of the production is intended for interior or protected uses. It is categorized by species and characteristics of face veneer, bond durability, and composition of center layers (veneer, lumber, particleboard, medium-density fiberboard, or hardboard).[5]

11.2.4.3 *Structural Composite Lumber and Timber Products*

Structural composite lumber (SCL) was developed in response to the increasing demand for high-quality lumber at a time when it was becoming difficult to obtain larger diameter, clear and straight grain trees needed for traditional beams and joists. Structural composite lumber products are characterized by smaller pieces of wood glued together into sizes common for solid-sawn lumber.

One type of SCL product is manufactured by laminating veneer with all plies parallel to the length. This product is called laminated veneer lumber (LVL) and consists of specially graded veneer. Another type of SCL product consists of strands of wood or strips of veneer glued together under high pressures and temperatures. Depending upon the component material, this product is called laminated strand lumber (LSL), parallel strand lumber (PSL), or oriented strand lumber (OSL). Different widths of lumber can be ripped from SCL for various uses. Compared with similar size solid-sawn lumber, SCL often provides a stronger, more reliable structural member that can often span longer distances and has less dimensional change.

Structural composite lumber is a growing segment of the engineered wood products industry. It is used as a replacement for lumber in various applications, such as prefabricated wood I-joists, which take advantage of engineering design values that can be greater than those commonly assigned to sawn lumber.

Structural glued-laminated timber (glulam) is an engineered, stress-rated product that consists of two or more boards glued together with the grain parallel to the length. The maximum board (lamination) thickness permitted is 50 mm (2 in.), and the laminations are typically made of standard 25- or 50-mm (nominal 1- or 2-in) thick lumber. The boards are joined end to end, edge to edge, and face to face, the size of glulam is limited only by the capabilities of the manufacturing plant and the transportation system. North American standards require that glulam be manufactured in an approved manufacturing plant.

Douglas Fir–Larch, Southern Pine, Hem–Fir, and Spruce–Pine–Fir (SPF) are commonly used for glulam in the United States. Industry standards cover many softwoods and hardwoods, and procedures are in place for including other species.

The ASTM D5456 [6] standard provides methods to develop design properties for SCL products as well as requirements for quality assurance during production. Each manufacturer of SCL products is responsible for developing the required information on properties and ensuring that the minimum levels of quality are maintained during production. An independent inspection agency is required to monitor the quality assurance program.

11.2.4.4   Particleboard

Particleboard is produced by mechanically reducing the wood raw material into small particles, applying adhesive to the particles, and consolidating a loose mat of the particles with heat and pressure into a panel product.

Particleboard is typically made in three layers. But unlike OSB, the faces of particleboard usually consist of fine wood particles, while the core is made of coarser material. The result is a smoother surface for laminating, overlaying, painting, or veneering. Particleboard is readily made from virtually any wood material and from a variety of agricultural residues. Low-density insulating or sound-absorbing particleboard can be made from kenaf core or jute stick. Low-, medium-, and high-density panels can be produced with cereal straw, which has begun to be used in North America. Rice husks are commercially manufactured into medium- and high-density products in the Middle East.

All other things being equal, reducing lignocellulosic materials to particles requires less energy than reducing the same material into fibers. However, particleboard is generally not as strong as fiberboard because the fibrous nature of lignocellulosics, *i.e.* their high aspect ratio, is not exploited as well. Particleboard is widely used in furniture and flooring system, where it is typically overlaid with other materials for decorative purposes. Since most applications are interior, particleboard is usually bonded with a UF resin, although PF and MF resins are sometimes used for applications requiring more moisture resistance.

All particleboard is currently made using a dry process, where air or mechanical formers are used to distribute the particles prior to pressing. The various steps involved in particleboard manufacturing include particle preparation, particle classification and drying, adhesive application, mat formation, pressing, and finishing.

Alternatively, a few particleboards are made by the extrusion process. In this system, formation and pressing occur in one operation. The particles are forced into a long, heated die (made of two sets of platens) by means of reciprocating pistons. The board is extruded between the platens. The particles are oriented in a plane perpendicular to the plane of the board, resulting in properties that differ from those obtained with flat pressing.

Particleboard that has been grade marked ensures that the product has been periodically tested for compliance with voluntary industry product performance standards. These inspection or certification programs also generally require that the quality control system of a production plant meets strict criteria. Particleboard panels conforming to these product performance standards are marked with grade stamps.

11.2.4.5 Fiberboard

The term fiberboard includes hardboard, medium-density fiberboard (MDF), and insulation board. Several things differentiate fiberboard from particleboard, most notably the physical configuration of the wood element. Because wood is fibrous by nature, fiberboard exploits the inherent strength of wood to a greater extent than does particleboard.

To make fibers for composites, the naturally occurring bonds between the fibers must be broken. Attrition milling, or refining, is the easiest way to accomplish this. During refining process, material is fed between two disks with radial grooves. As the material is forced through the preset gap between the disks, it is sheared, cut, and abraded into fibers and fiber bundles. Grain has been ground in this way for centuries. Refiners are available with single- or double-rotating disks, as well as steam-pressurized and unpressurized configurations.

Refining can be augmented by steaming or chemical treatments. Steaming the lignocellulosic under pressure raises the temperature to the point where the lignin bonds between the cellulosic fibers soften. As a result, fibers are more readily separated and usually are less damaged than fibers processed by lower-temperature processing methods. Chemical treatments, usually alkali, are also used to weaken the lignin bonds. All of these treatments help increase fiber quality and reduce energy requirements, but they may reduce yield and modify the chemistry as well. For MDF, steam-pressurized refining is typical.

Fiberboard is normally classified by density and can be made by either dry or wet forming processes. Dry processes are applicable to boards with high density (hardboard) and medium density (MDF). Wet processes are applicable to both high-density hardboard and low-density insulation board.

Dry-process fiberboard is made in a similar fashion to particleboard. Resin (UF or melamine-UF) and other additives may be applied to the fibers by spraying in short-retention blenders or introduced as the wet fibers are fed from the refiner into a blow-line dryer. The resinated fibers are then air-laid into a mat for subsequent pressing, much the same as mat formation for particleboard. ANSI A208.2 classifies MDF by physical and mechanical properties, and identifies dimensional tolerances and formaldehyde emission limits.[7]

Wet-process hardboards differ from dry-process fiberboards in several significant ways. First, water is used as the distribution medium for forming the fibers into a mat. The technology is really an extension of paper manufacturing technology. Secondly, some wet-process boards are made without additional binders. If the lignocellulosic contains sufficient lignin and if lignin is retained during the refining operation, lignin can serve as the binder. Under heat and pressure, lignin will flow and act as a thermosetting adhesive, enhancing the naturally occurring hydrogen bonds.

Refining is an important step for developing strength in wet-process hardboards. The refining operation must also yield a fiber of high "freeness;" that is, it must be easy to remove water from the fibrous mat. The mat is typically formed on a Fourdrinier wire, like papermaking, or on cylinder formers. The wet process employs a continuously traveling mesh screen, onto which the soupy pulp flows rapidly and smoothly. Water is drawn off through the screen and then through a series of press rolls, which compress the fiber mat to remove additional water.

Wet-process hardboards are pressed in multiopening presses heated by steam. The heating press cycle lasts 6 to 15 min. A maximum pressure of about 5 MPa (725 lb/in^2) is used during the press. Heat is essential during pressing to induce fiber-to-fiber bond. A high temperature of up to 210 °C (410 °F) is used to increase production by causing faster evaporation of the water. Lack of sufficient moisture removal during pressing adversely affects strength and may result in "springback" or blistering.

11.2.4.6   Cellulosic Board

Cellulosic boards are low-density, wet-laid panel products used for insulation, sound deadening, carpet underlayment, and similar applications. In the manufacture of cellulosic board, the need for refining and screening is a function of the raw material available, the equipment used, and the desired end-product. Cellulosic boards typically do not use a binder, and they rely on hydrogen bonds to hold the board components together. Sizing agents are usually added to the furnish (about 1%) to provide the finished board with a modest degree of water resistance and dimensional stability.

After drying, some boards are treated for various applications. Boards may be given tongue-and-groove or shiplap edges or can be grooved to produce a plank effect. Other boards are laminated by means of asphalt to produce roof insulation. A grade-mark stamp will be given for these cellulosic fiberboard products conforming to ASTM C208.[8]

11.3 Wood–Nonwood Composite Materials

Wood may be combined with inorganic materials and with plastics to produce composite products with unique properties to match end-use requirements. Wood–nonwood composites typically contain comminuted wood elements suspended in a matrix material (for example in fiber-reinforced gypsum board, or in thermoplastic material), in which the proportion of wood elements may account for appreciably less than 50% of product mass.

11.3.1 Inorganic-Bonded Composite Materials

Inorganic-bonded wood composites are molded products or boards that contain between 10% and 70% by weight wood particles or fibers and conversely 90% to 30% inorganic binder. Acceptable properties of an inorganic-bonded wood composite can be obtained only when the wood particles are fully encased within the binder to make a coherent material. Because hardened inorganic binders have a higher density than that of most thermosetting resins, the required amount of inorganic binder per unit volume of composite material is much higher than that of resin-bonded wood composites. The properties of inorganic-bonded wood composites are significantly influenced by the amount and nature of the inorganic binder and the woody material as well as the density of the composites.

Inorganic binders fall into two main categories: gypsum-bonded and cement-bonded. Magnesia and Portland cement are the most common cement binders. Gypsum and magnesia cement are sensitive to moisture, and their use is generally restricted to interior applications. Some inorganic-bonded composites are very resistant to deterioration by decay fungi, insects, and vermin. Most have appreciable fire resistance.

Paper-faced gypsum boards, widely used for the interior lining of walls and ceilings, have generically been called drywall because they commonly replace wet-plaster systems. These panels are critical for good fire ratings in walls and ceilings. Paper-faced gypsum boards (and glass fiber-faced gypsum panels), also find use as exterior wall sheathing. Gypsum sheathing panels are primarily used in commercial construction, usually over steel studding and are distinguished from regular gypsum wallboard by their water-repellent additives in the paper facings and gypsum core. The facings of drywall and of gypsum sheathing panels are adhered to the gypsum core, providing the panels with impact resistance, and bending strength and stiffness. The paper facings of gypsum panels are derived from recycled paper fiber.

The properties of cement-bonded composites are influenced by wood element characteristics (species, size, geometry, chemical composition), cement type, wood:water:cement ratio, environmental temperature, and cure time.[9] They are heavier than conventional wood-based composites, but lighter than concrete. Therefore they can replace concrete in construction, specifically in applications that are not subjected to loads. Wood-cement composites provide an option for using wood residues, or even agricultural residues. However species selection can be important as many species contain sugars and extractives that retard the cure of cement.[3]

In the last few years a new class of inorganic binders, nonsintered ceramic inorganic binders, has been developed. These nonsintered ceramic binders are formed by acid–base aqueous reaction between a divalent or trivalent oxide and an acid phosphate or phosphoric acid. The reaction slurry hardens rapidly, but the rate of setting can be controlled. With suitable selection of oxides and acid phosphates, a range of binders may be produced. Recent research suggests that phosphates may be used as adhesives, cements, or surface augmentation materials to manufacture wood-based composites.

11.3.2   Wood–Thermoplastic Composite Materials

Wood–thermoplastic composites have become a widely recognized commercial product in construction, automotive, furniture, and other consumer applications in the last decade. Commercialization has been primarily due to penetration into the construction industry, first as decking and window profiles, followed by railing, siding, and roofing. The automotive industry has been a leader in using wood–thermoplastic composites for interior panel parts.

The class of materials can include lignocellulosics derived from wood or other natural sources and different thermoplastics including virgin or recycled polypropylene, polystyrene, vinyls, and polyethylenes. Other materials can be added to affect processing and product performance of wood–thermoplastic composites. These additives can improve bonding between the thermoplastic and wood component (for example, coupling agents), product performance (impact modifiers, UV stabilizers, flame retardants), and processability (lubricants).

There are two main types of the wood-thermoplastic composite. In the first, the lignocellulosic component serves as a reinforcing agent or filler in a continuous thermoplastic matrix. In the second, the thermoplastic serves as a binder to the majority lignocellulosic component. The presence or absence of a continuous thermoplastic matrix may also determine the processability of the composite material.

The manufacture of thermoplastic composites is usually a two-step process. The raw materials are first mixed together, and the composite blend is then formed into a product. The combination of these steps is called inline processing, and the result is a single processing step that converts raw materials to end products. Inline processing can be very difficult because of control demands and processing trade-offs. As a result, it is often easier and more economical to separate the processing steps.[10]

References

1. J. E. Winandy, A. W. Rudie, R. S. Williams and T. H. Wegner., *Forest Prod. J.*, 2008, **58**(6), 6–16.
2. Forest Product Laboratory. *Wood handbook—Wood as an Engineering Material*. General Technical Report FPL-GTR-190. Madison, WI: USDA Forest Service, Forest Products Laboratory. 2010, Chapters 11 and 12.
3. J. L. Bowyer, R. Shmulsky and J. G. Haygreen, *Forest Products and Wood Science. Fifth Edition*. Blackwell Publishing Professional, Ames, Iowa, 2007, p. 558.
4. NIST. *Voluntary Product Standard PS 1–09 Structural Plywood. National Institute of Standards and Technology*. Gaithersburg, MD: United States Department of Commerce. 2010.
5. HPVA. *American National Standard for Hardwood and Decorative Plywood, ANSI/HPVA HP–1–2004*. Reston, VA: Hardwood Plywood & Veneer Association. 2004.
6. ASTM. ASTM D 5456-08. *Standard Specification for Structural Composite Lumber*. Annual Book of ASTM Standards. Philadelphia, PA: American Society for Testing and Materials. 2008.
7. CPA. *Medium density fiberboard (MDF), ANSI A208.2–2002*. Gaithersburg, MD: Composite Panel Association. 2002.
8. ASTM *ASTM C208–95(2001). Specification for Cellulosic Fiber Insulating Boar*. Annual Book of ASTM Standards. Philadelphia, PA: American Society for Testing and Materials. 2001.
9. F. C. Jorge, C. Pereira and J. M. F. Ferreira, Wood-cement composites: a review., *Holz Roh Werkst*, 2004, **62**, 370–377.
10. C. M. Clemons., *Forest Prod. J.*, 2002, **52**(6), 10–18.
11. NIST. *Voluntary Product Standard PS 2–04. Performance Standard for Wood-based Structural-use Panels. National Institute of Standards and Technology*. Gaithersburg, MD: United States Department of Commerce. 2004.
12. CPA. *Particleboard, ANSI A208.1–1999*. Gaithersburg, MD: Composite Panel Association. 1999.
13. CPA. *Basic Hardboard, ANSI A135.4–2004*. Gaithersburg, MD: Composite Panel Association. 2004a.
14. CPA. *Prefinished Hardboard Paneling, ANSI A135.5–2004*. Gaithersburg, MD: Composite Panel Association. 2004b.
15. CPA. *Hardboard Siding, ANSI A135.6–2006*. Gaithersburg, MD: Composite Panel Association. 2006.
16. AHA. Cellulosic fiberboard, ANSI/AHA A194.1-1985. Palatine, IL: American Hardboard Association. 1985.
17. AITC. *ANSI/AITC 190.1 Standard for Wood Products: Structural Glue-Laminated Timber*, Englewood, CO. 2007.

Subject Index